D0341286

The Woman Who Knew Too Much

Alice Stewart signing petition for Friends of the Earth

THE WOMAN WHO KNEW TOO MUCH

ALICE STEWART AND THE SECRETS OF RADIATION

Gayle Greene

Foreword by Helen Caldicott

Ann Arbor

THE UNIVERSITY OF MICHIGAN PRESS

Published in the United States of America by
The University of Michigan Press
Manufactured in the United States of America
∞ Printed on acid-free paper

2002 2001 2000 1999 4 3 2 1

*A CIP catalog record for this book is available
from the British Library.*

Library of Congress Cataloging-in-Publication Data applied for
ISBN 0-472-11107-8

In Memory

Agnes Elinor Paterson Greene
(May 22, 1907–May 20, 1997)
Lydia Paterson Greenspoon
(October 10, 1908–March 28, 1997)

My mother and my aunt,
who did not see the end of this project,
but were its sine qua non

My Father, Jack Greenberg
(January 10, 1907–October 6, 1993),
a good doctor,
whose stories I never collected

Curve

Using the language of science
she showed us the steep curve
mysterious on the graph
pointing to casualties of a certain kind.
Only over time
did the idea come to her, hidden
under this curve early deaths
of another sort produced a different but
invisible curve
making it
one long slope of dying.
All this proceeded
in a now to be calculated way
after the first explosion
began the chain of events
diminishment, loss
collapse, cancer
the disappearance of
family, friends . . .
The consequences
grim as they are
sterling clear.
There to see in the numbers,
one might easily infer
the stories,
the telling blank spaces.

—Susan Griffin (for Alice Stewart)

Foreword

Helen Caldicott

This is a fascinating account of the life of an Englishwoman, born into a medical family almost 100 years ago, whose discoveries have brought her into conflict with the powerful nuclear industry and the international regulatory bodies that set radiation safety guidelines. Alice Stewart is a pioneer worthy of a Nobel Prize. Her work has been largely unrecognized by the scientific mainstream because it challenges received wisdom that radiation in small doses poses no threat.

In the 1940s, Dr. Alice Stewart had a brilliant career in clinical medicine. She won honors and accolades few women have received, having been elected Fellow of the Royal College of Physicians (the ninth woman fellow ever and the first under forty to attain this distinction) and the first woman to be elected to the Association of Physicians. But World War II and a serendipitous turn took her to the new area of Social Medicine and an appointment at Oxford, where her work helped carve out the new field of epidemiology before it was even called epidemiology. While at Oxford, she initiated a survey on the etiology of childhood leukemia which led to the startling revelation that just one prenatal x-ray doubled a child's risk of developing the disease. This discovery won her few friends in the British medical establishment.

Twenty years later, she again lit a fire in the scientific world with the discovery that the nuclear industry is twenty times more dangerous than safety standards admit. Her work with Dr. George Kneale and Thomas Mancuso on the Hanford nuclear workers directly contradicted the Hiroshima data on which international radiation standards are based.

This book provides the most lucid account that I have ever read of the control and cover-up by the U.S. military of the data relating to the injuries and diseases induced by the atomic bombs dropped on Hiroshima and Nagasaki. It gives a compelling account of the personal story as well—of the scientist who carries on without funding or support; of the mother, wife, then divorced single mother, lover of Sir William Empson, and responsible and caring grandmother.

Alice Stewart at the age of 92 still dazzles an audience with her originality and penetrating clarity. Like other scientists whose discoveries were not fully appreciated in their time—Galileo, Pasteur, Semmelweiss—her work may receive the recognition and thanks of the future. "Truth is the daughter of time," as she's fond of saying, and her theories are even now finding validation as the health effects of radiation come to be better understood. That day when the value of her work is acknowledged will mark the beginning of the end of the nuclear age.

Contents

1. Introduction: Daughter of Time 1

Part 1. The Making of a Doctor

2. Dr. Lucy and Daddy Naish 19
3. School Days and Cambridge 35
4. Marriage, Motherhood, Medical Practice: Through the War Years 49

Part 2. Engendering Epidemiology

5. Changing Subjects 67
6. X-Rays and Childhood Cancer 78
7. Dr. Doolittle's Team for the Moon 94

Part 3. Through the Looking Glass, onto the International Nuclear Scene . . .

8. Up Against the Department of Energy 113
9. Taking on the International Nuclear Regulatory System 128
10. Rogue Scientists 147
11. Alice in Blunderland: Back in Britain 162
12. Fallout 177
13. The Invisibilizing of Alice 193

Part 4. A Message to the Planet

14. Epidemiology and Alice Stewart 213
15. The Good Doctor 232
16. Pioneer and Pariah 248
17. Endings 266

Notes 271
Selected Bibliography 301
Index 305

Illustrations *following pages 64 and 110*

Chapter 1

Introduction: Daughter of Time

In 1956, Dr. Alice Stewart discovered that a single exposure to a diagnostic x-ray shortly before birth will double the risk of an early cancer death. Her finding made a revolution in medical practice: on account of it, doctors have become very cautious about x-raying pregnant women. A few decades later, she produced a study showing that the nuclear weapons industry is about twenty times more dangerous than worker safety standards admit, a discovery that put her on a collision course not only with the U.S. Department of Energy but with the regulatory commissions that set international nuclear safety guidelines. If Alice Stewart had discovered that radiation was good for you, she might have won the Nobel Prize, as more than one of her admirers has commented. But since she is the bearer of bad news, there's been a tendency to ignore her.

Whereas no one disputes the dangers of radiation at high dose, Alice Stewart has been a lone voice warning of radiation risk at low dose. "In the old days, they killed the messenger who brought bad news," she's fond of saying; "a Cassandra is never popular in her time." But hers is a voice that is gaining power and credibility, as the biological effects of radiation come to be better understood.

I first met Alice Stewart in Berkeley, in May 1994. I knew her by reputation, as the woman who had discovered the link between fetal x-rays and childhood cancer and who had gone on, in her seventies, to ignite the controversy about nuclear worker safety and become a kind of guru to the anti-nuclear movement. I'd been a literary scholar who'd made a mid-career change to writing on health and the environment and was working, with Dr. Vicki Ratner, on a book on cancer. If you do any reading at all in the area of cancer and radiation, you come across the name of Dr. Alice Stewart—her work is a lodestone to the anti-nuclear movement. She is that rare thing in radiation research, an independent scientist who has found ways of surviving without institutional support, who has made her expertise available to activists and put her science to the service of society. The *New York Times* calls her "perhaps the Energy Department's most influential and feared scientific critic."[1]

Vicki and I felt honored to get an interview with her.

Alice was (I later realized) trying to enjoy one of her rare days off, spending the weekend with an old friend, Dr. Joyce Lashoff, who had worked with her on several projects and had recently retired from the University of California at Berkeley. But Alice had generously agreed to give up her afternoon to be interviewed by these two strangers, who came barging in on her at her friend's elegant Berkeley hills home. Lashoff was somewhat miffed, Vicki and I felt awkward, and everyone was slightly out of sorts—everyone except Alice, that is, who, though decades older than any of us (she was eighty-eight), warmed to the interview with an energy and enthusiasm that sparked ours.

Anyone who has met Alice Stewart knows what I mean by the Stewart charm. She seems a slight, granny-like presence, until you hear that strong, sculpted Oxford English and get a glimmer of her scientific acumen. She has fine deep-set eyes that sparkle with humor and curiosity and a gaze that holds yours. She is brisk, blunt, and to the point—one would not like to be on the wrong end of that wit—yet she is also amazingly patient. She has been over this material maybe a million times yet she takes pains to go over it again, carefully, precisely, until she's made sure you've got it—"got it?" she'll say. She has a smile that could melt stone.

Alice takes over and runs with our questions, putting Vicki and me at our ease. Though she is moving too rapidly and not chronologically, I begin to form a picture. It's a remarkable story of scientific discovery and its suppression by politics. It's a complicated story of a career in several stages. There are her early years in clinical medicine, when she gains honors that few women attained in this area, being elected—the youngest woman ever—to the British College of Physicians, while raising two children on her own. There are her years as head of Social Medicine at Oxford, when, on a grant of £1,000, she launches the landmark study that turns up the link between fetal x-rays and children's cancer. There's a remarkable post-retirement career when, at age sixty-eight, she ignites a firestorm in international scientific circles by suggesting that nuclear safety standards are too lax and wins, in her eightieth year, a grant for $2 million to study nuclear workers' records from the entire U.S. weapons complex. This is a woman who is courageous (and stubborn) enough to stick to her positions against the attempts of powerful authorities—the medical profession, the nuclear establishment—to discredit her. This is a scientist who has learned the cost of challenging mainstream opinions, in terms of funding and recognition, yet has kept, through it all, faith that "truth is the daughter of time," as she's fond of saying; "It's an old saying, but very true; it goes back to the classics, and earlier." And indeed, I know this saying from Shakespeare, only I'm surprised to hear it from this world-class radiation epidemiologist.

As the lines of her narrative become clearer, I am more and more intrigued. "This is a great story," I say; "somebody ought to write it up."

"Oh, they've tried," she replies, "but they never got anywhere."

I want to hear more and I sense that she is interested in my interest, and I sense that this is a dangerous moment. I tell myself to slow down. I am, after all, already writing one book and teaching full time. But my mind is racing ahead with questions and the irresistible feeling that this is a story that has to be told.

It turns out that not one, but several others have tried to write Alice's story, and I offer to look at the most recent effort. Over the summer I read it, and it is, as Alice has cautioned, a bit of a jumble. But there's a lot of information in it,[2] and it takes me to a world that seems oddly familiar, a world I know from English novels—from Evelyn Waugh, E. M. Forster, C. P. Snow.

Alice Stewart was born October 4, 1906, in the northern industrial city of Sheffield, to parents who were both pioneers in children's welfare at the turn of the century and who practiced medicine in conditions I recognize from the works of Dickens. So many of her family went into medicine that they once filled a whole page of the British medical registry. She was at Cambridge in the twenties and headed a department at Oxford in the forties and fifties. There are many names I recognize: her godmother was the daughter of Elizabeth Garrett Anderson, the first woman physician in England; her godfather was godfather to the poet W. H. Auden. She knew Geoffrey and John Maynard Keynes; she shared a house with the novelist Iris Murdoch.

There is a long relationship with William Empson, later Sir William Empson, a name that leaps out at me: he is one of the most important literary critics of the century, one of the "New Critics" who shaped the way literary studies were defined on both sides of the Atlantic in the forties, fifties, and sixties. He was a poet who had strong left-wing politics and a controversial personality; he was a powerful presence in the curriculum I encountered as a student at Berkeley and Columbia. He was

Dr. Alice Stewart's lover, and though they both married other people, their relationship lasted decades. ("How long did it go on?" I asked her; "from 1929 until"—she thought a moment—"1983.")

Other things strike me as I read through this manuscript. She has strong ties with people, and not just with family—her daughter and grandchildren and various in-laws—but with a kind of extended family of friends, colleagues, fellow scientists, activists. There are many people who are devoted to her, including the author of the manuscript I am reading. There are dozens of people she befriended in some way or other, gave a job to, loaned money to, made part of a project.

One story stands out. Once upon a time, Alice bought a house for Empson's son, Mogador, when he found himself in disgrace with his mother because he was about to marry the daughter of a Labour Party politician. The house was in Leeds, near the university, and it cost very little since it was slated for demolition. When Mogador left Leeds, the demolition date got postponed, and Alice tried to give it to the university, but the university didn't want it. Then comes a startling bit—a son's nervous breakdown and return from Canada with a wife and two children. At this point, the house becomes a lifeline, for Alice can offer her son and his family a place to live; after his suicide, she gives it to his widow, Jeanette. Eventually the demolition is rescinded, the house shoots up in value, and Jeanette is able to rent the other flats in the building and go back to school to complete her nurse's training.

There's tragedy there, and serendipity, and a generous act returning in time of need, for Alice felt keenly the need to see her grandchildren through after her son's death. It was partly this responsibility that kept her working.

The Woman

By the time I finish reading this manuscript, I am hooked, and at the end of that summer I take the first of several trips to England. We spend the time at Alice's flat in Birmingham, where she has a research appointment at the medical school (which provides an office and staff), and at Evenlode cottage, Fawler, about fifteen miles outside Oxford, which she's owned since 1949. Fawler, as the cottage is called, is a large, ramshackle structure with interesting nooks, crannies, corners, and wonderful vistas onto the hills of the English countryside. The grounds are ample, with large, established trees that Alice herself planted, a stone wall, and a statue of a French peasant woman, a figure with a wise, gnomic smile—

Alice calls her the "presiding genius." There is a vegetable garden kept by her daughter Anne, a doctor who lives and practices in London.

It is utterly unlike any place I've ever been. After I've wandered around and got thoroughly lost, wending my way up a hidden staircase to what seems a whole other cottage, Alice explains that Fawler is actually not one but several cottages. The original cottage was built in 1600, the second cottage a century or so later, the third a century after that. These are (she tells me) the characteristic Oxfordshire dwellings, made of stone with a steep tile roof and wooden beams like eyebrows above the windows. Alice has acquired them over the years and has combined them into one large structure that can accommodate the family and friends who converge there for holidays, summers, and birthdays. "It had to be done gradually because of financial straits," she says; "That's much the best way to do things, adding according to need, because you get a feel, you adjust, you keep the character of the place."

I am enchanted with Fawler, as is anyone who's been there. The place is, like Alice herself, warm, welcoming, multifaceted, unique. Alice speaks feelingly of it: "When I'm abroad sometimes and homesick, I think of the evening light at Fawler. It's my idea of what England's about." I am amused at the contrast between this costly piece of Cotswolds real estate and the modesty of the life lived within—the disregard of matching kitchenware or finery, the chipped crockery, the utter unpretentiousness. The microwave oven in the small kitchen is a recent acquisition, a Christmas present from her daughter.

I thoroughly enjoy these visits, and over the next few summers, as I come to know Alice better, my appreciation for her deepens. We work in a large, sunny room that is painted Venetian red and has windows opening onto the garden and fields. Our conversations range from Margaret Thatcher and Margaret Drabble through marriage, academia, and other institutions. She has a wonderfully barbed wit and a well-developed sense of the absurd. The accents are upperclass, the cadences formal, even nineteenth century. Her sentences are long and complex, and she actually finishes them, as Americans tend not to. They are filled with so many ideas—she gets so many clauses up in the air that you wonder how she'll manage to keep track of them all, but down they all come, in perfect grammatical order. Some of her expressions have a literary ring—"reluctant dragons though they were." *"Bible arithmetic!"* she mutters of a researcher's calculations from the Hiroshima data, the assumptions of which she dismisses as bogus. Yet she'll reach as easily for a homespun, housewifely expression. She named her method of calculating occupational radiation risk the *Ready Reckoner* after a book of conversion tables that children get in math classes at school.

I am intrigued by a reference to Thackeray she returns to more than once. "There's a story in Thackeray, of a godmother who doesn't get invited to the christening, and so she comes in bad temper and makes a gift of a little misfortune, and of course it's exactly the right gift to make. You must hope for a little misfortune," she says, "not too much, of course, and you've also got to be lucky. I've had my misfortunes and I think they were the making of me—but I was also lucky enough to have a steady salary."

Is this the source of the equanimity I sense in her?—for there's a marvelous calm about her, an assurance that things will work out, which is remarkable, considering the obstacles she's encountered. She has borne more than her share of slights, has had to scrape by with the barest of support. Only in the fall of 1996, as her last major grant ran out, was she given the title "professor"—an honorary title, conferring no pay. Yet there are no apparent scars—in fact, she sees her obscurity as having worked to her advantage because it's left her free to pursue her own ways. "Funnily enough, it was just right for me personally. If I'd landed a cushy job I'd have found myself sitting on committees on World Health and all sorts of things, but I stayed at my own drawing board in a way which, if I'd been a man, would never have happened."

She is an interesting text and full of complexity. She comes of good stock, of parents who both lived into their nineties, and her seven siblings all have, or had, the same amazing energy. But there's another gene there, too, for two siblings and a son have taken their own lives. She's a socialist, yet there's more than a touch of the aristocrat in her language, her background, the company she's kept. The old photographs show a woman who was stunning, with a mane of dark hair pulled back from her face, high cheekbones, a straight nose, lively eyes, and a fine figure. How did she make her way with those looks in professional circles in the thirties and forties? I wonder. The worlds she moved in—Cambridge, Oxford, the British medical profession and research establishment—were not welcoming to women, and she was a married woman, with children besides.

She enjoys company—she's such good company that I have to remind myself that this is a world-renowned scientist I'm chatting with, whose work has prevented untold numbers of malignancies and saved untold numbers of lives. Yet she's fiercely independent and protective of her time. The great love of her life has been research: "research is thrilling. . . . I love teasing out of figures new ideas. It's like producing a baby." Yet she has also raised two children and has helped raise four grandchildren, caring for them summers and holidays at Fawler, and has remained deeply involved in their lives. The generosity I sensed meeting her is real, and it returns to her in later life, in the form of friends who are there for her, ready to drive her about to pick up an item at the grocery store, or to meet a visiting American (me) at a bus or a train.

She is indeed a wonder. Friends have likened her to Alice in Wonderland and she herself enjoys the parallel: she sees things at odd angles, as though through a looking glass, and has found herself down more than one strange rabbit hole. She is also, I recognize, a fellow career-change artist, skilled at reinventing herself.

There are the intriguing subplots: the affair with Sir William Empson; the career of her mother, Dr. Lucy Naish, who became a physician at a time when this was barely a possibility for a woman. There's a lifelong rivalry with the esteemed Sir Richard Doll, the epidemiologist who made a reputation in the fifties by establishing the connection between lung cancer and smoking and who turns up in Alice's story during the Oxford years. His name is familiar to me from my research on the cancer book: he's well known in the United States for his 1981 studies asserting that at most 2 percent of cancers can be attributed to industrial pollution, a claim that lent authority to government agencies bent on deregulating industry during the Reagan administration.[3] Heralded by the *New York Times* as "one of Britain's foremost epidemiologists,"[4] he remains to this day at the heart of cancer research in England. His is a career with a very different trajectory from Alice's.

I begin to understand the role gender plays in Alice's story, both in marginalizing her and in making her the kind of thinker she is. She made her landmark discovery about fetal x-rays by devising a questionnaire for "the mums," asking questions that allowed them to recall what happened before the child's birth. It was a revolutionary approach: "ask the mothers? They don't know anything! To men, this would seem unscientific, whereas it made perfect sense to me that they might remember something that the doctors had forgotten." In our discussions of epidemiology, she refers to "a feeling for the data" that reminds me of the "feeling for the organism" described by Evelyn Fox Keller in relation to Barbara McClintock:[5] a willingness to keep questions open and let the material carve out its own shape. And she's not one of those women I am so tired of reading about, who breaks new ground for women but makes herself one of the old boys. She identifies strongly with women and with feminism.

The Work

Alice Stewart's findings about fetal x-rays, published in 1956 and expanded in 1958,[6] were not welcomed. Physicians didn't like being told they were killing their patients. Radiography was the new toy of the medical profession and was being used for everything from examining the position of the fetus to treating acne and menstrual disorders, to measuring foot size in shoe stores. Besides, it was the fifties, the height of the arms race, when the governments of England and America were pouring vast resources into weapons testing and building a powerful nuclear industry dependent on public trust of the friendly atom. In the United States, the Atomic Energy Commission (AEC) was waging a publicity campaign to assure the world it could survive all-out atomic war. Nuclear medicine was good publicity for nuclear power, nuclear power was a useful cover for the arms race, and there was little incentive to knowing that low-dose radiation could kill you.[7]

Alice was able to persist in her study of children's cancer, extending, refining, and elaborating her data in what became the Oxford Survey of Childhood Cancer, because she managed to scrounge together funding from America and was willing to work for a pittance. In the course of this project, she linked up with George Kneale, who became, as her collaborator, a brilliant statistician. They continued their work, collecting data on childhood cancer in relation to family history, parents' occupations and social class, illness, infection, inoculation, asking questions about cancer and the immune system that are on the cutting edge of cancer research today, until funding dried up altogether and they were made unwelcome at Oxford.

Then suddenly, in the fall of 1974, as Alice was winding up work on the Oxford Survey and relocating to Birmingham, she got a phone call from America. Dr. Thomas Mancuso, who had been appointed by the Atomic Energy Commission to do a study of U.S. nuclear workers, wanted her to "take a closer look" at his findings about nuclear workers at Hanford. Alice had barely heard of Hanford, the vast weapons complex in a remote corner of eastern Washington that had been built in 1943 to produce plutonium for the Manhattan Project; but she and George Kneale made the long trek to the United States to look at Mancuso's data. Their investigations indicated that "this industry is a good deal more dangerous than you are being told"—about twenty times more dangerous.

"That put the cat among the pigeons," she says. Mancuso was dismissed, the AEC attempted to seize his data, and she and Kneale returned to England, taking with them a copy of the data so that they could continue their analysis.

The report published by Stewart, Mancuso, and Kneale in 1977 had momentous implications. Once again, Alice found herself at odds with official assurances about the safety of low-dose radiation, only this time it was more than the medical profession she was up against: it was the nuclear industry and the international regulatory committees charged with setting safety standards. She found herself down a strange rabbit hole indeed.

At stake in this controversy are the worldwide guidelines for radiation exposure for workers and the general population. The international regulatory committees and the national committees as well—the International Commission on Radiation Protection (ICRP), the United Nations Scientific Committee on the Effects of Atomic Radiation (UNSCEAR), Britain's National Radiological Protection Board (NRPB), the U.S. Nuclear Regulatory Commission (NRC), and the prestigious U.S. National Academy of Sciences committee on the Biological Effects of Ionizing Radiation (BEIR)—all base their standards on the studies of the Hiroshima survivors. The Hiroshima studies were carried out by the Radiation Effects Research Foundation (RERF), which maintains that low-dose radiation is negligible, arriving at this position by extrapolating from high dose to low dose in linear fashion: radiation becomes less dangerous as dose diminishes, becoming negligible at very low dose. The RERF supports this position by arguing that since we're bombarded continually by background radiation that emanates naturally from the sun, from space, rocks, soil, and radon gas—and mostly we don't get cancer—the risks from exposure to low-dose radiation must be negligible. But the RERF is an organization that had close ties with the U.S. Atomic Energy Commission, and the AEC—created in 1946 to preside over nuclear research and development—was hardly a disinterested party.

The Hiroshima data are the basis not only of worker safety standards; they determine the standards that set risks and benefits of nuclear installations, settle liability and compensation claims, and establish classification of radioactive waste. Any acknowledgment that low-dose exposure is as dangerous as Alice suggests would have enormous consequences. "If we are correct, occupational safety standards will have to be changed and it will open the floodgates to claims from workers, veterans, and downwinders. If we are correct, radioactive waste is a bigger problem than anyone thought—you can't just dump it in the ocean or anywhere else and hope that as long as it comes off slowly to imitate background radiation, there's no effect. Because if you increase the world level of background radiation, you increase the numbers of mutations and cancer deaths. As one rises, so does the other. Inevitably."

Also at stake are the potential compensation claims of the million or more workers in U.S. and U.K. nuclear weapons facilities, of the hundreds of thousands of people living near nuclear installations or downwind from the Nevada test site, and the hundreds of thousands of U.S. and U.K. soldiers and veterans subjected to fallout from nuclear tests and operations.[8]

No wonder nobody wants to hear what Alice Stewart has to say.

The Nuclear Industrial Complex

Alice would need all the faith she could summon that "truth is the daughter of time," for there are few areas of scientific inquiry where truth has been so slow to come out.

Nuclear science, conceived in the dark days of World War II, was born in secrecy and was to continue in secrecy. The Manhattan Project, a vast, complex enterprise organized to develop the first atomic bombs, employed 150,000 men and women and had facilities scattered across more than 37 sites in nineteen states and Canada.[9] The people working in the plants were kept ignorant of what they were producing; even the scientists were kept in the dark, and those who knew anything were sworn to secrecy, their publications censored.[10] Only a few approved experts working at the highest levels were allowed to see the whole picture. The very existence of the project was kept out of the media, concealed from Congress, concealed even from Vice President Harry Truman until he was sworn in as president.[11]

The sophisticated equipment that produced the plutonium and uranium used in the Manhattan Project required the expertise of large corporations like Du Pont and Union Carbide. This meant that after the war, the United States was left with an extensive physical plant, a complex of factories, production facilities, equipment, trained personnel, and a far-flung network of vested interests.[12] Atomic bombs don't easily lend themselves to commercial spin-off, but the reactors that transform uranium into plutonium give off tremendous heat, heat that could be used to boil water to produce steam to turn the turbines that generate electricity. They could thus be adapted to existing technology, whereas alternative technologies could not; alternative sources of energy, such as solar, would have required a complete redesign of the energy system and so were of little interest to those making the decisions. "It's a hell of a way to boil water," comments author Karl Grossman. "But it did keep the machinery going."[13]

In 1953, President Dwight Eisenhower delivered his "Atoms for Peace" speech to the United Nations, promising that nuclear energy would transform life on earth. A massive and costly public relations campaign was launched that included brochures, films, literature, exhibits that went out to schools, featuring *Citizen Atom,* a friendly, smiling little fellow with a lightning bolt through his head.[14] The new technology required vast federal subsidies to make it commercially competitive: in subsequent years, the U.S. government poured $70 billion into the development of nuclear power, and the electric utilities invested an additional $125 billion—more than the cost of the entire space program or the war in Vietnam.[15]

Part of the reason nuclear energy got such large subsidies was that it linked domestic energy to the same technology as weapons production, thereby bringing commercial research and development in line with military goals. The United States began testing nuclear devices in the South Pacific almost immediately after the war. The Soviet Union exploded its first atomic bomb in September 1949. The United States responded by exploding a hydrogen bomb in 1952, with fifteen thousand times the power that devastated Hiroshima. Britain got into the race in 1952, when it exploded its first nuclear device in the Montebello Islands off the northwest coast of Australia, a device made from plutonium manufactured at Windscale (later renamed Sellafield), the soon-to-be-notorious facility built on England's scenic northwest coast. In 1956, Britain began operating its first electricity-generating reactor just across the Calder River from Windscale. In England the civilian program was even more directly linked to military goals, the production of plutonium for building bombs.[16]

The nuclear industry grew rapidly, becoming what may be the largest and most powerful business enterprise in history.[17] By the late fifties, when Alice Stewart's discoveries were making their way into the scientific literature, England and the United States were in every sense of the word *invested* in this technology and had no desire to hear the bad news.

But they had to hear, for an international anti-nuclear movement had grown up around the issue of weapons testing so strong that it succeeded, in 1963, in driving testing underground. The movement reemerged in full force in the mid-seventies, to protest the proliferation of nuclear power plants; it gathered momentum in the years the Mancuso scandal was breaking. It ultimately succeeded—thanks to the efforts of independent scientists like Alice and owing to the industry's dangerous inefficiencies—in halting the building of new reactors and the siting of nuclear waste dumps.[18]

Cold War Science

The Atomic Energy Commission, mandated with protecting atomic secrets and assuring U.S. monopoly on nuclear technology, presided uncontested over research and development at the vast system of national laboratories that grew out of the Manhattan Project—at Hanford, Oak Ridge, Los Alamos, the Savannah River Plant in South Carolina, the Lawrence Livermore Laboratory in California, and Brookhaven on Long Island.[19] Weapons production at these facilities went on for two decades before there were any inquiries into the health effects of radiation on the workers—and even when the Mancuso study was funded, in the sixties, many felt that it was more about public relations than public safety. That suspicion was corroborated when the government clamped down so fiercely on Mancuso, Stewart, and Kneale for turning up a cancer effect. The Atomic Energy Commission—which was being reconfigured during these years as the Energy Research and Development Administration (ERDA), subsequently the Department of Energy (DOE)—tried to seize the data that the three researchers had in their possession. It didn't succeed, but it did manage to deny them further access to the workers' health records.

So began a long battle on the part of activists to wrest radiation health research away from the Department of Energy.

Alice at this point—she is in her early seventies—becomes a major player in an international drama, in demand at conferences, hearings, inquiries throughout Europe and the United States. She becomes a familiar figure in Congressional hearings and addresses citizens' groups throughout the country. She testifies for nuclear workers seeking compensation, for American and British veterans of atomic testing, for women protesting the siting of cruise missiles at Greenham Common. Often she is the only expert witness willing to appear; often she receives standing ovations. Always she speaks out for scientific freedom.

Within anti-nuclear circles, she becomes a bit of a legend. In 1986, the year of Chernobyl, she is awarded the Right Livelihood Prize, the "Alternative Nobel," as it's called, a prestigious and well-known prize (better known in Europe than the United States), conferred in the Swedish

Parliament the day before the Nobel to honor those who have made contributions to the betterment of society. In 1991, in Carpi, she receives the Ramazzini Prize, the leading prize in Italy for epidemiology.[20]

In the mid-eighties, Alice is drawn into a wider public arena when she is awarded a $2 million grant from the Three Mile Island Public Health Fund. When activists win a $25 million class action suit against the Three Mile Island nuclear facility for the accident that occurred there in 1979 and a fund is designated to explore the effects of the radiation, they turn to Alice Stewart and the workers' records that the government still holds in its possession. Alice, then eighty, receives $2 million to study the Hanford workers' records, along with records from Oak Ridge, Los Alamos, and Savannah River. It takes several more years for the activists to pry the data away from the government, during which time she makes frequent appearances on their behalf. When they finally succeed in securing the workers' records, the event is hailed on the front page of the *New York Times* as an unprecedented victory against the Department of Energy.[21]

She and George Kneale have been at work on this data ever since.

Because radiation is invisible, imperceptible even to those exposed, and because nuclear technology is so complex, the anti-nuclear movement has been highly dependent on the support of experts.[22] But since most nuclear scientists are employed by or contracted to government or industry, few are willing to speak out, and those who do so usually find themselves without jobs.

"I speak out because there are not a lot of people who can," says Alice; "I have nothing to lose. A lot of people do."

The Shape of Things to Come

What Alice Stewart has to tell the world about the hazards of low-level radiation is, if anything, more urgent today than it was four decades ago, when she alerted the world to the hazards of fetal x-rays. The nuclear age is not over: in fact, since the end of the Cold War, nuclear technology has become more dangerous as it spreads across the globe and as radioactive waste piles up.

The nuclear industry continues to find markets for its reactors and equipment in the many countries that remain committed to nuclear energy—France, Japan, South Korea, India, Eastern Europe, the former Soviet Union. The public relations arm of the industry, with its annual budget of more than $20 million, has mounted a massive publicity campaign to promote nuclear power, pushing it as a "clean" energy solution to the problem of global warming.[23] Far from being "clean," nuclear technology has left us with waste accumulating at 435 reactors around the world, where millions of pounds of highly radioactive spent fuel sit in corroding storage tanks.[24] The cost of the cleanup is estimated to exceed the cost of the installations themselves—assuming that anyone can figure out how to clean them up.[25]

In the former Soviet Union, where fifteen Chernobyl-type reactors continue to operate, many areas are alive with contamination from nuclear waste dumps and storage facilities, from reactors and reprocessing plants, and from thousands of waste containers dumped into the sea.[26] In both east and west, hundreds of tons of plutonium and enriched uranium must be kept track of, and as black-market trade in these materials picks up, the threat of nuclear terrorism increases.

The industry's argument is that the waste will "dilute and disperse" and disappear; but, as Alice has warned and as experts are increasingly agreeing, it will not. Wherever it is dumped, it will be blown by the wind or carried by the tides or seep into the earth; it will be eaten by insects, birds, fish, mammals, and will make its way into us. It will add, is adding, to the sum total of cancers and birth defects. Its legacy will haunt us for longer than civilization has existed.[27] Plutonium, with its half life of 24,000 years, is, in human terms, forever.

And cancer is not the worst of it. "Even more than the cancer is the threat to future generations," Alice warns; "that's what you ought to be really afraid of. It's the genetic damage, the possibility of sowing bad seeds into the gene pool from which future generations are drawn. There will be a buildup of defective genes into the population. It won't be noticed until it's too late. Then we'll never root it out, never get rid of it. It will be totally irrevocable."

Alice's Story

Since that day in May 1994 when I first met Alice Stewart, we have had many conversations. I have had access to dozens of interviews and lectures taped by others.[28] I have drawn on and edited this material in a way that allows for the transposition from spoken to written word and the construction of a narrative. Wherever possible, I let Alice speak for herself, for her voice is eloquent and distinctive.

Alice's story begins at the turn of the century in Sheffield, then moves to Cambridge in the twenties and to Oxford in the forties. Chapters 2, 3, and 4 concern the personal story, the education, and affairs of the heart.

Chapter 5 finds her in Oxford, heading the Institute of Social Medicine and working with Dr. John Ryle, a visionary who hoped to inspire physicians to a greater sense of social responsibility. These are chapters not only in Alice's story but in the history of medicine, for the surveys Alice designed in these years helped shape the emerging field of epidemiology.

Chapter 6 describes the Oxford Survey of Childhood Cancer and the landmark discoveries that came out of it. Chapter 7, "Dr. Doolittle's Team for the Moon," describes the company Alice kept during these years, the unique quality of life and work on the Oxford Survey.

The next chapters follow her onto the international nuclear scene, where she has lived ever since. Chapter 8 concerns her work on the Hanford nuclear workers' records. Chapter 9 describes her revolutionary challenge of the A-bomb data, which several of her colleagues believe will be her most lasting contribution. Chapters 10 and 12 detail the long struggle to pry the nuclear workers' data away from the DOE, a victory that has had major repercussions in breaking the Energy Department's hold on radiation health research and getting it transferred to the public health branch of the government, the Department of Health and Human Services. Interspersed between these two chapters is a discussion of Alice Stewart's role in the anti-nuclear movement in England. Chapter 13 looks at the ways the governments of the United Kingdom and United States have stonewalled her findings.

We then move to two chapters on Alice's science, exploring what her work has meant for the field of epidemiology and for theories of childhood cancer and cancer and the immune system. Alice's ideas deserve to be better known—and they would be, if she were. The final chapter is Alice's (and my) speculations about her special qualities as a scientist, the role gender has played in her story, the way it's worked to her disadvantage—and advantage.

The Woman Who Knew Too Much is not a biography that tells every single detail of its subject's life. I am interested in the life story as it illuminates the making of this extraordinary woman, her mind and her work, and her role in this major scientific-political controversy. The book is a kind of collaborative memoir, since there's a lot of Alice's voice in it, but it is more than a memoir, since I've added much information that contextualizes her story historically and politically. I don't give a lot of time to the other side of the controversy, though I'll state briefly what it is: it holds that the Hiroshima studies are a satisfactory basis for radiation standards; that there is no late effect of radiation except cancer; that you can predict cancer risk by extrapolating from high dose to low dose in simple linear fashion; that risk from low-dose radiation is negligible. It maintains that current safety standards are adequate, that the risks from nuclear industry and weapons are well within those of other industrial hazards and the benefits are sufficient to warrant a certain number of cancer deaths, a number that can be arrived at according to risk-benefit calculations—calculations derived from the Hiroshima studies.[29]

The other side has on its side the power of national governments, the prestige and influence of the international regulatory committees, the wealth of the nuclear industry, and all the access to funding, publicity, and publication that money can buy. Alice Stewart has none of these. Not that she has been without access to publication—she has, despite her unpopular positions, succeeded in publishing nearly four hundred papers in refereed scientific journals, and her 1958 paper on childhood cancer and fetal x-rays is among the most quoted in the literature, after the famous DNA paper by Watson and Crick. But apart from a few pieces that have appeared in the *New York Times,* the *Times Higher Education Supplement, Ms. Magazine,* a *60 Minutes* segment, and a recent Channel 4 British television documentary,[30] she has had few opportunities to present her side.

This is Alice's story.

Part 1
The Making of a Doctor

Chapter 2

Dr. Lucy and Daddy Naish

"It takes two generations to make a person."

Alice Stewart is daughter to two physicians, Lucy (née Wellburn) and Albert Ernest Naish. Both were pioneers in pediatrics, and both became heroes in Sheffield for their dedication to children's welfare. He was known far and wide as "Daddy" and she was nicknamed "Granny."

Alice has qualities of both parents, combining Lucy's extraordinary intuition and gift for problem solving with Ernest's keen analytical intelligence and talent for diagnosis. She takes from both an idealism about medicine, a willingness to sacrifice financial gain to devote herself to the prevention rather than the cure of disease, and an ideal of medical science as committed to the betterment of society.

"I take no credit for having carved out my path in life," she says; "others paved the way for me. I had a flying start in life."

An Unusual Girl

"How on earth my mother ever came, from a remote Yorkshire village, to become a doctor, in almost the first generation of women doctors, is hard to imagine." Alice speculates that the absence of a father may have helped. Lucy Wellburn was raised by three women who had lost husbands while still young.

Lucy was born in 1876 in Scarborough, a thriving center of trade on the North Sea. Her great-grandfather John Mills was from a family of seafaring men and owned nine sailing ships himself. Her mother, Anne Matilda, married Henry Wellburn, champion weight lifter for Yorkshire, Town Councillor, owner of a wine-importing shop, a grocer's shop, and four farms. A few months before Lucy was born, he died suddenly, lifting a heavy sack of seed, leaving his wife Anne Matilda at age thirty-one, with five small children.

"It was in the days when a widow's property reverted back to the husband's family," Alice explains. "But my grandmother stood up to her husband's family and took over the family business. It was difficult for her financially and she had to send the two younger children away."

Lucy was sent to Robin Hood's Bay, an isolated fishing village, to live with her aunt and grandmother, both of whom had lost husbands to the sea. She was eight before she was brought back to live with her mother in a house above the grocery shop in Scarborough. When she was seventeen, in 1892, she was sent to a school in Mecklenburg-Schwerin, where her mother hoped she'd receive a more serious education than she'd be likely to get at the sort of French finishing school that girls of her class were customarily sent to. Returning to England, she sat the College of Preceptors examination, a national examination to qualify for university, and she came out first in the country. The prize she won for this, ten volumes of *Chambers Encyclopedia* bound in yellow gilded calf, she would cherish for the rest of her life. But after this distinction, she found herself living at home, working as a part-time governess.

Lucy at seventeen was five foot ten, large boned, and tended to be clumsy because always in a hurry. She had a frank gaze and a forthright way; she was extroverted and outspoken; a teacher described her as "extraordinarily uninhibited for a Victorian young lady." She was vigorous, restless, and so voracious a reader that she'd been advised to take long walks to cure her of this intemperate habit. And she wanted to be a doctor, an ambition her mother was leery of.

One day during the summer when Lucy was back from school, the family doctor, Dr. Everley Taylor, paid a call, and literally collided with her on the stairs. "The story goes," recalls Alice, "that Lucy sent his top hat rolling to the bottom of the stairs, and there was my grandmother apologizing, saying, 'what am I to do with this great boisterous girl of mine? She wants to be a doctor'; and Dr. Taylor replied, 'Well, why not? Why not?' This seemed to be what my grandmother needed to hear. Dr. Taylor was able to help, to give a bit of advice on what was available to a woman wanting to be a doctor in those days."

Becoming a Doctor

There was not a lot available to a young woman aspiring to study medicine in the last decade of the nineteenth century. Before 1877, there were only two women in Britain officially recognized and allowed to practice medicine—Elizabeth Blackwell, educated in America, and Elizabeth Garrett, who earned an apothecary's license before the rules were changed to bar women.

Elizabeth Garrett and Elizabeth Blackwell had succeeded in founding a women's medical school, the London School of Medicine for Women, having failed in their efforts to gain women admittance to coeducational institutions. Their school was formally placed on the list of registered medical schools in 1874, and agreement was reached with the Royal Free Hospital in 1877 for women to obtain their clinical training there. The Royal Free was the only teaching hospital in London to admit women. In 1893, the year before Lucy came to London to study medicine, women were still barred from the examinations of the Colleges of Physicians and Surgeons of England, and as late as 1914, the most prestigious schools at Oxford and Cambridge and the twelve medical schools of London remained closed to them. (England was slower to open coeducational medical training to women than any other major country.) Women began to be admitted during the war years and Oxford conferred its first medical degree upon a woman in 1922, but after the war, many of the schools and hospitals that had admitted women closed to them again. By the end of the twenties, when Alice got her medical degree from Cambridge, there were only two schools in London teaching women—the London School of Medicine for Women and the University College Hospital, which admitted twelve women a year.[1]

Lucy entered the London School of Medicine for Women in 1897. It took her three years to get in. She'd had no previous training in science, and at age eighteen, began to struggle with biology, chemistry, and physics for the first time. Once she got in, she continued to find the material difficult. She found chemistry incomprehensible and was overwhelmed by anatomy, stunned by the number of Latin names to memorize. Her health was suffering, and her mother insisted on taking her away for a summer holiday, at the end of which Anne Matilda moved to London and took a flat near her daughter. Finally Lucy passed her exams and started her clinical training on the wards at the Royal Free Hospital in 1899.

The Royal Free, known simply as "the Free" because its founder wanted a hospital for those who had no powerful friends to write referrals for them, as was the usual means of admission to hospitals those days, was a grim place in a poor district. Lucy felt lost and miserable. She didn't know how to help her patients and no one seemed inclined to teach her. She'd recall later how her long skirts, three yards in circumference, would catch and tear on railings and beds. In late 1901, she went to Queen Charlotte's to learn midwifery. During her first week, doctors, students, and nurses hardly spoke to her, and the atmosphere was so hostile that she was not sure she'd be able to stick it out.

Yet in old age, she'd look back to her days as a medical student as the happiest of her life. She loved London and could be seen on a Sunday out walking with a half dozen dachshund puppies waddling beside her. She bred dachshunds and showed them at Crufts, the leading dog show in England. She was one of the first members of the Ladies Branch of the Kennel Club, founded in 1886.

The British Kennel Club, like the British medical profession, was closed to women, comments Nora Naish, author of the unpublished biography "Dr. Lucy," from which many of these details are drawn.[2]

The Doctors Naish

While Lucy was doing surgical training at the Royal Free, in 1901, she met Dr. Albert Ernest Naish. Earnest was a tall, lean man, reserved and studious, with sensitive, finely drawn features, a gentle voice, keen blue eyes, and black hair. Lucy was smitten and arranged her time to be on duty whenever he was.

Albert Ernest Naish had taken his medical degree from Cambridge. Born in 1871, he was the youngest son of a large, well-to-do Bristol family. "They were Quakers, not hard up in the way my mother's family was, but not rich," recalls Alice. "Father's family had made money in cotton, but his father had sold the business and devoted himself to civic life."

Ernest had become interested in pediatrics while at Great Ormond Street Hospital, one of the first children's hospitals in England. There he met Dr. George Frederic Still, a pioneer in the field. (It was said in London medical circles that the only thing doctors knew about children's diseases in those days was Dr. Still's telephone number. Infantile rheumatoid arthritis was named *Still's Disease* after him.)[3]

Ernest and Lucy were drawn together through their work. She assisted him in operations and surgical procedures. They were married in July 1902, six weeks after her final exams. She was twenty-six and he was thirty-one.

The young couple had the revolutionary idea of practicing medicine together, and Ernest made arrangements with senior partners in Harrogate, a fashionable town in Yorkshire. Ernest put his nameplate on his door, and Lucy, who had been given, as a wedding present, a nameplate engraved *Dr. Lucy Naish,* put hers on the same door. Two days later, the senior member of the practice stormed in and began railing that he was not having his practice tainted by the presence of a female doctor.

The next day both nameplates came down and the Naishes left town. They moved to Sheffield, an industrial city with a large working-class population in the northwest midlands, where they bought a practice of their own in the not very fashionable east end of the town.

"My father was outraged that Lucy had been treated this way, and though it meant losing the money he'd paid for the practice, he determined that they should go elsewhere. My mother was very shaken," recalls Alice. "Years later, the experience could wake her with nightmares. She continued to work as a physician and worked closely with my father, but she never put the nameplate up again. It remained in the attic."

Lucy was from Yorkshire and found the move to Sheffield easy, but Ernest was never at home in the north: the winters were harsh compared to those in Bristol, and Sheffield was far from the mainstream of medicine. Ernest found family life demanding, and the family grew rapidly: Lucy gave birth to eight children within the next fifteen years. The first, Jean, was born in July 1903, and the last, Charles, was born in July 1918. Alice, the third child, was born in 1906. The baby, Charles, was born when Lucy was forty-two; Lucy worked right up until the time of her delivery, through the smells and sights of the dissecting room.

Infant Welfare in Sheffield

Like other industrial centers in England—Manchester, Liverpool, Birmingham—Sheffield had seen a population explosion during the nineteenth century that created dire living and working conditions. Entire families, sometimes several, slept in one room. Tenements were unventilated and refuse and excreta were thrown into yards and streets. The air was black with soot; heaps of horse dung lay about, swarming with flies. Epidemics of cholera, dysentery, and typhoid swept the cities, which the public health services were incapable of handling.

The cholera epidemic of 1848 killed fifty-three thousand people in England and Wales.[4] In fact, it was this epidemic that gave birth to the new science of *epidemiology,* the study of the etiology and spread of disease, the branch of medical science that Alice would later help shape. In 1848, Dr. John Snow, a London practitioner looking for the cause of the epidemic, mapped out the cases of cholera in his own neighborhood, Soho, and traced them to a single well where raw sewage was seeping back into the public well. (His paper "On the Mode of Communication of Cholera" became a landmark of epidemiology.) The well was closed and the spread of disease in that area was halted. Snow had difficulty persuading his colleagues of his findings; he published them at his own expense, and they were ignored, then attacked. It took decades before public officials would act on his discoveries: it was not until 1875 that the Public Health Act was passed, which saw to it that drains and sewers were installed that assured clean drinking water. Once they were built, there was a marked improvement in the health of urban populations.[5]

Yet the infant death rate did not fall. Infants died just as frequently at the beginning of the twentieth century as they had a hundred years before. Throughout the nineteenth century an average of 150 out of every 1,000 died, and in industrial cities, where factory work undermined the health of both mother and child, the death rate was higher. In some areas, it was actually rising—in Sheffield in 1902, it was up to 202.[6]

In 1875, the year before Lucy was born, William Farr, the founder of the science of vital statistics, wrote a "Report to the Registrar General" in which he noted a clear connection between social class and survival, pointing out that the babies who had the best chances of living were the infants of peers and Anglican clergymen. Unwanted infants were often allowed to starve slowly or languish from disease. It was a regular practice to feed them opiates to keep them quiet and cut down their appetite. They were rarely washed, since running water was not easily available. Illegitimate children were farmed out to foster parents, in a practice known as baby farming. It was estimated by the Infant Life Protection Society that 60 percent of farmed-out babies died.[7]

Toward the end of the nineteenth century, an infant welfare movement began, and by the early years of the twentieth century it had become a popular cause. Physicians and activists began creating special institutions for children, including orphanages, infant asylums, dispensaries, and the Hospital for Sick Children in Great Ormond Street. By the 1890s, a group of physicians had carved out a special concentration in pediatrics. But very little was known, there were no textbooks and few studies, and what the Doctors Naish knew, they learned for themselves.

In 1907, Lucy and Ernest began the first Infant Welfare Clinic in Sheffield, a "School for Mothers" that gave away free milk in the basement of the town hall. The local hospitals refused to allow them to use their premises, so they and a surgeon friend paid for the dried milk and, with the help of the Sheffield Medical Officer of Health, a Dr. Scurfield, got the use of the town hall. When the news spread that milk was being given away, mothers and babies flooded the town hall. The mothers then had to wait for their babies to be weighed, take instructions as to how to prepare the feedings, and agree to admit Health Inspectors into their homes, in exchange for the milk.

"So began health education in Sheffield," comments Alice. "My parents had quite modern notions of prevention—they realized that teaching people how to raise their children was more important than curing sick children. They got the idea that the medical profession must take some responsibility for this."

Lucy ran these infant welfare clinics, sometimes doing as many as three clinics a week. In one clinic in one afternoon, she saw 102 new infants. In 1902, the infant mortality rate in Sheffield was 202 per 1,000 births; in 1911, it was 127 per 1,000; and in 1917, when Lucy retired from her infant welfare clinic, it was 100 per 1,000. (Today it is approximately 10 per 1,000.)[8]

Dr. Lucy's Reforms

In 1907 Lucy was approached, by the Liberals, to stand for election to the Sheffield Board of Guardians of the Poor Law.[9] She refused—her family had always been Tories. Within a few days she was approached by the Conservative agent and agreed to stand. She was elected by an overwhelming majority.

Admission to any hospital at the turn of the century was practically a death sentence for a child, but the Poor Law institutions were the worst. At the Firvale Workhouse, where Lucy had jurisdiction, the infants were tied to commode chairs in order to save time on toilet training and nappie-washing and were never taken outside to the fresh air. Lucy got rid of this practice altogether and arranged for them to be put outdoors for part of each day. She tried—unsuccessfully—to increase allowances for those on the "out-relief" wards, destitute old women, widows, and deserted wives.

As she'd say, she'd been elected a guardian of the Poor Law institutions, but she soon came to see herself as a guardian of the poor.

Lucy was dismayed at the way children went about sniffling, dripping, and wiping their noses on sleeves, so she set up clinics to instruct them to blow their noses properly. "She became quite famous for these noseblowing clinics. It wasn't so easy—when the children were reprimanded, they'd tend to sniff in rather than blow out. So she'd produce a little bowl of water with walnut shells floating on it and instruct them how to make the shells move away." Alice demonstrates by holding her teacup to her nose and making snorting sounds, laughing as she recalls, "she had this comical way of teaching."

Another of Lucy's reforms stands out. The Firvale Hospital required that all drugs and treatments be approved by the Matron. The Matron had complete control over the physician's treatment, including the power to veto a treatment and keep the physician from seeing the patient again. Even Arthur Hall, the most prominent physician in Sheffield, was not allowed to prescribe without her approval.

Lucy thought something should be done about this and got hold of a copy of the "Special Orders Concerning Poor Law Institutions." There she found written in the regulations, "The Matron is responsible in all matters of hospital administration where the treatment of sick patients is concerned. The Medical Officer must in all cases be consulted." "She took this to bed with her and had one of her brainwaves—she was famous for her brainwaves, we always teased her about her brainwaves," Alice recalls. Her inspiration was that if the period came after the word *administration,* the orders would read very differently: "The Matron is responsible in all matters of hospital administration. Where the treatment of sick patients is concerned the Medical Officer must in all cases be consulted."

She pointed this out at the next meeting of the Board of Guardians, when the press happened to be present. Her Majesty's Inspectors from the local government were called and the original documents in the offices in London were unearthed. They found that Lucy had been right—the period in the original was after the word *administration.* So Lucy's brainwave made a major reform.

"This was so typical of my mother," says Alice. "She had this quick intuitive sense. She would sense a solution, though she wouldn't necessarily know why it was true or how she knew it. She'd be more often right than not, though not necessarily for the right reasons."

Alice was to experience similar brainwaves herself.

In 1920 there was pressure on Lucy to stand for Parliament, but she refused. "She might have been elected, too—she was a good public speaker. She knew how to make people laugh and had a great gift for laughter herself." She wholeheartedly supported the Suffragists and while studying in London had got to know Suffragist leader Dr. Flora Murray. She entertained Emmeline Pankhurst and her daughter Christabel, founders of the Women's Social and Political Union, the first and largest militant suffrage group in England, when they visited Sheffield. The Pankhurst group had chained themselves to the gates of Parliament, hurled rocks through windows, set fire to mailboxes to get themselves arrested and draw attention to their cause. Lucy was much impressed by their zeal but took no part in the movement herself.

"When a girl marries she must be prepared to lose her own individuality," she told Nora Naish. Luckily, it was advice she herself largely ignored.

Two-Career Family

"Two more different people than Lucy and Ernest you could hardly imagine," remarked John Emery, a friend and colleague. "Lucy was tactile, extroverted, bouncing and full of energy. She loved children and could get along with anybody. Ernest was timid, shy, reserved, had great difficulty in communicating with people he did not know. He was very critical. He lacked the common touch."[10]

"He was reserved," recalls Nora Naish, "and though he could be genial among friends and loved real conversation, he did not unbend easily. He was so little interested in money that a colleague once said of him, 'he talks as if it's wrong to want to make a living!' He was much admired for his integrity and goodness to patients. Women adored him. His children were in awe of him."

Alice remembers her father taking her, at age eight, for a walk on a winter's day. She was wearing a sailor's hat and its ribbon had worked loose—the snow clung to it, melted as it touched her face, and froze again in the wind, making an icy little whip that kept striking her ear. They walked seven miles, Ernest striding very fast and talking the whole time, expecting her to keep up with his pace and conversation. It never occurred to her to complain.

Ernest was not happy as a general practitioner. In England there is a sharp distinction between general practitioners and Consultant Physicians who have positions on the staffs of teaching hospitals, and he felt he was wasting his talents. "He might have had a real career in London," Alice explains. "He'd done medical studies without any sort of backing from his family and had proved to be very gifted. Then he left the whole scene and took up general practice in order to marry and found himself backwatered in Sheffield."

But he was spotted by Dr. Arthur Hall, a prominent Physician at the

Royal Hospital in Sheffield, as someone of promise. Hall persuaded him to take the exams to qualify as a Consultant Physician. In 1911, Ernest decided to sit the membership examination of the Royal College of Physicians, a difficult decision, since the exam had a 20 percent pass rate and he wasn't allowed to practice while taking it. This meant that he had to give up his livelihood and take the risk of not qualifying, at a time in his life when he was the sole support of five children. He passed easily and launched his new career as a Consultant Physician.

He began by taking an unpaid residency. Since he'd already been without income while taking the exam, this imposed further financial hardship on the family. His new career had necessitated a move to a new house—"the east end would no longer do," Alice explains, "so we took up residence in Sheffield's west end, at 5 Clarkehouse Road, a large Georgian house with a circular drive—posh as Sheffield went, though Sheffield didn't go very posh."

The family had to make do without such amenities as carpets, lampshades, and curtains (though, Alice speculates, this may also have been because Lucy didn't see the point of curtains—the windows had shutters and the Sheffield soot would have caused curtains to need constant washing.) Lucy furnished the new house with an eclectic assortment of second-hand items. The younger children went about dressed in hand-me-downs, faded and patched, and in the coarse linen Lucy would convert from flour bags to overalls and frocks. A patient was once heard to comment that the doctors' children were wearing sacks.

Factory Doctor and Lady Tutor in Anatomy

Once Ernest became a specialist, Lucy could no longer be a general practitioner. "There was a law intended to prevent conflict of interest—it wouldn't do to have her referring patients to him. The advancement of his career ended hers—yes, well, we all know the story," comments Alice; "from then on she did the sorts of things she could do without clashing with his career."

When the war broke out—*the Kaiser's War,* as Lucy called it—Sheffield, the center of steelworks, became the center of armaments, and Firth's Steel Works, expanded and renamed the National Projectile Factory, began to employ women. Six thousand women workers were employed. "It was dangerous work and there were many accidents and someone was needed to provide medical supervision, so Mother went to do this.

She loved it, though it meant traveling two hours each way to the factory. She once had to quell a riot—she came into the canteen and found the workers in an uproar because they thought they'd been fed maggots. It turned out some dried haricot beans had sprouted that had been left soaking overnight."

The armament factory wanted Lucy to work full time, but in early 1917, a domestic crisis forced her to resign. She lost her cook and was suddenly faced with the task of doing the cooking for a family of eight. When she finally found a replacement, she offered her services to the Royal Hospital. Hospitals became increasingly short of medical staff in the course of the war, so the services of women were for the first time welcomed. Lucy received no pay, but she was delighted to be doing "real medicine" again.

Ernest also happened to be practicing at the Royal. On one occasion Lucy burst into his office and when not finding him, demanded of the group of students, "Where's Daddy?"—which was how the reserved and formal Dr. Ernest Naish came to be known as *Daddy.*

Lucy was also, during the war, appointed Lady Tutor in Anatomy at the Sheffield Medical School, which had only recently opened to women students, and she was later promoted to Lecturer in Osteology, with a salary of £100 a year. Each week she gave two lectures in anatomy and spent the rest of the mornings demonstrating dissections to students. Alice recalls boxes of bones lying about the house. "She was famous for carrying her bones around in a battered brown suitcase, which had a tendency to fly open at inopportune moments, scattering the contents about."

Over the years, Lucy coached hundreds of students, mainly those who had begun the study of medicine late in life and ex-servicemen whose schooling had been interrupted by the war. "She made a point of never charging anybody. She called it her *war work,* though it went on long after the war. There'd be students coming to stay with us even after my parents had retired, even when she was old and ill. Many years later, when I was at Cambridge and Oxford, I'd run across people who'd tell me, 'I'd never have got through my exams without your mum.' "

Lucy also continued working with Ernest in the children's clinics. "With her organizational ability and his professional know-how, they were a powerful pair," Alice recalls. "He was up on all the new developments, particularly from abroad, but she was the one who knew how to get things done. She could have organized anything."

Lady Great Heart

"She'd been good-looking as a girl but became stout as a woman and never took much trouble with her clothes. She was portly and he was thin. But she had very good eyes—deep set with dark gray coloring. They gave her a thoughtful and intelligent look.

"She was known far and wide for her acts of generosity. There'd be waifs and strays about the place, lame dogs and other people's children. Some of these children would be foisted on us because they had eating troubles. From my point of view I could never see that *they* had the eating problems—*we* had the problem because as soon as they got with our family, they ate our food. Yet there was a certain gruffness to her, too—she'd never tolerate a fool, she could tell a person to buck up and get on with it. She was quick tempered but there'd be no residual—there'd be these flash rows but they'd be over and done with as fast as they'd start. But she'd tolerate vast amounts of foolishness where she thought people couldn't help themselves.

"She was the one you'd turn to for the small needs of life. My father was scrupulously fair—he took care to see that the girls got the same education as the boys, which was generous of him because we weren't well off. He was an important guiding principle, but he'd starve us of pocket money, send us to school without stockings and that sort of thing. She was the one you'd go to for the pennys and ha' pennys.[11]

"She'd miss no opportunity to instruct us in scientific principles. On one famous occasion she was dressing a turkey for Christmas and held up the neck to give a lesson on the vocal cords. She sent us to church, but I'm sure it was to get us out of the house—she never had any feeling for religion."

But she had boundless capacity to extend herself to anyone in need, a bounty that earned her the name *Lady Great Heart*. She breast-fed all her children and even on one occasion, breast-fed a friend's child who was languishing.

She had strong opinions on breast feeding, which she expressed at a conference in London on infant welfare and published in the British medical journal *Lancet* in 1913. In this article, Lucy writes as a physician—and as a woman who has nursed six children. She gives a vivid sense of the pains of nursing: the anguish caused by minute fissures in the nipples, the afterpains caused by the sucking, which are sometimes so strong that they make the mother feel quite faint. She offers women the understanding they are not likely to encounter elsewhere, for "few medical men appreciate these difficulties," and so few can give "the intelligent and sympathetic help which comes from such appreciation." She insists on the importance of her firsthand experience: "Some of the details may seem trivial enough, but my subjective experience . . . leads me to believe that neglect of such details will have a profound effect upon the after-course of the feeding." She writes in the hope that if the mother "understands that certain things which appear to her in her weak and aching condition to be unusual and unnatural, are really common and natural," she will persevere in breast feeding.[12]

Lucy would soon find her maternal energies taxed to the limit by the long nightmare of her son's illness. In 1924, as she leaned over David, age twelve, to tuck him in for the night, he asked, "Mummy, can you stop this arm moving?" Lucy sat in helpless vigil for the next several weeks, as he became increasingly unable to control the twitchings of his arms and legs. David was diagnosed with encephalitis, a virus that destroyed parts of the brain and left him with Parkinson's disease and behavioral disturbances. It was thought to be a residual of the 1918 influenza carried by soldiers returning from Europe, the Spanish influenza that swept across the globe, killing more people than the war had.

Sheffield had been particularly hard hit—there were over two thousand influenza deaths in the city during the winter of 1918–19.[13] All Lucy's children had it: Ernest, age seven, was ill for several days, and Alice, age twelve, nearly died. Lucy came down with it at the same time as her maidservant and had to crawl about the house caring for her while also feeding her newborn, Charles.

David was, by popular consent, the sweetest of the Naish children, and now the family had to watch him degenerate physically and mentally. He became prey to emotional disturbances verging on the psychopathic: he once drowned a kitten and, on another occasion, attacked Lucy, grasping her by the throat. In his later years, he became subject to painful seizures and was confined to a wheelchair. Ernest found the whole scene so unbearable that he withdrew emotionally.

Through this time, Lucy continued to work as a lecturer and tutor at Sheffield University. Sometimes, to attend a meeting of the Anatomical Society, she would take the overnight train to London, arrive in the morning, go to a day of meetings, take the evening train back and walk home from the station the following morning. As exhausting as it was, she saw such work as a lifeline.

When war was declared, on September 3, 1939—*Hitler's War,* as Lucy called it—she decided to stay in Anglesey, an island off the northwest tip of Wales where the family had been spending holidays since 1914. She resigned her position teaching anatomy and gave herself over to nursing David. David was then in his late twenties and becoming more crippled physically and more difficult in his behavior. Lucy, in her mid-sixties, had arthritis in her knees and back, yet managed to push the wheelchair and pull him in and out of bed, washing, feeding, and nursing him.

"Mother nursed him until he died, at age twenty-eight—for sixteen years. Why it didn't kill her, I don't know. This left me," says Alice, "with very strong ideas about how wrong it is to leave families with severely defective children. It is horrible for the siblings. Even with both parents doctors and all the backing we had, it was terrible."

For five years Lucy remained in Anglesey, and she and Ernest were separated. In some ways this arrangement suited him. It allowed him escape from domestic life, made extremely oppressive by David's situation and what seemed often to be Lucy's obsessive involvement. He was busy with new honors and distinctions. In 1932, at age sixty, he was made Professor of Medicine and Examiner in Medicine for the Royal College of Physicians and for the University of Cambridge. In 1935 he was elected President of the British Paediatric Association, of which he had been a founding member.

He ventured into artistic and philosophical circles. He founded the Literary and Poetry Society and was a prime mover in a philosophical society, a group comprised of clergy and doctors that met regularly in the Bishop's Palace. He confessed that the happiest years of his life were those after sixty. Nora Naish recalls that Ernest at seventy had "a light step and agile mind."

Family Stories

John Emery tells this story: "For several years, the Literary and Poetry Society met at the Naishes. These meetings would contain forty or so people and would be followed by tea. Tea was prepared in the kitchen and brought forth by Granny. If the lecture was a good one and Granny was interested, she would slip out at the end and the tea would quietly appear. If she thought the lecture had gone on long enough, she would leave and within a minute or so a loud strident whistle would come from the kitchen, announcing that it was over. If you asked about this she'd give one of her gleaming smiles and explain how, by regulating the height of the gas and the amount of water in the kettle, you could make a kettle sing. She played that kettle with wonderful skill."

Alice tells this one: "After my parents retired, in the late forties, to Anglesey, Mother got a dog, a wild, scrappy mongrel named Mickey. Dearer was Mickey to my mother than husband, children, anything. Once when my parents were traveling by train to Anglesey, my father, who was always a bit restless on trains, was walking about and wanting his lunch, so Mother gave him his sandwiches. But she gave him the wrong packet, she gave him the bread crusts intended for Mickey—which he ate without complaint. When, afterward, it was discovered that Father had eaten Mickey's crusts, it became a great family joke that he knew his place—he didn't question that he should be given Mickey's crusts!"

John Emery recalls how at Anglesey, the routine centered on reading, writing, and the post. "For a number of years I wrote to Daddy Naish, but I found him an almost impossible person to correspond with because every letter was answered by return post. When the post came at 9 A.M. one or the other of them would collect it and they would each take their piles of letters to their rooms, read them and answer them immediately, and this took the better part of the morning. In the afternoon Ernest would take the dog for a long walk by the sea, post letters, and return to tea. There followed the evening reading. These hours of reading round the fire in this isolated house on the Welsh coast are something I'll never forget."

Alice recalls, "Lucy had always been a reader—she'd read aloud to her children and to her grandchildren. We gave her a subscription to the London Library and every week a pack of books would be sent back to the library and more books be sent for. She loved biography—she was fascinated by Byron. I always wondered why she didn't disapprove of him, but she didn't—she loved the sounds of his words. I remember she discussed Byron with Empson."

In 1952, when Daddy and Granny celebrated their Golden Wedding anniversary, they received a congratulatory telegram from the Queen. He was eighty-one and she seventy-six. They were both mentally and physically vigorous. All the children and most of the twenty-seven grandchildren attended, along with the first of their forty-five great-grandchildren.

Shortly thereafter, Lucy developed shingles. She lost strength and progressively lost her sight and taste. During those years, she became dependent on others to read to her. She taught herself braille and turned her attention to the news on the radio. She had children and grandchildren and students in cities all over the globe, and these ties made her follow avidly what was happening throughout the world. She'd listen to reports on market trends because her grandson Michael was working as a rural auctioneer and valuer.

"People would flock around Granny because she had so many interesting stories to tell. They'd come to hear her stories and catch up with the news. She had this gift—she felt there's a story to be told about everything. She'd get the children to bring out the encyclopedia she kept under the bed, the yellow gilded volume she'd won as a girl, and she'd tell a story. Even when she was bedridden and couldn't see to read, she maintained this gift. She'd keep everybody's interest."

Ernest read to her every day. But in his last six years his own health failed. "He was operated on for prostate and something happened during the operation," John Emery recalls. "I had seen him a few weeks before and he had been his usual self, but when I visited him after the operation, the fire seemed to have gone right out of him. From that time onwards he became a very old man. All correspondence with him ceased."

He died in 1964. Lucy died in 1967.

Chapter 3

School Days and Cambridge

"There was no stinting on our education."

The first house Alice remembers is not 5 Clarkehouse Road, but the house in Sheffield's humbler east end, where she was born. It was built on a hill where a broad avenue split into two narrower roads. Facing the junction was a nursery, a semi-circular room with four windows. In each window was a window seat just right for a child to defend as a military post on the many occasions the Naish children played soldiers. The flat roof of the house was used as a battlement.

"We were a large, rambunctious bunch."

Alice, born in 1906, was the third of eight. Before her was Jean, born in 1903, then George, born in 1904, who, as eldest son, always had a special place in his mother's affection. Next after Alice came Charlotte, in 1908, the blond, blue-eyed sister who was Alice's torment. Then in 1910 came Ernest, whom everyone agreed was the genius of the family; a painter and inventor, he has done green farming in North Wales for the past forty years. In 1912 came Anthony (or David, as he was called). John, born in 1915, was Alice's special charge, and the two formed a close friendship that has lasted through life. Charles, the youngest, born in 1918, fell out of a tree and nearly died of cerebral hemorrhage at age six, in the same year that David developed encephalitis; he was left with damaged vision but managed to qualify in medicine at Cambridge. It became a family joke that it took only half a Naish brain to do medicine at Cambridge.

Alice, Charlotte, John, and Charles became doctors.

Country Freedoms

Alice was six when her father's career change necessitated the family's move to 5 Clarkehouse Road, in Sheffield's west end. 5 Clarkehouse

Road was an imposing Georgian brick house, built in 1795. It had a lawn that sloped away from the pillared portico of the main entrance and large windows commanding a view of the smoking chimneys of Sheffield. It had stabling for two horses, a carriage, and a circular drive that became a racing track for the children's bicycles. Its garden was large enough to include a tennis court, and—what was thrilling for the children—it had a grotto. "You know, one of those things pretending to be from an old monastery, a secret entrance to nowhere. It was dark and mysterious—we loved it."

One day when George was out front climbing on the sundial, a policeman came by and ordered him off the premises. "I always thought that was so typical of our family—we'd been mistaken for trespassers. Even the move to so stately a place as 5 Clarkehouse Road couldn't make us respectable. We had bare legs and sandals and knickerbockers, and I was terribly ashamed and wondered why didn't I have a pretty skirt and a petticoat. We were allowed more wildness than most children—for town children we had a lot of country freedom."

The house on Clarkehouse Road practically bordered on the Pennine Mountains, allowing easy access to the country. The Naishes were walking distance from their first cottage, in the Rivelin Valley, which they first discovered while out hiking. "It was a tiny stone structure. Water had to be fetched from the river, and it was lit by candles. But to us it meant freedom, adventure, mystery. At home we were always being hushed—we weren't supposed to make noise near my father's consulting rooms—but at the cottage we could run wild. And there were no servants, so we were expected to do things for ourselves. We cooked on an open fire. We learned to swim. I used to be very good at waking up early, and I'd be detailed off to wake everyone else so that we could go swimming before school."

In August 1914, Lucy discovered a retreat that was to serve the family even better. The month before World War I broke out, she traveled with her three girls to Anglesey, an island off the northwest end of Wales. She hired a horse drawn wagonette to drive them from the railway terminus in Holyhead to the village of Four Mile Bridge, at the far end of the island. For every summer thereafter, the family traveled this two hundred miles to Four Mile Bridge. Toward the end of 1919 Lucy found a cottage for sale and bought it.

"We were at the very tip of the island. It seemed so far away, another world. When we first started going there, the locals hardly spoke English. It was a paradise for children, an inland sea with an ancient stone bridge, all protected and child size. Mother let us run wild—we'd arrive and pour into the water; before we'd been there five minutes we'd all broken our shins and got ourselves soaked." The children climbed on the rocks, dived, swam, explored the tide pools, and shot the rapids made by the tide as it rushed through a hole in the bridge, in a dinghy named the *Suicide*.

The cottage, Derwen, is still in the family. So is the second cottage Lucy bought, Seaview, across the road: Lucy gave it to Alice when she married, and Alice gave it to her daughter Anne. Both places continue to be a center to the Naish and Stewart families: children and grandchildren and cousins and an extended network of friends, girlfriends, boyfriends, ex-boyfriends, ex-girlfriends descend on these two small dwellings each summer, pitching tents in the yards to take up the overflow. The inland sea is today as peaceful and gleaming and dotted with sailboats as it was in the old photos, though the road in front of the cottages is crowded with traffic.

Sibship Pressures

With seven siblings, two older and five younger, all with strong personalities, family life was, as Alice says, "a somewhat pressured situation." The older children bossed her and the younger ones needed tending. Alice's main memory of the first World War was "looking after young children after school. Mother was working in the factory and, being an enterprising person, made sure that her whole family was mobilized. We all became very independent, very fast. If the war hadn't happened, we might have stayed more like children."

Alice had a specially vexed relationship with her sister Charlotte. "She had blue eyes and fair curly hair whereas I was plain and mouse colored. We were always quarreling and down came the grown-ups on us, saying, 'Charlotte is younger than you, you ought to know better.' She had these bright blue eyes that would fill with tears, whereas I would refuse to cry, and this gave her an unfair advantage. I always thought she was taking advantage—I *knew* she was."

Alice learned a curiously passive, yet effective, way of fending for herself. "My way of dealing with sibship pressure was simply disappearing. I could never have won any open combat with that lot, I knew that, so I would just slip away. At moments of crisis I'd be found behind the door. I acquired a useful reputation for not being good and not being there." She feels that this skill has remained valuable throughout life—"It never occurred to me to go out and battle these impossible situations, compete with the men for the jobs. I'd look for the cracks in the system and slip through."

But she also remembers the first time she stood up against the family. It was on her brother David's behalf.

It happened one summer when she returned home from school to join the family in Anglesey. George, the eldest and her mother's favorite, had just returned, on leave from the navy, and "no sooner had he arrived than he started bossing everyone about. He was everyone's darling, and we all adored him. He had this charm and it lasted the whole of his life. As we sat down to dinner, there was Mother, hanging on his every word; when my little brother David interrupted to ask for a second helping of pudding and George cut him off, Mother took George's side and told David to wait—I was furious. I got up and announced that I'd had enough of George's lording it and stomped out. Mother came after me, trying to calm me down. I remember she said this was just the way men were, 'they think themselves lords of the earth'; 'well, they're not,' I said, 'not while I'm here!' "

Good at Science

"Our clothes might have been shabby, but we had a top level education. Father was rigorous in insisting that the girls be educated the same as the boys. There was a strict rule in our household that you were treated according to merit, not according to sex. My brothers thought this was a great waste."

Sheffield Girls' High School was a paid day school for girls, Sheffield's best, and walking distance from the Naish home. Alice won a full scholarship on the basis of examination and was delighted to see her name appear in gold script on a commemoration board and to learn that her fees would be waived.

When she was thirteen, her father sent her to St. Leonards School in St. Andrews, Scotland, which she attended from age thirteen to age seventeen. She loved it: "It was an escape from my family. It's not that I was unhappy at home—it would never have occurred to me to be unhappy—but at school people actually laughed at my jokes. At home I could never get a word in edgewise; nobody even listened to my jokes, much less laughed at them. It really was rather a relief to be away. I enjoyed it very much."

St. Leonards had been founded by Louisa Innes Lunsden, one of the first students at the new college at Hitchin, later to become Girton College, the first college for women at Cambridge, founded in 1877. It was Lunsden's intention to provide a serious education for girls, an unusual goal in an age when women's education was geared to teaching domestic skills.[1] "St. Leonards had a reputation—it was the first boarding school for girls established on the lines of the public school for boys. It aimed to give girls an education equal to their brothers." Alice adds, "It was no accident that you had to go to Scotland to find this."

St. Andrews had been a center of learning for centuries, home to one of Scotland's oldest and finest universities. St. Leonards School had a splendid setting overlooking St. Andrews Bay; the wind blew straight in from the North Sea, whistling dramatically through a ruined abbey on a promontory. Many of the students complained of the cold and got chilblains, but not Alice: the Sheffield winters had toughened her. Many of the girls objected to the drab wool uniforms, but these were the finest clothes Alice had ever worn.

"It was a serious place. The headmistress, Emily P. Storey—E. P. S. we called her—was a strict disciplinarian, very, very proper. She would not permit a girl who had been christened 'Betty' to be addressed by anything other than 'Elizabeth.' It was assumed we would work hard and do well. We were offered science, which was fairly modern for girls. It was only the 'elegant' girls who got shipped off into domestic economy."

Standards were higher than they had been at Sheffield, and Alice never quite reached the top, as she had at Sheffield—"except in math and science, which didn't count for much." Even a school as serious as St. Leonards neither expected nor encouraged the fairer sex to take an interest in science. "Games and English were what really mattered. You had to be a brilliant athlete or one of the literary people—this determined how you ranked in class. To be good at math and science was second class stuff."

Nevertheless, E. P. S. recognized in Alice someone of promise. So did the math mistress—"She thought I was rather good. I began to have thoughts about being a doctor. Truth to tell, I wasn't any good at English."

St. Leonards had a strong Cambridge tradition, on account of its founder's connection with Girton. Alice's family also had Cambridge ties—her father had gone there and so would two of her brothers, John and Charles. So when it came time to apply to college, "it never occurred to me to apply to Oxford. Father had no use for Oxford science, whereas he said Cambridge was buzzing with famous scientists."

St. Leonards attracted girls from intellectual and academic families throughout Britain, and while there, Alice got to know people she would stay in touch with through her years at Cambridge and throughout her life. By the time she got to Girton College, Cambridge, in 1925, she had a network of friends and connections outside the world of medicine.

She was going to need them.

Insiders, Outsiders

Alice will never forget her first physiology lecture at Cambridge.

"It was a large room, an auditorium you entered from the rear with a long set of steps descending to the speaker's podium in the front. I slipped in, hoping to take a seat as close to the back as possible. But when I stepped into the hall and took my first step, the students, all male, began stomping, slowly and deliberately, in time with my steps. As I took my first step into that room, *bang!* came the sound of two hundred men stomping their feet in unison. I took my second step and the stomp was repeated. Every step I took, there was this *stomp, stomp, stomp.* My first instinct was to duck into a seat and disappear, but no—every row was blocked by the men. I was forced down to the front row, where I found three other girls and a Nigerian. These medical students had managed to segregate us out—they weren't going to have anything to do with women or minority populations.

"I wasn't whipped. I was stomped.

"When you think of it, it was very cruel. I at least had brothers and was accustomed to being pushed about and had been out in the world a bit more than the average schoolgirl. But here was my first lesson in racism and sexism—I learnt all about minority groups in those few minutes. From then on, I had no illusions. I swore from that day that I would never make friends with any other medical student. They were hoydens. They moved in packs and were very rude."

Cambridge may have been "buzzing with famous scientists," as Alice's father said, but it was also the last university in Britain to admit women to full membership. Women at Cambridge were allowed to attend lectures and take examinations from the early 1870s and were officially admitted to examinations in 1881, but they were limited access to the university library and many laboratories and could not formally receive degrees. In 1922—three years before Alice arrived—Cambridge granted women "titles" to degrees, greater access to the library, and the right of admission to lectures and labs. However, the number of women students was limited to five hundred, a quota that was raised to one-sixth the student body in 1948, when women were granted the status of full members of the university. From 1926—the year after Alice enrolled—women faculty could be appointed to university posts, but they still had no voice in university government. Like faculty wives, the dons of the two women's colleges, Girton and Newnham, could attend social functions and ceremonials by courtesy only, as guests.[2]

Women had been allowed to study medicine for only a few years. They were still far from welcome.

Luckily for Alice, there were other things going on. Cambridge in the twenties was alive with interest and activity. It had been, since the sixteenth century, a hub of intellectual life, boasting such graduates as Isaac Newton, Francis Bacon, John Milton, Edmund Spenser, Christopher Marlowe, Alfred Lord Tennyson, Ludwig Wittgenstein; but after World War I, "the place went through a transformation as sudden and complete as that of a Roman spring," writes Michael Holroyd; "the sap began to flow again, and the exuberance of youth burst out . . . and it became possible to believe once more in the New Age of Civilization."[3] Alice loved Cambridge—the streets humming with bicycles, the black gowns of undergraduates flapping in the breeze; the long afternoons punting and picnicking on the Cam, the riverbanks lined with willows and lawns that sloped downwards from the backs of the spired colleges. She loved the musical and dramatic performances and the evenings spent discussing revolutions in poetic verse. "A lot of things were opening up in front of me—it was a whole new world—the concerts, plays, lectures." She was thrilled to be a part of it.

Girton was built two miles outside town to keep the girls separate from the boys, but there was a lounge that medical students had access to in the center of town, in a house just opposite the entrance to King's College. This made an enormous difference: "You'd bicycle in to the center in the morning and stay there the whole day—you could stay there for lunch, or wait between classes and labs."

Alice was especially drawn to the literary circles. Geoffrey Keynes, the brother of John Maynard Keynes, the economist, was a friend of a school chum from St. Leonards. Keynes was a doctor, but he'd edited a volume of William Blake's poetry, was active in the theater, and had married the lead ballerina in Diaghilev's Company, the major Russian ballet troupe. Knowing him, Alice got to go behind the scenes of theatrical and musical performances. Knowing him also brought her into the orbit of Bloomsbury, the group of writers, artists, and intellectuals that centered at the Woolf household in Bloomsbury and had so strong an influence on literature and the arts in the early twentieth century.

One event stands out particularly in her mind. "There was an organization called *One Damn Thing after Another,* they'd organized this lecture on women writers. It was quite a good talk—but who knew then . . . ?" "A woman must have money and a room of her own if she is to write fiction," intoned Virginia Woolf—tall, magisterial, austerely beautiful—to the women of Girton. She began with a description of how, in researching the topic of women writers, she had been ordered to get out of the library, to get off the grass, because she was not a Fellow of the College. This lecture was the beginning of *A Room of One's Own,* the literary-critical work that became inspiration to generations of women writers and critics.[4]

"I remember reading *The Waste Land* with T. S. Eliot, but then everybody seemed to be doing that. There were lectures by E. M. Forster and Gertrude Stein. I found that modern poets spoke to me, Auden especially." Asked if she felt a need to downplay her intelligence in order to have a social life, Alice replied, "well, no, but then I don't think I was all that intellectual—I was enjoying myself too much."

"We had punting races—I was quite keen on that. We did Scottish dancing—I'd been to school in Scotland and had a lot of Scottish friends. One member of our group could play the bagpipes, and there was a famous occasion where we were dancing in the court of Trinity College and were told to pipe down, as it were—we made a terrible racket. Officially, there were strict rules, but we were busy breaking them. Of course boys weren't allowed in your room in Girton, unless you had permission, in which case you were allowed to take tea in the presence of a chaperone. There were iron bars on the windows, but there were known places where if you were slim enough you could get through, and the owners of those rooms would allow you access at night.

"We lived dangerously, as people of that age did."

There were many of her fellow students who were to become well known: Henri Cartier-Bresson, who became one of the great photographers of the century; Michael Redgrave, one of the great Shakespearean actors; Malcolm Lowry, author of the twentieth-century classic *Under the Volcano;* Jacob Bronowski, the mathematician and humanist; several of the Trevelyans; Robin Darwin; Julian Bell, nephew to Virginia Woolf; Alistair Cooke, the journalist who became famous for his weekly radio talk, *Letter from America,* and for introducing the weekly episodes of *Masterpiece Theater;* Anthony Blunt, who became notorious as a spy. There was William Empson, the hub of the literary group, whom Alice would come to know very well.

"Of my good friends, not one was reading medicine, and I think it was really rather fortunate. My leisure time was spent with people who were reading English and had nothing to do with medicine. My feeling so angry about that first lecture gave me a real stimulus to look elsewhere for friends."

There was Kathleen Raine, the poet and critic who wrote of Cambridge in her memoirs: "My life in college seemed like a dream. Cambridge seemed to me paradise." It was a sense that Alice shared. But Raine wrote as one outside the charmed inner circle of "those beautiful well-groomed young women from Cheltenham and St. Leonards who all seemed to have come up already knowing one another, or with friends in common, with brothers and cousins in the men's colleges—they were of another race. They were merely continuing to live within a world which was already theirs; no metamorphosis was demanded of them. From afar I admired these proud creatures who came and went with ease and assurance. I would gladly have resembled them, but I did not; they knew, and I knew."[5]

Alice may have looked like an insider, with her St. Leonards friends and family connections, but she never felt like one: "There I was without a proper pair of stockings! No, I was hardly at the center of all this, I was just an ignorant medical student." She occupied the peculiar insider-outsider position that she was to keep throughout her life.

The Other Cambridge: Cavendish Laboratory

But something else was going on in Cambridge in those days, something that was to transform Alice's life in ways she could not have imagined. Scientists at the Cavendish Laboratory, the world center of experimental physics—a stone's throw from King's College, where the medical students hung out—were revolutionizing the structure of modern science. Their discoveries would lead, within the next two decades, to the making of the atomic bomb.

Ernest Rutherford, a Nobel–Prize-winning nuclear physicist, "the Newton of atomic physics," was formulating the modern conception of atomic structure.[6] Working with him at the Cavendish were Neils Bohr, "the Socrates of atomic science," and James Chadwick, along with other "English, American, German, Russians, who were in atomic physics at the time when the search was hottest," writes C. P. Snow, physicist turned novelist, in his novel *The Search*. "Research in Cambridge was on a different scale from anything I had seen. There were more great figures than junior lecturers in London. We were all asking ourselves: how soon are we going to be able to disintegrate atoms as we wish?"[7]

Rutherford described each atom as a tiny solar system, consisting of a small, heavy nucleus made up of a tightly bound bunch of protons and neutrons, surrounded by orbiting electrons. He found that some atoms are naturally unstable. These radionuclides, as they are called, are constantly disintegrating, and as they disintegrate, they give off energy. Uranium gives off several different types of radiation, which he named alpha, beta, and gamma. Alpha and beta are high-speed particles that behave much like x-rays. In 1919 he discovered that when alpha particles bombarded nitrogen nuclei, energetic protons were released; he had produced the first man-made nuclear transformation.

Rutherford was laying the groundwork for the development of nuclear power, though he did not predict this: in fact he did not believe it would be possible to harness the energy from splitting the atom, and he announced at a lecture in 1933 that "anyone who expects a source of power from the transformation of the atom is talking moonshine." It happened that living in a hotel in Bloomsbury, recently fled Germany, was Leo Szilard, a Hungarian-born physicist. Szilard read an account of Rutherford's talk in the *Times* and set out on a walk to think. "I remember that I stopped for a red light . . . I was pondering whether Lord Rutherford might be proved wrong," and in that moment he had a sudden insight into how a nuclear chain reaction might be sustained.[8]

"At the time I was there, this work was at its height—it was all going on, unbeknownst to me," reflects Alice. Robert Oppenheimer was there between 1921 and 1926. "We must have passed in the street."

Deft at Diagnosis

"I took as a matter of course that I had to choose a profession. I assumed a natural progression from leaving school to entering some profession, and it never occurred to me there was anything unusual about a woman being a doctor because there was my mum. Medicine was the family trade.

"But when I first got to Cambridge, my sister Jean was going into medicine and I thought, no, no, this is too dull, I must break away from the family tradition. I decided I would teach science—I thought I'd better do science since I wasn't any good at English. But Father said I should sit the medical exams, just in case I changed my mind, so I got those exams behind me my first year, and I was very glad I had; because later, when Jean decided to get married, I decided to do medicine. Marriage, you see, was a death sentence to a career."

Alice finished the Natural Sciences Tripos, consisting of chemistry, physiology, and anatomy, in 1928. She got a second ("not particularly distinguished") and stayed on an extra year to finish part II of the Science Tripos, specializing in comparative anatomy (this was the equivalent to a B.S. and M.S. in natural sciences). "I stayed four years at Cambridge, though I only needed three—I was having a marvelous time."

Alice was drawn to medicine less by the desire to heal than by an interest in science. "I wasn't attracted to it because I wanted to do good works or save the world but because it presented really interesting problems."

She had originally intended to do her residency in Sheffield, but her father discovered that the staff at Sheffield Hospital had voted against having women staff members, so she would have been allowed to study there but not do a residency. Ernest decided this was no place for a woman, and though it would have cost him far less if she'd lived at home, he came up with the necessary funds to send her to London.

It was as a clinical student at the Royal Free Hospital that Alice hit her stride. "This was the first time I became noticeably good at anything—clinical medicine seemed to bring out my special capabilities. I had been running along just above average, doing well but nothing exceptional, but this was the first time I put my head above the parapet: I found out I had what they call *green fingers.* I was especially intrigued by diagnosis and found I was rather good at it. In the final year there were prizes for the exams and I think I was the only one to take three or four. This turned out to be important later, because when I did get married, marriage took a huge chunk out of my career. But when I tried to get back into mainstream medicine there was this record people could point to."

The Royal Free was the hospital her mother had found so grim, and it was still the only teaching hospital in London for women. Alice was not as horrified by it as Lucy had been, though she was fairly shocked. She'd had the impression that her family was poor—"I can hear Mother muttering, 'we're very poor, very poor' "—but now she saw poverty.

"I didn't think I'd lived a sheltered life—I'd been brought up in Sheffield and I knew very well what these smart boys from Eton and Cambridge didn't know. I thought I knew about poverty, but I hadn't seen anything. You'd be in a basement, a girl on the floor giving birth—she'd have run away from home in Scotland or Ireland, because she was ashamed to let anybody know. She'd be alone, nothing but a dirty rag on the floor. Or you'd go up a horrible crowded staircase in the East End, lit by one skylight window way up in the highest story. The dust would be inches thick, you'd find a room with an entire family living in it, one water tap, no heating, just one bed. I had no idea how bad it was. It gave me very sharp, pro-Labour opinions. I've always wondered how anyone could do clinical medicine in the thirties and be a Conservative."

She found to her surprise that she loved clinical work and developed a new seriousness about her career. "Diagnosis is fascinating. Here's a patient in a bed complaining of something, what on earth is causing it? It calls into play your powers of observation, your ability to read signs, pick out from a history. It was a good time to be doing this—there was less by the way of tests and computerized help; it put more on your ingenuity. And I obviously had an appetite for it. I was powerless to cure people in most cases, but that didn't interfere with my passion for solving problems.

"I became quite famous for my bedside teaching. It was said that I had my mother's gift. I don't know if it was that or that I simply drew people's attention to the interest of the problems, trying to make a joke now and then. This is a good thing, even in the grimmest of situations, to make people laugh—it helps them remember. I think if you met any student of mine from those days they'd say I filled an important role there because I remember they came after me, wanting special coaching and all sorts of things."

Affairs of the Heart

But meanwhile she'd gone and got herself engaged. Alice met her future husband, Ludovick Stewart, through his sister Katharine, who was in the same year as Alice. He had been at Eton and was in the class behind her at Cambridge.

Alice isn't quite sure how she came to be engaged. "It was partly about Cambridge. Cambridge broadened the world for me, in an extraordinary way, and the Stewarts were a big part of this." They were a prominent Cambridge family, friends of the Darwins. Ludovick's father was a well-known French scholar and the dean of Trinity College, and his mother, Jessie Stewart, was a classicist and first-generation woman scholar at Newnham.[9] "I'd been taken in by the family and had this feeling of being approved of. Ludovick was slim, dark, very attractive. He was very talented musically, could improvise anything at the piano. He had a charming way about him, a funny dry humor—he was a great asset to any party. I remember a lovely last summer there, lots of parties and punting parties and madrigals, a Trinity ball. It was a kind of magical whirl, all part of this wonderful new world."

Alice and Ludovick knew they would have to wait until Alice completed her clinical work in London before they could marry. The engagement dragged on, but the more interested she became in her work, the more she realized how much she would be giving up. "Married women were not highly eligible for jobs. I nearly broke it off several times."

Meanwhile, she met William Empson.

Empson was a year ahead of Alice at Cambridge, though they were the same age. He was a magnetic presence, at the center of the Cambridge literary scene. "He was much grander than me," she recalls, and he was on his way to becoming grander still, to become one of the most dashing and controversial literary figures of the century.

Empson was born into the landed gentry, into an old Yorkshire family, the youngest of five children. He came to Cambridge in 1925, where he made a brilliant career. He was a published poet; he won a first in the Math Tripos in 1926 and a starred first in the English Tripos in 1929.

Julian Trevelyan recalls that "by far the most brilliant member of [our] group was William Empson, whom we all, to some degree, worshipped."[10] Kathleen Raine describes meeting him: "I remember the impression he made upon me—as upon all of us—of contained mental energy. . . . This impression of perpetual self-consuming mental intensity produced a kind of shock; through no intention or will to impress, for William was simply himself at all times . . . but in any company William was the one remembered. Never, I think, had he any wish to excel, lead, dominate, involve, or otherwise assert power; he was at all times, on the contrary, mild, impersonal, indifferent to the impression to the point of absent-mindedness. Nevertheless his presence spellbound us all. His shapely head, his fine features, his eyes, full lustrous poet's eyes but shortsighted behind glasses and nervously evading a direct look . . . was the head, in any gathering, that seemed the focus of all eyes."[11]

In 1929, Empson was elected a research fellow at Magdalen. He had nearly finished turning his dazzling undergraduate thesis into *Seven Types of Ambiguity,* which would become a classic of the so-called New Criticism, the critical approach that would define the terms of literary discourse for the next several decades. "One of the most influential works of modern criticism was produced in a fortnight by an undergraduate," wrote a colleague.[12] On the strength of it, he'd been offered a postgraduate job. "The Empson cultus is ubiquitous," wrote a contemporary.[13]

But when his bedmaker, Mrs. Tingey, reported to the college authorities that she had discovered "various birth control mechanisms" among his belongings and that he had been in the habit of having a woman in his rooms late in the evening, the college's governing body gave him twenty-four hours to leave town. Empson's spectacular success seems to have inspired animosity; his mentor, I. A. Richards, was away in China, and he was left without defenders. Empson handed in his resignation and left, "feeling very puzzled about my future," as he wrote, wondering whether to join the Civil Service or "live in a garret, as I should like to do . . . and take some small journalistic jobs, if I can get them."

Alice found Empson electrifying. They spent a whirlwind few weeks together. Knowing him was a revelation—"He taught me to see things differently, to look at the world in a new way." But she cut off the affair because "I found it too dangerous." It was 1929, she was twenty-three, and it was risky to be this deeply involved.

She ended it on account of what seems to have been a misunderstanding. "He walked by me in the street and I thought he'd cut me—I think now he just didn't see me because he was terribly shortsighted and was always thinking about other things. I resolved to end it then. It all seemed—too complicated. I suppose I was afraid, and then too there was my engagement.

"I didn't know what to think," she concludes. "I was a stupidly ignorant girl."

Their friendship would resume in 1953.

"I was very reluctant, very slow to get married. I hadn't anticipated how much I'd care about my work—and I could see it all about to be snatched from me. And I thought, what on earth am I doing, what have I committed myself to? Marriage put tremendous demands on a woman in those days; it simply wasn't compatible with a career. I wasn't seeing much of Ludovick—he was teaching at a posh public school in Uppingham. And then the thing with Empson showed I was thoroughly muddled.

"There were many moments when I wished I could have unstuck the whole thing, and in fact I came near to it. I'd got to the point where I'd decided to break it off, and I really think I would have, but then Ludovick got the sack from Uppingham and I hadn't the heart to. I think if he hadn't, I would have—but it would have meant he would have to go home and say I've lost my job and I've lost my girlfriend too, and it was too much."

Alice decided she could make a go of it, if she set her mind to it.

Chapter 4

Marriage, Motherhood, Medical Practice: Through the War Years

"You could say I got into Social Medicine because of the war—it was an accident of war."

Alice Naish and Ludovick Stewart were married in June 1933 and moved to Manchester, where Ludovick became master at Manchester Grammar School. "It was a very good job—Manchester was said to be the only school in England to rival Eton. So I upped stakes to follow him."

Alice feared that this move was going to put an end to her career. She found a position as a locum—a stand-in or substitute—for a doctor in Tyldesley, a nearby cotton town. "I can remember waking in the morning to the clatter of the women's clogs going down the street. My most vivid memory was visiting the poor people's houses—how the smell would come at me out of the letter boxes. I had no idea I was pregnant at the time, and this gave me a very heightened sense of smell." (Anne was born in May 1934.)

Their time in Manchester was limited, however, for Ludovick was offered a post as French master at Harrow, one of England's finest public schools, located just outside London. Alice was delighted—the move back to the London area meant the possibility of doing real work again. "Well, it wasn't *real* work, it was assisting in an experiment at the London School of Hygiene and Tropical Medicine. It was feeding the monkeys, actually."

Harrow

The house she and Ludovick moved to, Kennett House, was an old mansion with high-ceilinged rooms and an elegant staircase leading down to an enormous hall. It was at the top of a hill and had French doors that opened out on a large garden and a lawn sloping down to a lake. They shared the house with two other masters and began using the hall and garden for musical and dramatic performances. They did orchestral performances and productions of Shakespeare's *Tempest* and *Twelfth Night* and Milton's *Comus*.

"It was an idyllic setting and an idyllic time. Ludovick was delighted with these performances; he was excellent at them. He'd always wanted to do music, and he would have done—he was a very good musician—except that his father had talked him out of it, thinking he'd never make a living. Our friends came from London and Cambridge to join in—it was a rather unusual mixture of medical people with people in the arts. Things jelled beautifully for us, for two or three years."

Kennett House became a center of activity for the school and community. Alice combined working at the London School of Hygiene and Tropical Medicine with running the household. Though she had the help of a cook and a nurse, the management of this large, complex household fell to her. Dinner parties in those days demanded tails and long dresses; people still paid formal visits, leaving visiting cards in silver trays. (On one occasion Mrs. Oldroyd, the cook, answered the door holding a colander in her hand, and the caller put the card into it, mistaking it for the tray.) When Alice returned from work in the evening, she'd find dozens of calling cards awaiting her. Courtesy required that she return the visits, so she hit on the idea of returning them the day of the rugby match, when she knew she'd find no one at home.

One of the master's wives called Alice the luckiest woman in Harrow because she had managed to break through "all that." Alice recalls being "the only wife connected with the school who had a job to go to, though I think if it had been anything but medicine, I'd have been run out of town." She also recalls driving to London in "a little open motor car but being ashamed because nobody in that area had a car, so I started commuting by train. I did a lot of commuting by train.

"It was all in all a very interesting period. Life was full of excitement. It was extraordinary, really, to combine this life with medicine—though I did very nearly derail my career. It was a happy period for Ludovick. He was very musical, very inventive—he had a knack for getting people to give their best, while convincing them it was their own idea. My son Hughie was born in February 1937. He and Anne grew up in this background. But I can remember one night talking to Ronnie Watkins, one of the teachers there, about how the war would soon come and break it all up. . . . " (Ronnie Watkins was to become famous for organizing the restoration of Shakespeare's Globe Theater. Alice would come across his name years later when she was in Denver, doing work around the Rocky Flats nuclear facility.)

Back to Work

Alice's real concern at this time was that her career was going nowhere. "If Ludovick hadn't taken a position so close to London I'd never have been able to get back into medicine at all, but it wasn't much of a job." She was helping a biochemist friend at the London School of Hygiene and Tropical Medicine do research into anemias and blood diseases. It was Alice's task to take blood samples, keep the records—and feed the monkeys.

At about this time there came an opening for Registrar at the Royal Free Hospital, London. This was 1935, and the Royal Free was still one of few hospitals that admitted women, so it was a rare opportunity and a position with considerable responsibility. The Registrar keeps the hospital records and notes, besides doing medicine and teaching.

Alice applied and, on the basis of the distinctions she'd won during residency, she got the job. "There I was, twenty-nine, with a marvelous job. Being a Registrar at a teaching hospital put me right on the road to becoming a Consultant. But I was at a disadvantage in that I hadn't sat the Membership of the Royal College of Physicians exam." This exam was still, as in her father's day, the basic requirement for being a Consultant, and it still, as in his day, had only a 20 percent pass rate. So she set about preparing for it. She was struck by how much more difficult it was to study now that she had children.

She sat the exam in January 1936. After it was over, candidates assembled in the hall at the College of Physicians, which was located in Trafalgar Square, to be told who had passed. First the names of those who had passed were read out, then the names of those who had failed. Alice's name was in neither category, so she sat and waited, wondering if this meant she was a borderline case. But it turned out that the delay—which went on for a very long time—was because the committee was translating the Latin document signifying a pass into the feminine gender. The committee needed to change *magister* into *magistra*.

Alice had passed on the first try. "I was in the right job for it—day in and day out I was doing clinical medicine, getting marvelous experience. I got that job and it was a turning point. I got back into mainstream medicine and was on my way to becoming a Consultant Physician at a teaching hospital, which was very rare for a woman. Women were getting into medicine as obstetricians, but to be a Consultant Physician was to get into the elite. There were very few women Consultant Physicians in those days, fewer than ten, and virtually none in a London teaching hospital."

Alice remained at the Royal Free for three years. It was a part-time appointment, but she was working full time and under enormous pressure. Besides the teaching and tending of patients there was the administrative work. "The Registrar did all the general medicine, heart, blood, bone—it all got channeled through me. I also covered for other physicians when they were away on holiday.

"I wore many hats and worked hard, but I was back with my first love, clinical medicine. I was teaching bedside medicine. I'd send one of the students ahead to attend a patient; she'd make the diagnosis and I'd come along with the rest of the students and she'd deliver it to me. The job was a gift for someone of my interests—I loved a difficult diagnosis. I got to practice diagnosis without being responsible for the cure.

"I got a bit of a reputation—if you have a diagnostic problem, take it to Stewart. (If you want prolonged treatment, find someone else!) Whether it was an inheritance from my father, who was also a very good diagnostician, I don't know. I began to share notes and experiences with him, and I think he was very pleased.

"I stayed in that job from 1936 to 1940 and learned an enormous amount. I loved having students—they'd see what was interesting in a case; they'd feed me ideas. I got no holidays and my salary was absurdly low—something like £200 a year, not enough to cover the domestic help. There was great competition for these jobs, even though a lot of people couldn't even try for them because they couldn't afford to be so badly paid. I remember getting maternity leave when Hughie was born, but I had to pay for my own replacement.

"When I'd come home in the evening, there'd be some crisis. There were not only the children, there were the dramatic and musical events. I had domestic help, but the moment I walked in the door, I was expected to take over. I'd get back after a day of work, and everyone—family, the help, other residents—ran to me with their problems. They never did that to my husband."

When an appointment as Consultant at the Elizabeth Garrett Anderson Hospital came through in 1939, she started her new job in the hospital while finishing the Registrarship at the Royal Free—which meant that for a time, she was doing two full-time jobs.

She took the new position in September. September 1939.

Alice remembers the newspaper headline, "German army sent to Polish Corridor." Such were the distractions of these days that what went through her head was, " 'isn't that absurd of the Germans, to be so tidy at this moment'—I had this vision of the German army sweeping up the corridor, and then I suddenly realized what I'd done. Fortunately I didn't say anything out loud—the whole thing was so serious that it would have been unthinkable to think anything like this."

The War Years

Harrow was not evacuated, nor was it far enough from London to receive evacuees, so except for the changes in staff when the younger masters joined up, the school underwent few disturbances. But, as Ronnie Watkins had predicted, the war put an end to that life of culture and elegance. For a time the musical and dramatic performances went on, but they too came to a close. Kennett House was soon to burn down.

Ludovick was sent to Bletchley for work on decoding. Early in the war, the Government Code and Cypher School was moved from London to an old country house in Bletchley, where experts from all over England set to work breaking Germany's military code.[1] They succeeded, and on account of their work, Churchill had information about the military and naval maneuvers of the Nazis. Ludovick's work made it an interesting war for him, though he was never able to talk about it with his family. He and Alice began to drift apart.

For Alice, also, it was an interesting war.

The outbreak of World War II found her as a Consultant at the Elizabeth Garrett Anderson Hospital in London. Medical facilities in London had been reorganized in anticipation of bombing—"Everyone thought there would be air raids at the outset, so the major London hospitals were emptied out in readiness for the casualties, and the women staff were sent away to the outlying hospitals. The staff of Garrett Anderson was transferred to a Poor Law hospital called Oster House, a low-grade hospital in St. Albans, twenty miles to the north of London. Oster House was full—teeming, in fact—not only with patients who'd been there already but with others who'd been transferred from the London hospitals. Meanwhile all the big Consultants sat in the main London hospitals twiddling their thumbs while nothing happened, while the women who'd been transferred to Oster House got all the real medicine.

"Into this hospital poured the most extraordinary, marvelous medical problems. At first, it was just London being ill as usual, though we did later get the air raids, and I'd see these extraordinary cases brought in mornings after the bombings. I had better medical experience there than before or since.

"It was a hive of interest. I tipped off some of the students I knew that this was the place to be if they wanted to get on with their education."

Among those she tipped off were Honor Smith and Lavinia Allington. Honor was the daughter of Lord Bicester, who founded what is today National Westminster Bank. She was a redheaded, fast-driving, high-spirited, witty woman who loved steeplechasing and aspired to winning the Grand National on one of the family's horses. (Dick Francis was one of her father's jockeys.) Lavinia was Honor's cousin; she did medicine after "reading the Greats." Others soon followed Alice to Oster House—Ruth Bowden, Eleanor Davis Jones—who were to become distinguished practitioners in their fields.

Honor and Lavinia were older than the other medical students and became close friends with Alice, who was only two years their senior. Alice helped Honor to qualify and got to know her family well. "Through her I became close with a level of society that I never would have known otherwise. She lived in a house where a liveried footman served lunch. There was a tennis court and a cricket pitch, horses, and a private petrol pump. There were houses in town and country, full of priceless artifacts. When Honor traveled to America she stayed with J. P. Morgan. It was fascinating."

When Alice's children were evacuated to Cambridge to stay with Ludovick's parents, she set up household with Honor and Lavinia in St. Albans. "We formed a sort of camaraderie." It was said of this place,

There was once a wonderful flat
the centre of this and that,
a lot of high thinking,
a lot of hard drinking,
a doctor, a dog and a cat.

"We were having, professionally speaking, the time of our lives. It would never have happened if it hadn't been for the war. I was earning the enormous sum of £800 a year—quite a boost over the £200 I'd been making. To me, this was wealth untold."

Serendipity: To Oxford and Social Medicine

"Then another opportunity presented itself. It was 1941, and I happened to meet an old colleague, Molly Newhouse, at a sherry party in London. She was doing a residency at the Radcliffe Infirmary in Oxford, and she said they were in urgent need of a replacement in Oxford because somebody had fallen ill. I was just due for a leave and said, 'well, I can come and do it instead of having a holiday.' I thought springtime in Oxford might be pleasant and that the children might like it.

"So I packed myself and two small children off to Oxford, and it was a disaster. The weather was bitter cold, it never stopped raining and there was nobody to look after the children. I'd engaged someone I'd been told was 'very reliable' who never turned up, and I had to put them in school because there was no other means of looking after them. They hated the school, and in desperation I had to send them back to St. Albans, to stay with my friends, while I finished my stint at the Radcliffe.

"I came away and was rather thankful it was over, thinking that was the end of that. But it turns out I had made a good impression on the professor there, Dr. Leslie Witts, and he wrote me a bit later and said that his assistant had been called up and would I come and work for him for the duration—meaning the war. It was the sort of job that was normally never open to women—to be a senior assistant in the Nuffield Hospital in Oxford. It was rather a plum."

The Nuffield Department of Clinical Medicine had become a teaching hospital of some importance. Oxford was flooded with more work than it could accommodate, on account of the war, and it had had to expand its clinical facilities. Alice suddenly found herself with the necessary qualifications for the job: she couldn't be called up for service. "A man could be called up or a woman without children could be called up, but a woman with children could not. So there I was, with two small children, and suddenly in demand.

"The second time I went to Oxford, I returned in rather better order—the children were in school and the mother of the landlady took it for granted that she would help me look after them. Once the war came, we got active support from everybody. The whole of society was on our side—'don't worry, we'll help you look after your children.' It was quite astonishing. The war enabled me to leap over barriers that would otherwise have blocked my way as a woman. It tells you what women could do if society would change its attitude.

"I was given the job of looking after patients and teaching, and I was thoroughly enjoying it. Leslie Witts was bright, young, successful, and ran an excellent department. But suddenly it was announced that the person whose place I had taken had come back again—he'd been found to be ineligible for service, and I was told more or less politely that I'd have to go back to where I'd come from, which I really couldn't do because somebody had taken my place and my return would have knocked her out.

"But it turned out there was one research project that I might do. It concerned an outbreak of anemia, liver disease, and jaundice in some shell-filling factories, involving TNT. The upshot is—this is how I came to switch from general medicine to epidemiology, or Social Medicine as it was then called.

"So you could say I got into Social Medicine because of the war—it was an accident of war. I never would have got such a position otherwise. And the project I was put on was the sort that could only crop up during the war."

Alice's War Work: TNT and Poison Gas

Alice's boss, Dr. Witts, had been approached by the Medical Research Council to investigate a problem in the munitions industry. (The Medical Research Council, or MRC, is, the major source of funding for medical research in the United Kingdom, occupying a position like that of the National Institutes of Health in the United States.)

"What had happened was this: during the first World War there had been an outbreak of fatal jaundice and aplastic anemia among people filling shells with TNT. There had been some deaths, and people were worried they might have a massive outbreak on their hands. There had nearly been strikes and demonstrations—it had threatened the armaments industry at a very crucial point in the war. But the interest stopped when the war ended, so now they had this problem on their hands the second time round because they hadn't solved it. They wanted to know what would be the health effects of reviving this industry. I was to assess the risk of developing jaundice or aplastic anemia from filling shells with TNT.

"I had, of course, never heard of any of this. I pulled out all the books I could about TNT poisoning, but it was a subject right outside my experience. I had difficulty even understanding the chemical formula for trinitrotoluene.

"So I decided the best way to tackle this was to go to the shell-filling place myself, since nobody seemed to know what was going on. It was assumed that I'd go about solving the problem in the laboratory way, but instead of heading for the laboratory, I headed for the factory. I went on a tour of the factory, and I spotted straightaway that you could never get an experiment going from this. There wasn't a hope of setting up a standard from the men filling the shells—they were the ragtag of the population who hadn't qualified for the ordinary call-up. I decided I could get a much better grip on it if I went and filled some shells myself. From there it was a short step to saying, 'well, if I go, I'd better have some medical tests done on me before and after exposure.' Then I hit on the idea of taking volunteers with me, doing the necessary tests on them, and getting some proper control data.

"We had no problem finding volunteers—it was early in the war and everyone was eager to pitch in. And I—I took advantage of the situation. I recruited a batch of healthy undergraduates. The only problem was the shortage of time because they had to get back to college in October.

"So there I was in charge of this sort of human experiment, with forty-odd undergraduates. Students came to Oxford for tests, and we all set off to Kirby, where they were monitored for blood tests and diet. We got the project going even before the Medical Research Council money came through for it. Our volunteers were to work for at least four weeks in a job where they were required to use TNT. They were to live in the workmen's hostels and be treated exactly like the workers."

The problem Alice needed to solve was whether the response to TNT was idiosyncratic or dose related: was this an idiosyncratic effect that required the exclusion of certain types of people from the industry, or was it a dose effect?

"And we did it—within a short time we were able to say that it was dose related and identify the dangerous places in the industry.[2] The students were marvelous. They submitted to a lot of screening tests, blood counts, and even a sternum puncture to look inside the marrow and see if damage was being done. And sure enough, there was damage. They came out of the experience with slightly lower red blood counts than what they went in with—the hemoglobin level fell off. They were still within normal, but we realized we had an agent capable of destroying marrow cells on our hands. They were young and healthy and quickly overcame it, with no ill effects."

In the course of the experiment, Alice had to devise projects to keep the students occupied. "While they were in the factory they'd get rather bored and tend to slip away, and I wanted to give them an incentive to stay around. So I offered a prize for the best essay on their experience of factory life. They were delighted with this—it's just the sort of thing students like—and a lot of them wrote essays. A historian friend of mine judged them and wrote an article of her own. We later combined the essays and published them as a book, *Impressions of Factory Life*. It was a great success.

"In the meantime, the man who was supposed to come back for his position at Oxford hadn't come back, so instead of losing my job, I had two. I had to do the teaching and clinical medicine and I had to run this study. And having completed this study, I was asked to do another study, and another—I was acquiring a bit of a reputation. And I ran through about three occupational surveys."

Alice was doing epidemiology before it existed as an academic discipline. Little did she imagine where this experience would lead her: the questions she was asking of shell-filling workers, she would later ask of U.S. nuclear weapons workers on a much larger scale.

The second study was equally successful, and it too was related to the war. It was an investigation of a factory that produced gasproof clothing for the army.

"The threat of the use of poison gas in the second World War had led to the idea that you could impregnate clothes with antigas protection. The chemical used involved carbon tetrachloride, a substance which was liquid at room temperature but very volatile. Workers complained of nausea and lethargy and the labor turnover was very high—in fact, it was worse than in any other industry. But there was pressure to get this work done, so the MRC initiated the inquiry. It was, like the first study, classified as secret—we weren't allowed to publish it until after the end of the war. I have no idea why.

"There was a factory in Yorkshire. The workers would arrive on Monday feeling fit, but as the week wore on they would become increasingly ill, and often by Friday, they were vomiting. But then they'd go off and get a bit of air for the weekend and come back feeling better. So I was called in to solve this. We wanted to know what it was that was making them sick." Alice realized that these symptoms were like those induced by anesthesia, which similarly creates sickness and dopiness and then dissipates as the dose wears off. "When I heard all these symptoms, I was reminded—carbon tetrachloride was known to be closely related to chloroform, which was once used as an anesthetic but was found to be too dangerous. The work was being done in a closed space because of the blackout restrictions, and there was a fair amount of evaporation and the workers were getting a big dose."

The results confirmed her suspicions.

One of the more challenging moments in this study involved an encounter with a deaf radiologist in a dark room. "Somehow he'd got it into his head that we were stealing his data or doing him down—at any rate, there was some misunderstanding, and I caught the full blast of it. I blundered into the room where he was working, which was kept completely dark because radiologists need to have dark, and he began shouting at me, ordering me to leave—and I'd gladly have left, but I couldn't find the door and he couldn't hear my appeals for light because he was stone deaf. So I had to stand my ground while he heaped abuse on me. Finally someone turned on the lights, and I managed to quieten him down.

"On the way out I remember an assistant muttering to me, 'I have always heard that genius is an infinite capacity for taking pains, but I now see it is an infinite capacity for receiving pains.' I think we could have stood for a Marx Brothers routine. But I was there to make a go of things, to see things through."

The investigation became a landmark for Alice. "It was very dramatic from my point of view because Witts arranged for me to present the findings to the Association of Physicians, a prestigious organization indeed. They thought it was absolutely wonderful—it was a whodunit, you see, and I didn't lay upon it the dead hand of industrial medicine, which is usually very dull. I managed to turn it into an interesting detective story. And it had a sort of favorable message, unlike my poor radiation stories—I was able to say that probably the doses were too low for kidney and liver damage. I gave the whole talk without once looking at my notes. I think it was that more than anything else that got me elected as a fellow of the Royal College of Physicians. I'd been a member, but now I was made a fellow." It was 1946 and Alice was the ninth woman fellow, and the first under forty, to attain this distinction. More than this, she was the first woman to be elected to the Association of Physicians, the crème de la crème of the British medical profession.

The study was published in the first volume of the new *British Journal of Industrial Medicine,* of which Alice was a founding member.[3]

Alice's next assignment was to set up a study of pneumoconiosis among coal miners in Wales. The Medical Research Council, under the influence of the newly elected Labour government after the war, was encouraging research into occupational health problems of workers, and one of the groups that needed urgent attention was the Welsh coal miners. She was assigned to follow up those workers who had left the mines with sickness compensation to discover whether the often fatal disease, pneumoconiosis, continued to progress after they ceased breathing the dust.

She loved working with the miners and had a keen sense of the injustices they had suffered—"they were dead right, they'd been a much neglected group, they'd been very badly used."[4] She worked with Arthur Horner, the head of the union and an ardent Communist, to set up meetings with the miners. "Whenever I'd address them in groups I'd be aware of the unusual amount of coughing in the audience." She was appalled to discover that those dying between ages fifty-five and fifty-nine were thought to have "succumbed to old age." (When Alice came across one miner who had, at age eighty-two, died of drink she commented that this might seem to be an excellent advertisement for drink.)

"I got direct experience of what it means to set up a survey and picked up a bit of courage that would help me later on. I was acquiring a reputation for having a good nose for this sort of thing. That's how I came to be a part of the new department of Social Medicine when it was founded."

Family Matters

The end of the war brought marital problems. Ludovick went back to Harrow to teach French. Alice in the meantime had settled in Oxford and suggested that he come and look for a job there. There was an opening for a French don at Corpus Christi, one of the smaller colleges, and with his Cambridge degree and their connections, he'd have had a good crack at it. But somehow he could not be persuaded to apply—he did not want to come to Oxford and he did not want to teach French.

He wanted to pursue his interest in music. He had gone into French because his father advised him he'd never make a living in music, but after the war, there were new opportunities in music. Orchestras became part of schooling everywhere, and they required conductors. A position came up in Cambridge and Ludovick took it.

Since he wouldn't come to Oxford, Alice made an effort to get to Cambridge. She approached the newly appointed head of Social Medicine at Cambridge, applied for a position in his department, and got "very badly snubbed." By this time, she was involved in the study that would become the Oxford Survey of Childhood Cancer, and she would have been a credit to any department, but the prejudice against women was strong. As she made efforts to find employment in other departments that might bring her closer to Ludovick, she had a strong instinct to keep hold of the survey—"I was right to hang on to it, I knew it was important. There was no way I could go back to being a housewife just because the war was over."

Ludovick went to live with his mother in Cambridge and became involved with a woman he met during the war. "He'd been surrounded by adoring young women at Bletchley, and of this one he said, 'she needs me more than you do—there's so much more I can do for her than for you.' " A divorce in 1952–53 gave Alice custody of the children. "They were left free to visit their father, and he was left free from contributing to their upkeep," Alice comments. She became their sole support. She felt, at times bitterly, that her husband was looked after by a mother and a wife while she was left to look after herself and the children.

Ludovick wanted to send Hughie to Harrow, the posh public school where he had taught, but Alice had reservations. She was disturbed by certain weaknesses in the boy. Hughie had enormous charm but was well aware of its manipulative value. Alice was worried that at Harrow he would come into contact with the sons of the rich and famous and begin to fancy himself one of them. But Ludovick favored Harrow, and Alice gave in.

At Harrow, Hughie became close friends with the son of a wealthy paper manufacturer, and the boy's father offered him a job in the firm when he graduated. But he lost patience with the work and quit. He went to McGill University in Canada, hoping to become a journalist. While working as a reporter in Aberdeen, he met Jeanette Johnston, who was studying to become a nurse, and the two of them married. Their children, Charles and Christabel, were born in Canada. The career in journalism never materialized; he became a teacher instead and seemed to be doing well, until one day he arrived back in England in the midst of a breakdown.

"There followed a very troubled time during which he cut himself off from everyone. It went on for a long while," Alice recalls; "he began by antagonizing me, then every member of the family, then finally Jeanette. He once attacked her while she was driving, which was very dangerous in a car with children. She started to threaten divorce, and he became very lost. He'd drift off to London; you never knew when he'd turn up. He drifted off on his own and one day overdosed on lithium." He died in November 1977, age forty.

The official diagnosis was schizophrenia, though Alice suspects it was manic depression, "because that is the family trait."

"I always had a conscience about him. You have to have that feeling, as a parent—you feel something went wrong that you ought to have been able to fix. Well, we probably did a lot of things wrong. You could argue that it had something to do with my being away from home. But now it became a very strong incentive for me to go on working. He'd dragged Jeanette through a terrible time and left her and their children destitute."

After his death, Alice was in a position to help Jeanette because she was able to give her the flat she'd bought for William Empson's son, Mogador. "It was a lifeline—it was not only a place she and the children could live, but a source of income, with the rentals. When Jeanette decided to return to university for more nurse training it was beautifully located, near the university. There were two years after Hughie died when I took the children, Charles and Christabel, on their holidays up to Shetland so she could continue doing her degree. During the year that she took her exams, I took the children as much as possible. Jeanette was a very good partner in this—it's not easy to be the recipient of help."

Charles and Christabel, born in May 1970 and February 1972, were near the ages of Alice's daughter Anne's two daughters, Eleanor, born in November 1970, and Harriet, born in August 1973, and Alice's cottage in Fawler provided them a haven for weekends and holidays. "Anne's children had a flying start and I was anxious that Jeanette's shouldn't feel themselves disadvantaged. They were both clever children but it was a very dicey business. They were to be given everything they wanted, according to their brains, but they were not to be pushed. Now Charles is finishing his doctorate at Oxford, and Christabel has graduated from Glasgow, and we're only just out of the woods from my point of view."

Alice's daughter Anne, in the meantime, had started medical school at the Royal Free Hospital in 1952, much to everyone's surprise because she'd always been good at English and bad at math. She then married a man ten years her junior and put him through medical school. "They were a good couple for awhile, though after seventeen years marriage, they drifted apart and he remarried a childhood friend. She's getting along fine now—you know how you can tell when people are doing all right, when they're all right in their lives."

Anne is a successful and popular neuropathologist at a hospital in Essex. She works one day a week as a general practitioner.

On Alice's eightieth birthday celebration, held on the lawn outside Fawler, all four grandchildren and several grandnephews and nieces donned wigs and costumes and put on a play called *Granny Saves the World*. The text of this masterpiece is lost, but it was about some impending nuclear disaster, with Granny trying to warn the world and no one heeding her. "Everyone was going about their usual business, bored, filing their nails," recalls Eleanor.

Charles, under the table sawing wood, represented the sound of the world filing its nails.

At Alice's ninetieth birthday party, also at Fawler, all four grandchildren were thriving. As this book goes to press, Charles has just taken a position as research associate in theoretical computer sciences at Brandeis University. Eleanor is working as a printer in London, having done a degree in history at Manchester. Christabel is doing a Masters in twentieth-century arts. Harriet did a degree in ecology at Sheffield, then completed a Masters in London at Wye, "with distinction," and is doing research on sustainable agriculture in Almeria, Spain.

Lucy Wellburn and Albert Ernest Naish met in 1901 as medical students on the wards of the Royal Free Hospital.

Lucy and Ernest married in July 1902. "The young couple had a revolutionary idea of practicing medicine together."

The Doctors Naish became pioneers in pediatrics and children's welfare in Sheffield at the turn of the century. Baby Alice is in the carriage.

Ernest's advancement to Consultant Physician required a move to a new house, consulting rooms in front, at 5 Clarkehouse Road—"Posh, as Sheffield went!"

1913. "For town children, we had a lot of country freedom."
Alice is *top left,* David is the baby.

"We were a large, rambunctious bunch." Jean, born 1903; George, 1904; Alice, 1906; Charlotte, 1908; Ernest, 1910; Anthony (David), 1912; John, 1915; Charles, 1918.
Jean, Alice, Charlotte, John, and Charles became doctors.

Four Mile Bridge, at the far end of Anglesey, off the northwest end of Wales, has been a summer retreat for the Naish and Stewart families since 1914. "A paradise for children—an inland sea with an ancient stone bridge, all protected and child size."

Alice Mary Stewart—graduated from St. Leonards, St. Andrews, Scotland—began Cambridge in 1926.

Picnics and punting on the River Cam—Cambridge "was a whole new world." (Alice is *second from right.*)

Alice's first job, as House Physician at the Royal Free Hospital, 1932. (Alice is standing, *second from right.*)
"I hadn't anticipated how much I'd care about my work."

Alice Mary Naish and Ludovick Stewart were married 17 June 1933. (Alice had kept in touch with her Scottish friends.)

Alice and Anne, born 1934

Anne and Hughie, Anglesey. Photo by Dr. Leslie Witts.

Alice at Oxford. Alice found herself much in demand on account of the war, with several jobs to choose from.

Dr. John Ryle, chair of the Institute of Social Medicine at Oxford, in 1945.

Fawler, Alice's home since 1949.

"When I'm abroad sometimes and feel homesick, I think of the evening light at Fawler. It's my idea of what England's about."

In 1952 Ernest (age eighty-one) and Lucy (age seventy-six) celebrated their golden wedding anniversary. All their children and most of their twenty-seven grandchildren attended, along with the first of their forty-five great-grandchildren (with Mickey, the dog).

Part 2
Engendering Epidemiology

Chapter 5

Changing Subjects

"I was beginning to feel that more interesting medical problems lay outside hospitals than ever came into them."

Social Medicine came into existence as an academic discipline in 1943, with the appointment of Dr. John Ryle as chair of the Institute of Social Medicine at Oxford.[1]

Ryle had been Regius Professor at Cambridge, senior physician at Guys Hospital in London, and personal physician to the king. "He would have lent luster to any enterprise," says Alice. He was tall, blond, charismatic, with a presence people described as radiant, luminous. "There was, allied to a first-class brain, unfailing courtesy, sympathy, kindness, and tolerance," wrote Dr. Leslie Witts, Alice's former boss at Oxford; "His magnetic personality was an inspiration, and every member of his staff loved him."[2] Colleagues admired him for his diagnostic skills; patients loved him for the reassurance in his manner.

"He was at the top of the profession; you could get no higher," recalls Alice; "He had achieved every possible honor and distinction. People thought he was mad to give it all up. But he turned to Social Medicine because he was dissatisfied with medicine as it was being practiced."

The idea of a department of Social Medicine at Oxford came out of the newly awakened concern for public health created by the war, the same concern that generated the National Health Service. The Civilian Medical Board records, the medical examination required of men for recruitment, had revealed widespread ill health. A 1938 report of the Registrar-General on Occupational Mortality had drawn attention to the class basis of illness. Using census returns, the report divided the population into five social and occupational classes and found startling discrepancies: infant mortality rates, diseases of the ear, mastoid, and respiratory system, gastric and duodenal ulcer, and valvular heart disease increased steadily with descending social scale, becoming approximately twice as high in the lowest class as in the highest class.[3]

There was a growing feeling that the medical profession ought to be addressing this.

In 1939, Sir Arthur MacNalty, chief medical officer to the Ministry of Health and lecturer on public health at Oxford, submitted a report to Oxford University,[4] and in 1942, the Nuffield Provincial Hospitals Trust decided to finance an institute and a chair, for which it chose Ryle. If it proved successful, the university would take it over and continue funding. Other chairs in Social Medicine followed—at Birmingham and Edinburgh in 1944, at Manchester in 1947, and at Sheffield in 1948.

Ryle hoped to rally the medical profession to a sense of mission. As his first assistant, he chose Alice Stewart.

A Man of Vision

"He went down in salary, down in status," recalls Alice; "it was considered very eccentric of him to make this move—it was very brave. He had enlightened views—he had vision. He wanted to awaken physicians to a new sense of social responsibility."

"Some of my friends have rebuked me for leaving the clinical field," Ryle wrote in 1947. "I reply that I have merely taken the necessary steps to enlarge my field of vision and to increase my opportunities for etiological study."[5]

Ryle was appalled by the disparity of sickness and health at the two ends of the social scale. "We now know from indisputable statistical evidence, as well as from the evidence of daily experience in city hospitals and slums and elsewhere, that the opportunities for health and the chances of death in the several social stratifications of the community and between occupations are very different."[6] "Before the war, the infant death rate in some of our northern industrial cities was three times that in certain suburban districts," he wrote, describing the conditions Alice's parents had worked all their lives to ameliorate.[7] No wonder his ideals struck so deep a chord in Alice.

His writings are a ringing cry for an end to social injustice: "Human justice—by the standards of vital statistics—is still a shoddy affair," he wrote in *Changing Disciplines,* the 1948 manifesto that explained why he was changing disciplines and why he felt that medicine itself needed to be changed. "We cannot in a democratic age accept that the mortality at the two ends of the socio-economic scale should be so sharply differentiated." He rallied physicians to "a more inspired sense of duty to our communities and to the race." He urged them to take a public role in the world and assume responsibility for the prevention of disease.[8]

He chose the name *Social Medicine* because he saw the new discipline as linking up with the social sciences and addressing social conditions. The term *epidemiology* was rejected because it was thought to involve only infectious diseases, the science that began with John Snow's landmark work linking cholera with contaminated water. "There was a department of epidemiology in the school of bacteriology at Oxford and Ryle did not wish to be confused with it: he was interested in diseases that have their etiology in social ills, among which he included cancer, cardiovascular conditions, rheumatic disease, certain psychoneuroses." (Alice thought the name *Social Medicine* was "an unfortunate choice, because it had too many political implications. *Survey Medicine* would have been more accurate. In America, *Social Medicine* tended to be confused with socialized medicine or with social work, and the name never caught on there. Today the term has been abandoned by nearly everyone in favor of *epidemiology,* which is the right term because epidemiology is concerned with etiology and it was etiology that Ryle was interested in.")

Ryle acknowledged that medicine had made great scientific and technical advancements but lamented that it had lost sight of "the broader social needs of the group or community." Physicians place too strong a belief in the potency of treatments and pay too little attention to the fact that many illnesses "have discoverable origins in social, domestic, or industrial maladjustment—in fatigue, economic insecurity, or dietary insufficiency."[9] Looking back on his years as a student and teacher in clinical medicine, he had seen medicine become more mechanized and myopic, more focused on treatment than prevention, on the end stages rather than the etiology of disease. "In these thirty years I have watched disease in the ward being studied more and more thoroughly through the high power of the microscope. . . . Man as a person and a member of a family and of much larger social groups, with his health and sickness intimately bound up with the conditions of his life and work . . . and economic opportunity, has been inadequately considered."[10]

Medicine had become the study of "end-result conditions," Ryle said; "with etiology—the first essential of prevention—and with prevention itself, the majority of physicians and surgeons have curiously little concern."[11] "As he put it," explains Alice, "we were tackling disease from the wrong direction—why focus on the end stage when you should be looking for the cause? This made the physician a sort of premature mortician. So much more could be accomplished, so many more lives could be saved, by going for the causes of disease. But to do that, you have to address social, economic, occupational, and educational conditions, and that's not so simple."

Ryle hoped to bridge the gap between academic medicine and public health departments and services, and to bring together the physician, the public health officer, and the practitioner, in an enterprise directed to both research and practical ends. To do that required the reorganization of existing departments and a curriculum that would combine medical research with the social sciences and with the practical aspects of medicine.[12]

"He hoped to get his finest students interested," Alice explains. "He thought they would flock to his banner. He genuinely thought that the war would make the medical profession want to change.

"He couldn't have been more wrong. All people wanted to do was return to the way things had been before the war, to business as usual. The doctors all hugged their practices in Harley Street.

"His concerns weren't shared by the doctors, and they aren't to this day. Though he did catch a few fish—he caught me. My discovery about prenatal x-rays and childhood cancer was exactly the sort of thing he envisioned. On a tiny grant we made this discovery, and it was precisely because we were able to hook up with the public health services, as he urged. Just imagine how many other discoveries are out there waiting to be pulled in if you just ask the right questions, make the right sorts of connections, tap into the right sorts of information.

"How I wish he'd lived to see it—he'd have loved it."

Alice and Social Medicine

"I don't know why Ryle wanted me as his first assistant. I suspect he couldn't get anyone else. I can't imagine I was first on his list. I had no public health training. I had no statistics, no degree in mathematics—I had to sit down and learn statistics.

"But I did have firsthand experience of how to run a survey, and I was much in demand in those days. There were all sort of positions offered me. I was asked to head a new department of Occupational Health in London Hospital, which I would have liked very much. But I was put off when a colleague told me there was a notorious bias against women there. I was quite torn, really; I would have liked industrial medicine—my work on the pneumoconiosis study had given me a taste for it. [In fact industrial medicine at the London Hospital subsequently became very technical, concentrating on the chemical nature of industrial poisons, and might have taken a different turn with Alice as head.] Or I could have returned to my old position at the Elizabeth Garrett Anderson Hospital, but that would have ultimately led me to a practice in Harley Street and all sorts of social things that I wasn't prepared for. Or I could have stayed on with Witts."

In career terms, the turn to Social Medicine was as hazardous for Alice as it had been for Ryle.

"My friends thought I'd committed professional suicide. There I was on a runaway career, being offered positions and all sorts of things that had never come the way of a woman before. My friends thought Social Medicine had something to do either with the Labour Party or old ladies. I myself wasn't sure what it was, but I was beginning to feel that more interesting medical problems lay outside hospitals than ever come into them. I got rather tired of treating patients, though I loved solving problems. I knew how to go about diagnosis in clinical medicine, there were guidelines for that, but here suddenly were all sorts of new problems for which the rules had not been worked out."

Epidemiological investigation engaged her like a piece of detective work. "I'd been a body doctor—now here was a much larger body to work on."

So Alice came to work for John Ryle at the Institute of Social Medicine at Oxford in October 1945. She "crossed the Rubicon," as she says, between clinical and Social Medicine.

Alice wasn't particularly interested in Ryle's main project, the Child Health Survey. "He wanted to chart what happened to the child from age one to school age. There was data on children to age one, collected by infant welfare clinics, and there was documentation collected by the schools on children of five to fifteen years, but very little attention had been paid to the child's development in the intermediate years. Ryle wanted to look at that. I didn't much care to."

What did grab Alice's interest was a short article in the *Times* that described a survey in Northamptonshire of workers in the shoe industry. "These mass radiography units were lent to the army during the war, and after the war, any medical officer could have one for the asking. One medical officer had done an x-ray survey of workers in the shoe industry in Northamptonshire, where the shoe industry makes up 50 percent or so of jobs, and had turned up an excess of active cases of tuberculosis. This was puzzling, since most industries that had a tuberculosis problem had other health problems as well, whereas the shoemaking industry had an unusually good health record.

"The *Times* reported on this study, I read it, went in to Ryle, and he let me have my head with it, though he wasn't mad keen on it. He was an excellent boss, the best I ever had."

The question was, why the occurrence of tuberculosis in an otherwise healthy industry? Was there some sort of selective recruitment into this trade, or something that went on in the workplace? The orthodox view was that tuberculosis was spread primarily within the family, but Alice suspected that in this instance, the workplace might be the more likely source.

Alice turned to the Civilian Medical Board records, the records of the examinations used for recruitment, to see if she could discover something about the kind of workers who became shoemakers and boot makers. "These records were, in effect, a health record for the recruitment-age population. But after the war, nobody saw any particular value in them, and they were about to be destroyed—we got to them just in the nick of time.

"We concluded that there was selective recruitment of less healthy, more susceptible people into the shoe and boot industry. People with chronic tuberculosis found it a sort of favorable industry to go into since the work is so light. Whereas in more physically challenging jobs, the man with tuberculosis will not come back to work, he is likely to come back to this industry once he's feeling better. So the old ones spread it to the youngsters. We were able to document that the flow of infection was from the workplace to the home rather than the other way around. And that was new because nobody thought it could happen.

"We could trace the spread of infection, we could chart its movement: it wasn't just people working side by side that infected one another, the infection could move around the whole room, and the bigger the workroom, the greater the risk. We were able to correlate the number of people who caught it with the number of people working in the room and the ventilation. It was ingenious, it was fascinating—I loved it."[13]

Working with the Civilian Medical Board records made Alice realize what a missed opportunity these records were. "Their only function had been to determine the health of the male population, to call up men for the war. With a little attentiveness, they could have told so much more, yet there they were, about to be destroyed. I think that they finally weren't destroyed because of our interest in them, but neither did anyone make further use of them.

"Looking at these records, we turned up some amazing things."

They found that tendency to varicose veins was related to height, that eczema was related to slight build, and that men with peptic ulcers included an undue proportion of painters, decorators, and people in sports and entertainment industries. They discovered that children in top income groups tend to get more acne. They found an association of hernias with stoutness and spinal curvature with thinness.

"We turned up quite a few one-eyed lorry drivers—looking at the medical fitness of drivers of commercial vehicles, we found some dangerous lack of declarations of drivers applying for licenses. These records were a gold mine of information.

"There was much more that could have been learned—for example, when a person was bedridden, like my poor brother David, the family would send in a certificate stating the reason why he hadn't appeared for the recruitment exam and why he was bedridden. These records should have been kept side by side with the Civilian Medical Board records, but they weren't, they were tossed aside, so we found, when we came to look at them, that we had only one side of the story—about who was fit. Nobody thought of keeping the invalids' records alongside the records of the healthy so as to complete the story. And this is because the medical profession was nowhere around to point it out. These are the sorts of things Ryle hoped to change by linking up academic research with public health services—you see the kinds of things Social Medicine might have taught us, had he succeeded."

Alice and the Social Medicine Department published a report for the Medical Research Council that pointed out how useful vital statistics could be in identifying medical problems, but it sank like a stone.[14]

The moral Alice draws from this is a ringing indictment of the medical profession. "Practicing medicine without asking these larger questions is like selling groceries across the counter. You go in with an illness; the doctor sells you a pill. It's no more responsible than that. Nobody goes out and asks, 'who didn't come in because he was too sick to come? because he couldn't afford to? Why are so many people coming in with this, and so few with that?'

"A profession is defined as an occupation which has given a sufficiently good and prolonged account of itself to be self-regulating, self-governing; but can the medical profession make this claim? The medical profession has been very irresponsible."

Ryle's Death

Ryle had known, even as he accepted the chair in Social Medicine, that his days were numbered. An attack of coronary thrombosis in 1942 had given him clear warning. In his years at Oxford, as his health deteriorated, he worked unstintingly for the cause of Social Medicine, trying to engage the medical profession in this mission. Though he did not accomplish the reforms he'd envisioned, he loved his work at the institute and thought of these years as the happiest of his life. He turned in his final years to writing poetry, even as he continued to write and publicize on behalf of Social Medicine. In the summer of 1949, he suffered a heart attack from which he never recovered. He died in February 1950, at the age of sixty-one. He was hailed widely by colleagues and friends, for his idealism, humanism, and humanitarian approach to medicine.

Alice felt his loss deeply, not only for herself but for the profession. She took over as head of the institute. "Anyone in their right mind would have fled, faced with this situation, but I felt, like Ryle, quite committed to this subject by now, quite determined. I thought it would only be a matter of time before people would see the light. I now know I was overly optimistic."

In the years since the war, the tide had turned against Social Medicine. "I thought we were doing rather nicely and were making a bit of headway, considering how new we were and how small we were. But Ryle was the one important person in the subject, and when he was gone, there was no one to defend us. The faculty was expanding in other directions and there was fierce competition: the clinical school was seeking money and so was the preclinical school ["preclinical" is all the subjects you do before getting to medicine, such as physiology and biochemistry]. The clinical and preclinical were vying with one another, and we didn't belong to either. This new subject was nobody's friend. It didn't belong to the laboratory studies and it didn't belong to the clinical studies and Ryle was dead, and what's more, the department was conveniently headed by a woman. What easier thing than to squash us out of existence.

"So it was announced that the whole thing was going to stop. They'd never have done that if Ryle had lived."

But Alice feels there were deeper reasons for the disinterest in Social Medicine. "Social Medicine is difficult. It's multidisciplinary; it requires knowing something about sociology and statistics. Medical training doesn't give you this sort of background. Social Medicine is best done in teams, physicians joining up with public health workers to help with the collection of data and with statisticians to help with the analysis of the data. But physicians don't want to work in teams—they enjoy being autonomous, they enjoy being better paid than others. Physicians in England enjoy great superiority. If you're successful as a physician, you can end up in the House of Lords—that's where some of them do end up. Why should they play second fiddle to anyone?

"Besides, why should they want to associate with public health? Public health has always been a kind of second-class citizen: as far as physicians are concerned, nobody goes into it unless they've failed in medicine. Public health is traditionally easy to get into. It asks practical questions and administers rules and regulations over local conditions."

After Ryle died, Alice wrote "A Report on the Work of the Institute of Social Medicine, Oxford 1943–1950" to defend the institute, describing its research on the relation of disease to social circumstances and showing how important it is to improve social conditions to prevent illness.[15] She described Ryle's studies of infant mortality, of disease and occupational conditions, his investigations of children's development in relation to family history and social environment, his work on the elderly.

"My defense of the new discipline was sent around to the heads of all clinical and preclinical departments," Alice explains. "A few read the report and came round, reluctant dragons though they were. But most were not interested.

"I found the records of Ryle's Child Health Study in a terrible mess, everyone with their own style of recording, and I said, 'look, we've got to get this systematized. We've got to make the information manageable. We've got to invent a way of keeping the records systemized until something better comes along.' " She knew it was only a matter of time before computers would revolutionize this sort of data keeping.

She devised a system of ledgers in which data could be entered horizontally and read both horizontally and vertically, which she called *visible tape*. "On each page there are rows and columns with a numbering system—in Ryle's Child Health Survey, the columns listed the names of the children and the rows showed birth weight, weight at six months, weight at twelve months, etc.—and so on to eight years. You have 25 rows on the page, starting with 00 and ending with 24, then on page 2, you begin at 25 and go to 49, and the next one goes 75 to 99, and so forth. So that when you turn up any page, you know you've got the same information recorded on it. It meant you could look at a page and get a quick impression, and I think I turned up quite a few ideas that way. To increase the number of columns, just add more volumes—you can keep filling the book as the information comes in, so you can go vertically and horizontally into infinity."

This system became the cornerstone of the Oxford Survey of Childhood Cancer. "When we did the Oxford Survey we were covering all of Britain, with volunteers, and we had to have a way of systematizing the data. We were at the mercy of two hundred data collecting places—records came in from all over, collected by people with different interests, social workers, radiologists, pediatricians. Our system meant you could handle the information in a routinized fashion. You could fill in information early or late in the story and begin to see what was going on. It was invaluable. It lasted for eight years—it enabled us to manage without computers for eight years."[16]

Diminished Circumstances

With Ryle's death, Social Medicine at Oxford fell on evil days. Alice, who had been at the end of the war in the position of choosing among several jobs, found herself without any job or future prospects.

"I was rung up and told I'd been sacked, and I took off with Honor Smith to her favorite pub for a drink. Then someone rang back to tell me this wasn't happening—apparently it was realized it would be a slap in the face to one of the university's great benefactors to shut down the Institute—the donor, Lord Nuffield, an automobile manufacturer who was giving a lot of money to the university, was still alive. Somebody had pointed out, 'you can't do that to a living patron.' It also happened that I had tenure, so they'd have had to find me another job at the same rank.

"But I didn't get the second message because I was at the pub. In theory I hadn't heard either story, and then came a letter to me from the Registrar of the university saying they'd decided to replace the 'Institute' with a 'Unit,' so it would be the 'Social Medicine Unit,' and that its head would bear the title of 'Reader,' whether in clinical or preclinical, it didn't say—one had a higher salary than the other. When I got a letter saying I was to have a salary of £2,000, I assumed this meant I was a reader in clinical medicine. But in fact I hadn't been appointed to either group—I'd been made a reader *ad hominem,* as an individual rather than part of a group, which meant that my salary didn't go up when other people's salaries went up. If I'd had a proper appointment in clinical or preclinical I'd have gone up on the salary scale, but this way, they could do with my salary what they liked.

"It was only long afterward that I found this out. When I needed some information for a form and sent an inquiry asking what category I was in, a nice young man rang back asking, 'may I come in and see you?' He said he'd been worried about me for some years—each time there were increments of salary they simply picked a figure out of the hat for me. I do think I could have sued them.

"But I chose to lie low. I had of course been excluded from all committees, which suited me fine. But perhaps"—and here Alice has second thoughts—"perhaps I should have looked to my own interest and seen that they paid me properly. Maybe I should have said, 'look here, you can't do this to the department, you must give it a professorship.' If I had stood up for it, if I'd got myself more clout, people might have had to pay us more attention. As things developed, they were able to ignore us."

The Institute of Social Medicine was renamed the Social Medicine Unit and Alice was made head of it, but there was nothing to head. "They had to pay my salary and they had to give us some rooms, but we had no staff, hardly enough for a half-time secretary—barely enough to light a gas fire! We lost the original building given to Ryle and were given a room where there was no space for our books. I had to put them in a medical library until our situation improved. It was impossible.

"There was no money for research, so what was I to do? The terms of the job were that I was to do such teaching as the clinical professors required, but they required nothing. If I'd been that sort of person I'd have been able to sit with my feet on the mantelpiece until I retired. I think the university hoped I'd just go away.

"So here I am in this curious position, asking, how can I make the best of it? The good thing was that I had no work to do—they had to keep paying me until retirement age, and I had nothing to do. But I thought, I can do better than that."

Chapter 6

X-Rays and Childhood Cancer

"It was almost an accident, this discovery."

"I said, 'well, let's see if we can't show the world that this subject is much more important than they think.' "

Alice's situation meant that any funding had to come from outside sources. There was fierce competition for grants, but with the help of Leslie Witts, she got Radcliffe money to hire a statistician. "I got David Hewitt—he'd got a first in statistics; he's since made quite a name for himself in Canada, working with health statistics. And I got Josephine Webb—she'd worked for the Medical Research Council, had a bachelor of medicine. David was very young, Webb was near retirement. We were a good team."

This, then, was the situation of Social Medicine at Oxford. They called themselves the "Three Musketeers." Alice still believed that the university might somehow be persuaded to come to its senses and resume support of their department.

Designing the Survey

"The three of us sat down and asked, where are we going to get money to do research? We needed to get on to a subject that would attract outside funding. And the obvious way to do that was to think of a disease which was on the increase for reasons unknown and which lent itself to epidemiological study. So we went down the list." There were four possible candidates: lung cancer, cardiovascular disease, poliomyelitis, and leukemia. Three of these were already being studied at the School of Tropical Medicine. (Lung cancer was being studied by Richard Doll and A. Bradford Hill.) That left leukemia, which was on the rise but not yet under any epidemiological investigation because it was rare and epidemiology was not thought to be sufficiently sensitive to deal with so subtle an effect. But it was becoming less rare. In 1951 it was rising at such a rate that there was talk of a leukemia epidemic.[1]

"I think one of the things that drew me to this project was the death of a child I knew, the daughter of a good friend. She was my godchild, she died of leukemia at age three, and her mother asked me to find out why—'You know, Alice, you ought to be looking at this disease,' she said.

"So we went to the official stats to see if we could find any point of entry into this subject. I had Hewitt work up the statistics and he found that leukemia was mainly affecting people over fifty, which was no surprise, and children between ages two and four, which *was* a surprise.[2] What was odd about this was that age two to four is usually fairly safe—children that age tend to be healthy because they've survived the traumas of birth and have not yet gone to school. Yet here they were, leading the band. Why?

"That did seem to be something worth going after.

"Hewitt also showed that the type of leukemia affecting the children, lymphatic, was different from the type that adults generally got, myeloid. And he showed that children were getting leukemia much more in countries with better medical care and lower death rates—England and America—than in countries like Ireland and Scotland.

"It was at this point that I suggested we propose a project.

"I said, how about if we go back to the mothers of the children who died and find out whether something had happened before birth. I got it into my head that it might be something that had happened before birth, with a latent period expressing itself that much later. Nobody took this seriously since it was assumed that the latent period for childhood leukemia was a matter of a few months. But I said, let's go interview the mothers: they might have a memory of something prenatal that the doctors might have forgotten.

"The only clever thing I did was to remember that life begins at conception, not at birth, and to frame the questionnaire in such a way that what happened before birth was featured systematically in the form of questions to the mother. Like, 'did you have any illnesses during pregnancy? What happened at birth? Were you x-rayed and, if so, why?' and so on. What I did was to devise questions that encouraged the mother to describe what had happened to her from the moment her child was conceived."

David Hewitt had the idea of setting up a control group, of studying another group of dead children, matched as far as possible by age, sex, and region. But it was difficult to find a comparable group of dead children between two and four, except in those who had died from other forms of cancer. So the study broadened out to include all types of children's cancer, not just leukemia.

The researchers drew up a proposal to implement a nationwide interviewing of mothers of children who'd died of leukemia and other forms of cancer between 1953 and 1955. Alice had been involved with the Medical Research Council (MRC) committee on leukemia, having been called in on its survey on ankylosing spondylitis, headed by William Court-Brown and Richard Doll. A Dutch radiotherapist who had been using radiotherapy to relieve the symptoms of ankylosing spondylitis, a painful condition of the spine, discovered that no fewer than five of his patients had died of or were suffering from leukemia. Meanwhile, the MRC had been requested to report to Parliament on the hazards of nuclear radiation: the Atomic Bomb Casualty Commission, which began work in 1950, had been turning up leukemia in the Hiroshima survivors. So Court-Brown and Doll were asked to look into the relation of x-rays and leukemia in ankylosing spondylitis patients.

Alice and Hewitt and Webb were helping with data collection, and so they were there on the Medical Research Council committee on leukemia, and able to put in an informal request—there was no committee on epidemiology at the time, so this would have been the logical committee to ask. They were also present to hear the responses. It is Alice's recollection that "they turned us down flat. The committee insisted that one person would have to do all the interviews, for the sake of consistency, and our survey was far too extensive for that. David Hewitt pointed out that it might be preferable to have more than one interviewer, as a check against bias, as long as you had the same person doing both the case and control interviews. But the committee would have none of it."[3]

Alice hypothesizes that a major prejudice against her study would have been that it was *retrospective,* beginning from the effect and then looking back. "The new school of epidemiologists were trying to imitate the laboratory sciences, which analyze data *prospectively*—you enroll a number of people, document their exposures to whatever risk, and then follow them forward in time to see what happens. Retrospective data, collected after the event, was considered unreliable; it was thought if you depend on the memory of the subject after an effect is known, you pick up a prejudiced story. As a practicing physician, I was used to this method: a patient comes in, you ask them their history, you look backward to find the cause of the problem. There seemed nothing wrong with this. But it wasn't the fashion in epidemiology.

"But there was a Dr. Green on the committee who approached me and offered me £1,000 from the Lady Tata Memorial Fund for Leukemia Research. And I said, 'well, half a loaf is better than none.' We had something to start with. It wasn't much, but there was this to be said—it left us free to go our own way. So with all the nerve in the world I said, 'we'll go it alone.' And the Three Musketeers set off on this expedition."

Alice and her team set out to get copies of the death certificates of all the children who'd died of leukemia and other malignant diseases, and Alice designed a questionnaire to be submitted to the mothers. They decided to pick a live control group from the birth register, a random sample of living children from the same region, the same age, the same sex, so they'd have *case control pairs.*

They hit another snag when they tried to get the birth registrations (called birth certificates in the United States) and were told this was confidential information. "But I had a bit of good luck when someone at the general registrar office said, 'We can't give you the birth registrations, but you as a doctor can get the birth notifications.' There's a difference—you can take up to six weeks to register a birth, but someone present at the birth must immediately, within twenty-four hours, notify the public health department, because it has the responsibility, in Britain, to make sure that the mother gets help—and this is birth notification. This can be signed by anyone who attends the birth—the midwife, the husband, a family member, the bus driver. The birth notifications then become part of the public health records and are open to physicians. The birth registration becomes part of vital statistics.

"It made all the difference, this person who let me know that. It was very kind of him, really—he didn't just say, 'no, you can't have this information, now go away'; he said, 'no, I can't let you have it, but this is how you can get it.'

"But then Webb suddenly fell ill and I thought, this is impossible."

The Length and Breadth of England

"That's when I got the idea of going to the local public health departments. And what better place to start than my own hometown, where my family was well known."

Alice turned to the Medical Officer of Health in Sheffield, as her parents had fifty years before turned to a Medical Officer of Health in that same city when they began their Infant Welfare Clinic. "Dr. Roberts knew me and knew my family, and I came to him for help—'Here are the cases; here is a questionnaire we'd like you to distribute for us.' He listened to my idea and he was thrilled. He thought it was beautifully exciting and interesting. He said, 'I've got a young woman who was just appointed, and she'll be very interested, and I'll put her at your disposal for one day a week.' That was how we started the survey.

"Dr. Roberts was kind enough to get me introductions to the medical officer in Bradford, and from Bradford I went to Leeds and from Leeds, and so forth—and that's how I covered the country. Each place I'd go, I arrived as an honored guest. I'd go each time to the head man and would give him a list of cases with the questionnaires and tell him to send someone to interview the mothers and to send the same person to interview the controls [the mothers of the live children whose cases were used as comparison]. He was to send the questionnaires back to us and we'd do the rest. It worked like a charm. The public health departments were in a good position to help and they were more than eager to do so. They felt rather sad that their work had been taken away by National Health, and they welcomed an interesting project. And the public had good will toward the local health departments because they'd been administering rations and services to pregnant women during the war.

"So the disaster of not getting the grant, the disaster of Webb falling ill at the critical moment—they all turned to good advantage. When Webb got better she was quite cross with me for carrying on so successfully but I placated her by saying, 'well, you go and do London.' (And it turned out Hewitt was right in what he'd told that committee—you shouldn't have just one interviewer. She knew too much, she didn't fill in the records sufficiently because she knew she could fill them in later, and we had a terrible time with her records.)

"I eventually visited every County and County Borough Health Department in the country—all 203 of them. I'm probably the only person in history ever to have done so. I knocked on the door of every one of these medical officers. I spent those £1,000 on railway fares traveling the length and breadth of England, going to each public health official, saying, 'here are the questionnaires, will you help?' I spent £1,000, and I got in return something incalculably valuable.

"Later when the Americans tried to do a study like ours they gave up because it cost too much. But it cost us so little because I was making use of the existing records. I always said, 'I'm putting it on the rates' [*rates* are a local tax]—I was putting the cost of this survey on taxes. It's amazing to think now, we covered the whole country and we hadn't even a card sorting machine. The first questionnaires were actually in my handwriting—I think they must be historical documents by now; they're still in the file. Of course we didn't have photostat machines; we made carbon copies. One thousand pounds was the total cost of this discovery—plus the salary Oxford was giving me for doing nothing so I was free to go bumbling about the country.

"I wish Ryle had lived to see it because he would have seen the beauty of it; he would have felt, 'yes, that's what I meant, that's what we ought to be doing.' This was his vision, this linking of academic inquiry with public health. This was what he wanted to see happen—you keep these nice steady public health departments going and have academic medicine feeding them interesting questions. If you know how to use the existing records, there's no telling what you'll turn up. But there has to be somebody there being paid to do it, the way I was. It was such a fluke that I was there, really—a marvelous fluke.

"We started out with approximately five hundred leukemia deaths, matching them up with five hundred deaths from other forms of cancer and one thousand live children of the same age, sex, and region. And when we came to tally up the findings, we made an astounding discovery: both groups of dead children—those who died from leukemia and those who died from solid tumors—had been x-rayed before birth twice as often as the live children.

"We could see it quite early on, from the first thirty-five pairs: *yes* was turning up three times for every dead child to once for every live child, for the question, 'had you had an obstetric x-ray?' *Yes was running three to one.* It was an astonishing difference. It was a shocker. They were as like as two peas in a pod, the living and the dead; they were alike in all respects except on that score. And the dose was very small, very brief, a single diagnostic x-ray, a tiny fraction of the radiation exposure considered safe, and it wasn't repeated. It was enough to almost double the risk of an early cancer death.

"That was what we found and that finding has determined the course of my life ever since."

After Shocks

"But I was very, very nervous about the returns that were coming in. What if we were wrong? On the other hand, I had that sort of uncontrollable excitement that you get when you come across anything that is out of the ordinary and thought, was it possible that we had, by accident, come across something that might tell us something really important? Could it be that this one thing, this small dose of radiation, was going to tell us something about the cause of cancer?"

It was natural, in 1955, to ask a question about x-rays. But Alice's questionnaire asked about everything from fish and chips to colored sweets, tinned fruits, aerosols, and luminous paints; it included questions about family health history, exposure to automobiles, buses, hens, rabbits, and dogs. "None of us had the faintest clue that this tiny dose of x-ray occurring before birth might have this powerful effect initiating childhood cancer.

"I knew I was onto something important and I knew it was going to cause trouble. I thought I should give some warning to the Medical Research Council before we went public. So I went to the MRC Secretary, Sir Harold Himsworth, to say, 'look, I've got this report, I think you should know about it.' I knew Himsworth, I'd done a research project with him, and I knew Joan Faulkner, who was his secretary at that time, so I went round to their office. [Joan Faulkner had been one of Alice's medical students at the Royal Free and is now married to Sir Richard Doll.] And we chatted away in her office and the message came back that Himsworth was too busy to see me, so I left the manuscript and came away. I then set off to Anglesey for a holiday with the children and no sooner had I arrived than I received the message: I had, if you please, to return to London immediately! It was very inconvenient—I had to arrange child care and so forth.

"The MRC said they thought Bradford Hill ought to look at the thing, and I said, 'yes, that's fine,' and he took a look and asked me all sorts of questions about the research and didn't find anything wrong.

"What we published in *Lancet,* in 1956, rippled across the country.[4] It was just a preliminary report, but it created quite a stir. At first the praise was terrific—'this is wonderful, this is Nobel Prize stuff,' and so on. The Medical Research Council became interested and decided they wanted to have Scotland covered and offered me three times as much for surveying Scotland as it cost me to do the whole of England. We weren't above accepting the money.

"But then a reaction set in and the mood changed. It was as though I'd trodden on somebody's corns. The medical profession didn't like it. The obstetricians came down on me like a ton of bricks—how dare I say that x-rays are dangerous? They saw me as interfering with their practices. The radiologists were oddly divided. Some were very pleased—they said, thank goodness you've come along, we've been forced to work with this deficient machinery all these years, now National Health will have to give us new equipment.' But most thought I was taking the bread out of their mouths—they were afraid people would stop using x-rays.

"All sorts of people rushed in on the scene agreeing or disagreeing, trying to prove I was wrong or right.

"I became notorious. One radiobiologist commented, 'Stewart used to do good work, but now she's gone senile.' There were letters in the paper saying, 'she doesn't know what she's doing, we're saving lives with x-rays.' And I daresay there was a lot of backstreet gossip. Nothing faintly resembling what later happened over the nuclear issue, though—no character assassination. I kept out of earshot so as not to get upset.

"We were criticized for trusting the mothers' memories. Our study was retrospective, beginning with the effect and looking backward—a child has died, we were looking for the cause. This methodology was suspect because of the so-called recall bias: a woman interviewed after her child's death might be suspected of searching her memory for some explanation that could account for the tragedy. But we consistently checked the mothers' claims against the hospital records and found that their memories were reliable. Rarely did they contradict the hospital record, and whether the child had died or not didn't influence this.

"I'm convinced that the only way you could find what we found was to look backwards—you'd never have spotted it looking forward. If you begin with the effect—you start with a group of ill or dead persons and set out to discover what went wrong—your study can be much smaller and of shorter duration. What we'd found was a very small effect—only about one in two thousand of these x-rays resulted in cancer—too small to have left any mark on vital statistics. You'd have to monitor thousands of children for at least ten years after birth to pick up so small an effect."

Over the next eighteen months Alice and her team continued their interviews, gradually adding to their original series of case/control pairs. They published a fuller report in the *British Medical Journal* in July 1958.[5] There, with an expanded database, they were able to conclude definitively that a fetus exposed to x-ray was twice as likely to develop cancer within the next ten years as a fetus that had not been exposed—and not only leukemia, but all types of cancer. "We succeeded, within a three-year period, in tracing more than 80 percent of all childhood cancer deaths that had occurred in England between 1953 and 1955, which was a miracle considering that we had no money.

"We reckoned that a child a week was dying from this practice, which isn't all that many—though any death caused by a medical practice is very much the wrong side of the tally. We thought that doctors would stop x-raying on the mere suspicion that we were right, and we felt we must hurry to cover all the deaths that occurred in the next ten years, because once they stopped x-raying, there would be no further cases. We needn't have worried; they went right on x-raying, so we went right on monitoring. We went on and on and managed to include all children who died from 1953 onwards. It was a full-time job and kept me close to the data collecting. We spent the next twenty years proving we were right, and we did prove it—that a single x-ray, a fraction of a permissible exposure, was enough to double the chance of an early cancer. We emerged after twenty years with a genuine finding—there could be no mistake.

"But it was a very small effect that we'd picked up, and if we hadn't stumbled on it, I doubt that anyone would have."

Distant Thunder

Alice had no idea how large this small effect would turn out to be.

In 1956, the year she published her preliminary warnings about fetal x-rays, nuclear testing was at its height. In the United States, Democratic presidential nominee Adlai Stevenson called for a unilateral end to weapons testing on the grounds that fallout threatened the future of the species. Stevenson was overwhelmingly defeated, and the arms race went on: twice as many nuclear weapons were tested in 1957 as in 1956, and twice as many were tested in 1958 as in 1957. The U.S. Atomic Energy Commission was issuing instructions for the building of bomb shelters, by way of reassuring the U.S. public it could survive nuclear war; schoolchildren were drilled in "duck and cover" exercises, being taught to duck under their desk and cover their heads to ward off the effects of thermonuclear attack. The U.S. government was waging an enthusiastic public relations campaign to win trust in the friendly atom.

But an anti-nuclear movement, supported by a small but vocal group of scientists, was gathering strength.

In 1958, the Campaign for Nuclear Disarmament formed, with a program to rid Britain of nuclear weapons. On Easter 1958 the first Aldermaston march was held, in which 4,000 citizens marched from London to the weapons research facility in Aldermaston, calling for a halt to nuclear testing. (This was the first of several: later marches started in Aldermaston and finished with big rallies in London.) In Germany that year, a protest rally in Frankfurt attracted 20,000 people, and in Hamburg, 120,000 marched.[6]

During the years 1954 to 1957, several eminent scientists came forward to challenge the Atomic Energy Commission's position that hazards from fallout were insignificant. Among them were Hermann Muller, the Nobel Prize winner who demonstrated in 1926 that radiation produced mutation, and Linus Pauling, winner of the 1954 Nobel Prize in chemistry. In 1958, the year Alice's study of childhood cancer was published in the *British Medical Journal,* Pauling warned that weapons testing would produce millions of birth defects, embryonic and neonatal deaths, and cancers. "Each nuclear test spreads the added burden of radioactive elements over every part of the world," he warned. "Each added amount of radiation causes damage to the health of human beings all over the world and causes damage to the pool of human germ plasm such as to lead to an increase in the number of seriously defective children that will be born in future generations." He sent out letters to scientists around the world, asking them to join him in an appeal, and got signatures from more than eleven thousand scientists from forty-eight countries calling for an end to testing, which he presented as a petition to the UN. His petition was a major influence in persuading the United States and USSR to agree to the 1963 moratorium on testing.

For his efforts, Pauling was hauled before the House Un-American Activities Committee (HUAC) and made so uncomfortable at Cal Tech that he resigned.[7] Senator Joe McCarthy and HUAC were whipping up anti-Communist hysteria that made it dangerous to express even mildly liberal opinions. Robert Oppenheimer had his security clearance withdrawn and his career ruined for opposing the development of the hydrogen bomb.

It was many years before Alice would see the connection between these large political events and her findings. She was not one of the scientists asked to sign Pauling's petition: her work had not yet established her as someone whose opinion mattered. She didn't even take part in the Aldermaston marches. "I was always against the bomb, of course. I lined up automatically on the side of people like Pauling. But I never got round to the Aldermaston marches—they were held at Easter and I usually had the grandchildren for the holidays." Later, as it became clear where her radiation research was leading, there were other reasons she held back: "I wasn't big on marching in demonstrations; I felt it was important that I not be an activist myself. I didn't want my feelings about the nuclear issue to confuse my science, and I didn't want to do anything that would jeopardize my credibility or discredit my position. I don't think I do the cause any good by waving my flag in the street."

Besides, she had her hands full, defending her findings.

Attacks and Defenses

The major setback to the Oxford Survey was a 1960 study by Richard Doll and William Court-Brown, in association with A. Bradford Hill.

Court-Brown was secretary of the MRC committee to report on the hazards of radiation. The studies he and Doll had produced, of patients treated with x-rays for ankylosing spondylitis, as well as their 1956 report on leukemia in the A-bomb population, came to conclusions that corroborated the findings of the A-bomb studies—cancer risk could be extrapolated from high to low dose and there was effectively no risk at low dose.[8] Doll and Court-Brown had been among the first scientists invited to Hiroshima to evaluate the A-Bomb Commission's studies; Doll reported that he "was satisfied with the Commission's procedures and conclusions."[9] Now along comes Alice, suggesting a much higher risk than the commission—and Doll—was estimating. Court-Brown and Richard Doll responded by launching a study of their own, using the more accepted prospective approach.[10]

"The Court-Brown–Doll study was wrong on many counts," explains Alice. "It looked only at children who had been x-rayed and surveyed only eight hospitals, which was too small a sample. It followed the children forward in time, but it didn't follow them as long as it needed to, a full ten years. Besides, it looked only for leukemia. Everyone was sure we were wrong about the other cancers because there was nothing in the A-bomb studies about any cancer other than leukemia, so they assumed they could limit their investigations to leukemia. But if you limit the field this way you're looking at only half the story, cutting down your chances of finding anything by 50 percent. If you study only x-rayed children and then don't follow them the full ten years, you've limited yourself even further.

"So small and truncated a study as this was bound to have negative findings. It was outrageous how much influence it had! It got top billing—the *British Medical Journal* made it the lead article and give it an editorial. It shaped the way people perceived us in the coming years. Now everyone breathed a sigh of relief and returned to their usual practices. Doctors went back to using prenatal x-rays.

"We knew this was dangerous and decided to carry on. We felt we had to, at least until we could convince them to stop doing these x-rays—though after the Court-Brown–Doll study, we never got any support from Britain again. If funding hadn't come through from America, we'd have been finished."

Doll later admitted that his survey was "not very good" and that the results were "unreliable." "Unfortunately, our study was too small. . . . I never really thought it carried much weight. . . . I've never really been happy with that study."[11] But this was no help to Alice—the damage had been done.[12]

"What saved us was a study by Brian MacMahon, which came along two years after Doll's and rehabilitated us."

MacMahon, an epidemiologist from Birmingham who had taken a post at the Harvard University School of Public Health, looked at a much larger population of children, from thirty-seven maternity hospitals in New England. He compared the children of seventy thousand mothers who had received pelvic x-rays during pregnancy with the children of mothers who had not been x-rayed and found that cancer mortality was 40 percent higher among the children whose mothers had been x-rayed.[13]

"The funny thing was, MacMahon nearly published his study with a negative finding. He'd followed the children for eight or nine years and had found no sign of cancer, and had even gone public with these findings and had sent them in as a paper to the *Journal of the National Cancer Institute*. But then in the last year, the returns started coming in positive, and he had to withdraw his paper and rewrite it.

"And in the meantime, Court-Brown and Doll had gone ahead with their survey, in the confidence that the American study was coming in with negative findings—when at the eleventh hour, it turned around and corroborated *us*.

"It was much better designed than the British study. It took a larger sample and didn't confine itself to leukemia, and sure enough, it showed that the fetus was as vulnerable to x-rays as we'd said. And it was a prospective study, so it got official approval. But MacMahon could only find this prospectively because he knew what to look for—*because our study had told him what to look for.* Otherwise, a prospective study would have had to have been *huge* to find what we found."

Thanks to MacMahon's confirmation and the U.S. funding it generated, the Oxford Survey was able to continue. Other studies then began to corroborate Alice's findings. In the early sixties, the Tri-State Leukemia Survey was begun. One of the largest radiation-related studies ever undertaken, it covered some six million x-rayed subjects in New York,

Maryland, and Minnesota; by 1972, results published by Dr. Irwin Bross and Nachimuthu Natarajan indicated that children of mothers x-rayed during pregnancy suffered 1.5 times the leukemia rate of children of mothers not x-rayed. In certain subcategories, exposed groups are 5 or 25 times as likely to develop leukemia as the general population.[14]

What Alice could not have anticipated was that new methods of statistical analysis, the so-called Mantel-Haenzel procedures developed between 1957 and 1962, would reinstate the value of retrospective data and restore faith in the retrospective approach. These procedures would provide epidemiologists with new tools that were suited to the needs of a large database consisting of cases and matching controls.

"So we tested our theories again and again, using increasingly sophisticated statistical methods and a growing collection of data. We stuck to it, and in the end we were better for the opposition—we had the strength of our data and it was too strong for people to contradict. If everyone had accepted our initial findings, we'd have had no reason to go on and collect more data. We wouldn't know one-quarter of what we now know about childhood cancer. As it was, we spent years testing and retesting our hypotheses and expanding our database.

"The Oxford Survey, like the tortoise of the fable, was never designed for sprinting, but like the tortoise, it has shown great endurance and ability to outlast rival projects.

"And we gradually, over the years, made the profession and the public uncomfortable about medical x-rays."

Long Term

Alice believes "it takes about twenty years. It usually takes that long for an unpopular discovery to be digested, and you're lucky if it takes *only* that long."

Dr. Karl Morgan was the first U.S. government scientist to understand the implications of the Oxford Survey, and his understanding made for a conversion in his thinking about radiation risk. Morgan, a physicist who had worked on the Manhattan Project, is one of the most respected men in the field of nuclear health—in fact, he is the founder of the field of Health Physics. Morgan saw early on that Alice's work challenged the universally accepted "threshold hypothesis" that claimed there was a threshold below which radiation was safe. Her data "were so far ahead of their time that few scientists accepted them at first," he recalls. "However, as she collected more and more data it has become evident that. . . *there is no safe level of radiation.*"[15]

It took until 1980 before the major American medical groups recommended that doctors not routinely x-ray pregnant women—and even so, 266,000 pregnant women were x-rayed in that year.[16]

In that same year, Morgan told a congressional hearing for radiation victims that he and others had "fought for years to pass a recommendation that women in the childbearing age should not be given x-rays in the pelvic and abdominal region except in emergency," and the failure of the x-ray industry to comply had been "one of the biggest problems."[17]

"The medical profession was slow to alter its practices," Alice recalls. "The first thing to go was those x-ray machines in shoe stores. The French made them illegal, then the English followed. Many people ceased the practice of fetal x-raying before the official recommendations were made. In some countries there were codes of practice, if not actually laws, that grew up against x-raying pregnant women. I would meet doctors from Europe, Hungary, and so forth, who built such codes into their practice. There was a period when in utero x-raying fell off, but then it went up again, so that by 1980, between 7 and 10 percent of fetuses were x-rayed, the same as in the early fifties—even though the practice was now known to be dangerous. It's true that x-rays had become safer by this time, but one of our papers showed that though they had become safer, they were still having a small effect.[18] Yet the practice persisted.

"The place that held out the longest was England. That was because the Doll and Court-Brown study was so influential."

As late as 1977 there was a last gasp of official opposition to Alice's findings when a report from the National Council of Radiation Protection (NCRP report no. 53) claimed that obstetricians had x-rayed those fetuses which they somehow *knew* would get cancer, which explained why the x-rayed fetuses went on to develop childhood cancer. "God knows how the obstetricians were supposed to have known which fetuses these would be—they had somehow intuited the cancer. It was all nonsense."

Winding Down

In 1980, the year that major U.S. medical associations made official recommendations against routine x-rays of the fetus, the Oxford Survey of Childhood Cancer came to an end. Over a period of thirty years it had surveyed more than twenty-two thousand pairs of children, becoming the world's largest and longest-running study of childhood cancer. It had built an enormous database, including information on cancer in relation to maternal age, pregnancy illnesses, drugs, postnatal infections, inoculations, parents' occupations and social class, family histories, and later, ultrasound. It had demonstrated changes in radiation sensitivity during successive stages of fetal development, showing that the fetus is supersensitive during the first trimester of pregnancy. It had collected data that had the capacity for testing dozens of hypotheses.

George Kneale, who had been a schoolboy when the Oxford Survey began and had been working with Alice since 1962, used the database to demonstrate that by the time the child is on the brink of manifesting leukemia, the risk of a pneumonia death is over three hundred times greater than the normal risk, which points to a link between cancer and immune system suppression.[19] He and Alice were beginning to test connections between inoculations and resistance to cancer. One question that had never been answered was why leukemia in children was almost always lymphatic rather than myeloid, and why it tended to peak between ages two and four—a question Alice later addressed with a hypothesis about Sudden Infant Death Syndrome (SIDS) as the manifestation of a latent form of cancer, as we'll see in chapter 15.

Alice was quite clear that what she and her colleagues had come upon was not a simple story of the dangers of x-rays. She did not attribute the rise of childhood cancer only or even primarily to radiation, but she realized that it was also related to the advent of antibiotics that, by reducing the number of deaths from infection, had eliminated a major competing cause of death. Since the child incubating leukemia becomes more infection sensitive, in the days before antibiotics, the child wouldn't have lived long enough to have manifested leukemia.

"The survey ought to have continued," Alice insists. "It ought to have been put on an ongoing basis. It's exactly this sort of monitoring of the population that you need if you're going to get to the causes of cancer. If you are trying to prevent a disease like cancer, you need to understand the cause of the disease, its etiology—the rest, as Ryle said, is just end-stage treatment. Why is that so difficult to see? You need this sort of monitoring. We'd shown it could be done cheaply and easily, just by using the existing network.

"There should have been people flocking to our office, asking how we did it, instead of the obscurity we worked in."

A Bombshell in the A-Bomb Data

"There was, of course, one major study that said we were wrong—or rather, we were saying *it* was wrong. That was the A-bomb survivor study. People kept invoking it to refute us, saying, 'there's no evidence in the A-bomb data of any in utero damage, no evidence of any cancer other than leukemia; therefore Stewart is wrong.'

"I didn't see a direct route between the Oxford Survey and the A-bomb studies until 1970, when Jablon and Kato published a paper in *Lancet* attacking us. We'd said that if you give one million children one rad of ionizing radiation shortly before birth, you can expect in the next one to ten years to get about six hundred cancer deaths. But if we were right, the A-bomb studies should have turned up *twenty-six cases* of childhood cancer—and they found *one*. Therefore, they said we'd exaggerated the risk by a factor of twenty-six. There had been 1,270 bomb survivors exposed *in utero* and they'd followed them for ten years and found no cases of leukemia and only one case of cancer; therefore, they concluded that we must be wrong.[20]

"I said of course they hadn't found any cancer in the children because those children would have died off. They couldn't have survived those catastrophic years after the Hiroshima blast—they hadn't a chance of surviving, since leukemia would have made them hundreds of times more infection sensitive. I said that our differences weren't evidence that we were wrong but that the A-bomb data were wrong. This was the beginning of the quarrel I was to have with these studies, a quarrel which was to move to stage center of my life.

"But everyone assumed the A-bomb studies were right and so we must be wrong, and this gave a reason for continuing the attack on us."

Chapter 7

Dr. Doolittle's Team for the Moon

"We kept the show going on a sort of repertory theater budget."

"We were known as *Dr. Doolittle's Team for the Moon*—it was a phrase coined by some children of a friend of mine, and it stuck. A funny little team of misfits with big dreams, like Dr. Doolittle's team with their spade and their bucket, determined to get to the moon. There we were surveying the whole of England, without even a card sorter."

There were not many women at Oxford in the fifties, especially not many who headed science departments, even if it was only a "unit," not a department, and an impoverished unit at that. There were certainly not many women like Alice. Slim and vivacious, with intense blue eyes and a disarming smile and a disregard of formality, Alice stood out amid the general dowdiness of English academic life. Then, too, there was her propensity for hiring unusual people.

"There were retired and semiretired people and others who needed work on a temporary or part-time basis. There were young people who found themselves in some sort of trouble—there was the son of a former Oxford don who came to do work for us while getting through a religious crisis. There was a butcher, Mr. Stanley, who told us in interview that he needed the job because he had leg trouble and needed to get off his feet; he turned out to have beautiful writing and became an excellent record keeper. There were women trying to balance their work with raising children. There was a German refugee."

She and her team were given a wide berth.

"There was no very sharp delineation of jobs, and people might be asked to do coding one day, to check references in the library the next, or to look after the children, with me there to fill in the gaps where needed. Everyone was given the hours they wanted. I could do this because I had complete authority within the Social Medicine Unit—no one cared what we did."

In the end this approach assured Alice a staff that was devoted and dependable, grateful to be able to work at tasks and schedules that suited them. "There are countless beneficiaries of Alice's help," recalls co-worker and friend Renate Barber; "her offers were always extended in a no strings attached sort of way, whether it was a temporary loan or the offer of a chat." "I think I got a rich reward," says Alice. "We had virtually no turnover. Much of it was very dull work, but nobody seemed to mind—everybody felt that they belonged to the project. It was partly a way of treating people, giving them responsibility."

Alice's experience juggling single parenting with work gave her a keen sense of the difficulties women face. "I always believed that women should be given all possible help to return to work, even if this meant they had to have time off or bring their children to the office. There was Jesse Parfit, a very clever girl who found herself in Oxford on account of her husband's job as medical officer. There was Molly Newhouse, who was temporarily between jobs—she'd been abroad and would go on to make a name for herself in industrial medicine.[1] There were Winifred Pennybacker, the wife of a local surgeon, and Celia Westropp, who'd been employed by Ryle and was at a loose end after he died. And there was Maggie Kinnear Smith-Wilson, who stayed on in charge of the Oxford Survey after I left."

Accommodations for the demoted unit were spare. "First we had a room then we got a second room, with a carpet—it had been designed for an American professor of physics, a Nobel Prize winner, who took one look at it and didn't want it. One by one the rest of the rooms in No. 8 Keble Road fell to us, as people left. So from having a room, we gradually got the whole of this house, five floors, with a billiard room at the back which was useful for storing records. Very convenient it was, too, just opposite the Natural History Museum. It worked out rather well."

American Friends

"After the Court-Brown–Doll paper rebutting us, our support came from America.

"A visiting American, a professor of public health who was in Oxford for a meeting, decided to visit me. He arrived late and apologizing—he'd got lost—and pointed to the building across the way, saying, 'I thought *that* was you.' He was pointing at a grand building, the museum, and I laughed, 'oh, no, we don't rate so high, we're very small fry here, as you see.' We had two small rooms at the time and were about to run out of money again.

"So we got to talking about financing and he said, 'why don't you apply to America? You're very well known there.' That was news to me. I had no idea how to apply for an American grant and to the present day I have no idea—the efficiency with which you people apply for grants is unknown in England. But he was nice enough to return home and send me the forms by way of thank you. I remember, they were big thick papers, not in triplicate but more like in six, big red and pink forms, and I put them on the top of my desk tray, meaning to get to them. And they move gradually from the top of the tray down to the bottom; then I put them back to the top again and they move down again, and they do this several times and I still cannot bring myself to complete them. And I say to my secretary, 'Look, we must do something about these or throw them away.'

"That was the time of the Marshall Plan and America was turning swords into plowshares and giving money all over Europe. I applied to the National Cancer Institute, asking for something like $5,000 per annum, and got it with no difficulty. Later I realized how silly of me, I should have added another zero and nobody would have noticed. But actually, when the crunch came later I think we were better off, not having a big grant. Also, of course, America had tried the Tri-State Leukemia Study and found it was expensive, so they funded us for several years, right up until 1970. They sent us the money regularly every year, no questions asked. From then on we were free of the university; we had our patron.

"When the National Cancer Institute money dried up, we tapped into an organization called the Bureau of Radiological Health. Then Reagan withdrew support from that, and at that point we stopped getting money from America."

Alice felt that there were advantages to the obscurity—it left her alone to get on with her work. She was a fellow at Lady Margaret Hall, which gave her a residence and contacts outside Social Medicine, and she had friends. In 1953 she and a few other women started the 53 Club. The 53 Club met on Thursday evenings and gave women a chance to get together. The life of a faculty wife at Oxford can be lonely, since social life takes place exclusively at college. Dons were not even allowed to marry until the end of the nineteenth century, and until recently, wives were excluded from most university functions and not allowed to dine in Hall. (The 53 Club went on well into the 1990s, though it became a luncheon club as its elderly members began to find it difficult to get about at night.)

Alice was not lacking friends, though she did find herself alarmingly cut off from professional contacts.

Once when a Medical Research Council conference was held in Oxford, on the role of mathematics in medical research, she only became aware of it because an invitation to a Dr. Hewitt was wrongly delivered to David Hewitt. "It was outrageous that we weren't invited to that, simply outrageous—leaving us out like that. So we gate-crashed."

George Kneale

George Kneale, who has become the mainstay of Alice's work, her key statistician, came to the Oxford Survey in an extraordinary way.

"His mother had heard I'd befriended a boy, the son of a friend—you know, one of those clever boys reading classics who's suddenly all at sea. I'd given him a temporary home and provided him work at the office and helped him out. George's mother got wind of this and one day she cornered me and asked, could I help with George? She'd been an Oxford don for many years and so had her husband, and their son was a fixture around Oxford; everyone knew he'd been a sort of problem child. He was exceptionally clever, but he could never make friends. They'd sent him to Cambridge, where he'd got a degree, and now he was home and refusing to come out of the back room, and could I help?

"I thought about it and, remembering the trouble with my son Hughie, realized I must not channel anything through the parents, and so I rang him up when I knew his parents would be out. I got George on the phone and said, 'we're urgently in need of help on the survey, could you come round?' He got on his bike and was instantly there and we put him to work addressing envelopes. He was working in my room, where a lot of rather complicated things were going on, chat about what we were doing and so forth, and every now and then he'd put in some remark—he has this funny, blunt way of talking—that would set me wondering. I can remember Lavinia, one of our temporary people, saying, 'that boy is no fool, he's got something there.'

"David Hewitt had left by this time, and I needed some figures checked so I asked George to check them, and he came back with a sort of table. I took this document next door to the statistics department and said, 'can you tell me what he's done?' and they explained that what he'd done was read into the figures certain sorts of relationships and estimated a probable future. He'd anticipated something that I thought was going to take another two years' data collection, and when I asked him about it, he just said, 'it's obvious.' Well, it wasn't obvious, it was quite extraordinary, and I realized that we ought to make better use of him. What he needed was a degree in statistics, so I went to his parents and told them if he could get qualification in statistics, I think he might go far.

"So it was agreed that he should do a degree, though he continued to come to us during his holidays—he liked being with us. Well, he cuts through this degree like a knife through butter and passes with high honors. By this time his parents think they can see their way to a career for George, so he applied for a job in government, decoding secret information—intelligence, that sort of thing. Someone came to me about security clearance and I said, 'oh yes, he'd be wonderful for intelligence; he never talks to anyone,' but the fact that he'd once attended a mental hospital weighed against him and he was turned down. Meanwhile he'd drifted back to his parents' back room again, and I thought, oh no, we can't let this happen; so I called again and said, 'we need help at the Survey,' and he immediately came round. He's never left me since.

"I was right, too—he's a genius. He became the key to my project, the making of the whole show. He's difficult, he's a loner, he doesn't like working with people, but he found something with me that was all right, that made him feel all right.

"And I found—the oyster had a pearl. He wasn't an obvious prize, but a prize he was. And we—we are a pair. We require one another."

Misalliance

Alice was due to retire in 1974 at age sixty-eight. "I could in the normal course of events have stayed on at Oxford. I could have remained as a fellow of Lady Margaret Hall and have carried on with what money I had. I wanted the Oxford Survey to go on. It had tremendous potential—we were only beginning to see its uses.

"And it would have been easy enough to keep it going."

But in 1969, Dr. Richard Doll came to Oxford as Regius Professor. "I was glad when I heard about the appointment, I thought it would be a good thing. We'd never had a Regius Professor who was an epidemiologist, and a cancer epidemiologist, at that. I thought he would give us a ground for flourishing. I couldn't have been more wrong.

"Doll got a chair established—he changed the name, calling it Preventive Medicine (it later became Public Health and Primary Care); but it was our field, Social Medicine, all the same. He came into my room and said, 'isn't this marvelous, getting this chair?' and I looked at him rather quietly because we'd known each other all these years and said, 'you can't really expect me to be pleased, Richard.' He was taken by surprise. And I said, 'look what you've done—all the time I was here, I wasn't fit to be—you know, you might at least have asked that the chair be given to me.' He said, 'Oh, they'd never have given it to you.' And I said, 'I didn't say you should succeed, I said that you should have asked.' He could have arranged it—at least until I retired. It would have made a difference to my pension, but that was the least of it—there was my pride. It would have said, the only reason I hadn't had it before was because they'd been short of money.

"Doll made it quite clear that I wouldn't be welcome if I stayed one day beyond my retirement. And I had no supporters—I was neither fish nor flesh nor fowl. And I was a woman. The college where I lived, Lady Margaret Hall, was keen on my staying, but I knew I needed space more than geographical; I wouldn't have been allowed to flourish. So I sent out word that I wanted to move. Since I had a grant, I wouldn't cost anything—funding was still coming from America. Thomas McKewon, head of Social Medicine at Birmingham, got wind of this and went back to Birmingham and made inquiries and got me invited as a senior honorary research fellow.[2]

"Doll insisted on giving a retirement party for me. I didn't want it. But when I was asked to speak I said, 'here you have an example of someone who's fallen between two stools—or should I say, two chairs?' And it was so—there was a chair got for Ryle and a new chair to come, but never one for me. Doll didn't know at this time that I was due the next day to go to Birmingham; probably he thought I was going to stay about the place and make a nuisance of myself. I think he was surprised that I was able to move on without any reference or testimonial from him."

Alice had made arrangements to keep the Oxford Survey going—she felt there was no reason for the data collection to stop just because she was no longer in charge. She got the Marie Curie Cancer Foundation to house the records, since she had been told that Oxford didn't want them, and left Maggie Kinnear Smith-Wilson to carry on with data analysis. But there was a glitch—"when Maggie applied to the National Cancer Institute to continue the survey and they asked her who was now sponsoring it, she gave Doll's name; his endorsement never came. (Alice heard from a colleague that the word in London, at the Medical Research Council, was that the Survey was not worth carrying on.)

"We applied to the Medical Research Council repeatedly—we required very little, salaries for a doctor, two statisticians, and three or four clerks. The MRC gave us money to write up the results of the data we'd collected, but after 1980 we encountered all sorts of obstacles to getting access to more data. The last year for which we could get full data was 1980."

(When I asked Sir Richard Doll whether he'd opposed the continuation of the Oxford Survey, he said, "I think that's possibly fair [to say]." He added, "I thought that when a person retired they ought to leave the data to the people that stayed behind." Alice claims she would have been more than willing to leave it.)

"Look then what happened once I left—Doll was Regius Professor, a cancer epidemiologist; George and I were doing interesting work in his same field, being invited all over the world but never once invited back to Oxford to speak. We did get one invitation to return—to attend the leaving party for our charlady. It was known that Doll had a poor opinion of our work, so no one else asked us either. When the MRC put together a committee in epidemiology, he was made chairman, which gave him enormous influence. After that, every department in the country was called in to consult with the MRC—except us." Alice can't shake the feeling that she's been "swimming upstream" against some strong invisible current all these years—that "a word from Doll might have set it all right, but that word was never spoken. It's as though he can't bear to let me into the story."

Sir Richard Doll has another version of events. "I've done nothing but try to help her," he says; "she was a splendid teacher; my wife was a student of hers and said she was one of the best and most stimulating teachers at the Royal Free. But there was no way I could get her made a professor; there were very few professors in those days." To my comment, "she says you could have asked," he replied, "well, how does she know what I did?"[3] (However, in a 1993 interview, Doll speaks as though nothing of importance had gone on in epidemiological research at Oxford prior to his arrival: "There was very little there in the way of epidemiological research," which is what made the offer "very attractive": "I was able to bring several people with me.")[4]

The contrast between the prosperous career of Sir Richard Doll and Alice's threadbare professional existence is stark. Dr. Richard Doll was knighted for his discovery of the link between lung cancer and smoking. He has had the finest researchers working with him and remains, since his retirement, an integral part of cancer research at Oxford. Alice made her discovery about x-rays and childhood cancer on a grant of £1,000—a nonrenewable grant—and has kept her research going practically on pocket change.

Alice is not one to begrudge a person good fortune, but she does feel that Doll was exceptionally lucky to be in the right place at the right time when the study of smoking and lung cancer came his way—and that this shouldn't entitle him to have the power over the work of others that it has. As she recalls it, "Dr. Percy Stocks had been studying the rising rate of lung cancer and had called a meeting of the Medical Research Council; he had a hunch that the cause was smoking. Ryle had been invited to this meeting but he was ill—in fact, he was dying. I went instead. The table was full of grandees. Sir Austin Bradford Hill was at one end. We went around the table and all these experts gave their reasons why they didn't think smoking was the problem. Nobody wanted to do this survey, and everyone was saying why it wasn't necessary. If Ryle had been there he'd have jumped at it and nobody would have fought him for it. It came to Bradford Hill as a sort of last resort, who said, 'Right, well it looks like we've got to do something. I've got a young man in my office,' and he gave the study to Doll."[5]

Alice doesn't see why she and Doll couldn't have been working together all these years, and it is difficult to understand why they weren't, since they had such similar backgrounds and concerns. Both started out as physicians; both changed subjects after the war, moving into epidemiology before it was called epidemiology; both had left-wing political views that drew them to Social Medicine. Both made major discoveries in the fifties that helped shape epidemiology so it came to include chronic as well as infectious diseases. They both moved in Oxbridge circles, attended the same meetings, were on the same editorial boards. But one went on to fame and the other to obscurity.[6]

Whatever his reasons, it seems to have been Doll's refusal to endorse the Oxford Survey that made it impossible for it to carry on. "The American money dried up on account of Reagan, but that's not what stopped us—we were cut off at the British end. We were building a database capable of testing several hypotheses about cancer that would have been strengthened had we been allowed to carry on even a few more years; but not only was it made impossible for us to keep the survey going, it was impossible even to leave the records.

"So we had to bring them to Birmingham. Since my office in Birmingham was a trailer, a sort of hut, and the records were prodigious by this time, it became a real problem, where to put them." Alice solved this by finding an empty corridor connecting two temporary buildings, "out of the way, with plenty of space." There they sit to this day.

"I had managed to find new jobs for everyone except George. So I arranged for him to come with me. Within three months of our arrival in Birmingham comes this request from Mancuso to look over his study of the Hanford workers. It's interesting—if I'd still been at Oxford, it might have fallen to Doll's department. One wonders what they would have made of it."

Within a few years of Alice's retirement, the announcement came of a cancer epidemiology unit at Oxford funded at £200,000 a year for an unlimited period, which represented almost half the MRC's total funding for cancer epidemiology.[7] Several years after that, a major study of the causes of cancer in children was launched, under the auspices of Sir Richard Doll. "A unique nationwide investigation into the causes of cancer in children is to start on 1st April, 1992," proclaimed a press release from the U.K. Co-Ordinating Committee on Cancer Research (March 12, 1992): "The United Kingdom Childhood Cancer Survey will take five years to complete and aims to gather detailed information on each of the 1,000 children diagnosed with cancer every year in England and Wales and from twice as many comparable healthy children." It announced itself as "the largest study of its kind ever undertaken anywhere in the world," and as "new" and "unique."

"This study gets launched with great publicity and purports to be the first of its kind," comments Alice. "It's astonishing. It's as though the Oxford Survey had never existed, as though *we* had never existed, though the differences between Doll's work and ours are hard to see—except for the funding, which is staggering." "The estimated cost of £6 million will be met by donations from leukemia and cancer charities and the electrical and nuclear industries," said the press release. "Exposure to natural or man-made radiation will be investigated." One of the hypotheses to be tested is "that the amount of ionizing radiation to which a child is exposed either before birth *in utero* or after birth causes a material proportion of all cases."

What Alice finds dispiriting is that "there was no attempt to link up with our data, though we had a huge database by then. There was no effort to maintain continuity of data collection and similar methods of controls, though the new items in this survey could easily have been factored into methods that maintained continuity with ours, and an enormous amount would have been gained by doing this." (Doll claims that if she had left the survey and had been willing to relinquish control of it, he might have made use of it; Alice insists that she would have been perfectly willing to do so.)

There is something mysterious about the relationship between these two pioneer physician-researchers who started out so similarly. Whatever else was going on, major differences in their scientific viewpoints were to develop in the course of their research that may help explain the rift. Doll accepts the A-bomb studies on radiation risk whereas Alice—as we'll see in chapter 9—is very critical of those studies: thus they have fundamental disagreements about low-level radiation risk. Doll has announced on more than one occasion that the vast majority of cancers is caused by genetic bad luck, diet, and smoking—that radiation, pollution, pesticides, and food additives "have very little to do with the majority of cancers." (The month the 1992 study was announced, he said he was leaning toward the viral hypothesis.)[8] Alice's work has led her to believe that the majority of children's cancers are caused by radiation—as we'll see in chapter 15.

Alice and William

But there were other, more rewarding relationships in her life during these years. Throughout this time, there was an important, steadying presence in the man she reconnected with after more than twenty years, at a party in London.

Alice Stewart and William Empson remet in 1952 at a celebration held around an exhibition of Chinese art. He saw her from across the room and approached. " 'Let's dance,' he said, and we went right on dancing, as it were. I realized he wasn't holding anything against me. We picked up from where we'd left off more than twenty years before."

They began to meet often, sometimes at Alice's cottage, sometimes in London, and sometimes in Sheffield, where he had a basement flat.

"I found him living in these extraordinarily awful digs in Sheffield. I'd try to tidy up a bit and get some furniture and so forth. We'd spend the time walking on the Yorkshire moors. Or in the country around Fawler when he came down to Oxford. He had a large house in Hampstead which he'd acquired after the bombing. Hetta was mostly away, off with someone in the Far East; it was assumed she wasn't coming back. There were about ten years when we met frequently, and after that we more or less continued, on and off again, until he died.

"Wherever we were, we'd spend the weekend walking. We had marvelous talks. It was wonderful, really."

Empson had, by the time he and Alice got together again, acquired a brilliant reputation, not only in England but in America. He was known as a dashing, impulsive, iconoclastic, eccentric character. He was slender, intense, hawk nosed, with a fine forehead and sensitive mouth. He sported a bushy moustache in his youth but in later life grew a beard that hugged the rim of his chin and cascaded down his chest, making him look like a mandarin or an Old Testament prophet. He was notoriously shortsighted, "usually in a state of abstraction"; "his mind was so much its own place" that "environment was indifferent to him," recalls Kathleen Raine.[9] He had a penetrating gaze, tending to a scowl (or a squint) above thick glasses. He was full of energy and restlessness, had difficulty sitting still. He had his head in a book at every possible moment.

Empson's dismissal from Cambridge had by no means ended his academic career, though it had pushed it into some unusual forms. Since he and Alice had last been together, he had taught in Japan, at the Tokyo National University. He'd traveled to Japan on the Trans-Siberian Railway and stayed there for three years, lecturing from 1931 to 1935. He'd then come back to London for a few years before returning to another post in the Far East, at Peking National University. He arrived on a Japanese troop train in the autumn of 1937, just as the Sino-Japanese War broke out and Peking fell.

Fortunately, the Peking universities managed to evacuate in time and had instructed their students, then on summer vacation, to reassemble at Changsha, a high mountain one thousand miles to the southwest. Three major universities took to the road. "People got there . . . with the clothes they stood up in and maybe some lecture notes," Empson recalls; "a fairly dangerous business and you certainly couldn't take a library. . . . The lectures went on sturdily from memory."[10] At the end of 1937, the combined universities moved again, one hundred miles further southeast. Empson committed himself to this adventure, accompanying students and faculty into exile, showing unexpectedly good-humored tolerance for physical hardship, at one point sleeping on his blackboard. He soon became famous for being able to reassemble texts from his head. One of his students recalls that he simply "typed out Shakespeare's *Othello* from memory. . . . His typewriter provided us, totally out of 'nothing,' with Swift's *Modest Proposal.*"[11]

During his years in Japan and China, Empson became a serious student of Eastern culture and religion. He traveled widely, in search of the meaning of the Buddha—he believed Buddhism was "much better than Christianity"—and wrote a book, *The Faces of Buddha,* which was lost during the war and never retrieved.

At the outbreak of the war in Europe, Empson spent three months traveling in the United States, where he became convinced of the need for a propaganda effort to recruit U.S. aid. He played his role by joining the BBC in 1940, a job that assured him an "interesting war," as he said. He was made BBC Chinese editor, a post that involved organizing news broadcasts and talks in Chinese. He wrote to I. A. Richards in January 1943: "my office is next door to George Orwell's and I find him excellent company. I have written a ballet," he reveals in the same letter, "[but] I am afraid nothing more will come of it."[12]

Empson married Hester Henrietta Crouse and had two sons, Mogador and Jacob, during the war. He and his family returned to Peking National University in 1947 and were there during the Communist takeover. He met this situation with characteristic aplomb, seeing the barbed wire barricade round the university as a nuisance rather than evidence of danger. He returned to England in 1952, and at the time he and Alice resumed their relationship, he had taken the chair of English literature at the University of Sheffield, a position he occupied from 1953 to 1971.

The *New York Times* called him "the meteor of modern criticism in English, a living figure of almost legendary brilliance."[13]

He was reputed to be difficult and uncompromising, yet was known also for his generosity to students and younger writers. He was famous for his doodling: a colleague recalls, "when there was nothing else to do, he would work at a [mathematical] problem on a scrap of paper. It was always the same problem. I saw the diagram of it on at least two hundred different bits of paper, accompanied in some cases with a page and a half of algebraic signs. It had something to do with proving that a certain circle touches a triangle at nine points."[14]

But he was known mainly for his literary criticism, which played a key role in defining the way literature was perceived for the next several decades in England and America. The purpose of literary studies had been called into question by the First World War and the havoc that followed during the years that saw the rise of totalitarian regimes in Europe and preparation for a second major war; the value of studying literature was hotly debated at Cambridge in the years Empson was there. F. R. Leavis and I. A. Richards, Empson's mentor, were busily forging a role for it: far from being irrelevant to the modern world, they argued, literary criticism would save it.

"In the early 1920s it was desperately unclear why English was worth studying at all; by the early 1930s it had become a question of why it was worth wasting your time on anything else. English was not only a subject worth studying, but *the* supremely civilizing pursuit," writes critic Terry Eagleton. Leavis and Richards argued that the social upheavals and scientific discoveries of the century had so devalued traditional mythologies that religion could no longer fill the void—so poetry must.[15] Literary studies became a kind of substitute Christianity.

In subsequent years, this banner would be taken up by several influential American poet-critics, the so-called New Critics—T. S. Eliot, John Crowe Ransom, W. K. Wimsatt, Cleanth Brooks, Allen Tate, and R. P. Blackmur. And Empson's works would become a sort of bible of the movement. Empsonian influence extended to the United States and even Japan, where people extracted precepts from his critical essays and tried to implement them in readings of their own. An outbreak of "neo-Empsonianism" "raged like a low fever among undergraduate poets in Oxford and Cambridge."[16]

But Empson found this alliance with New Criticism uncomfortable.

He was especially at odds with what he saw as the smug "neo-Christianity" of T. S. Eliot and F. R. Leavis. He was also uncomfortable with the apolitical stance of the New Critics. A typical New Critical reading focused on the formalistic perfection of a poem, exploring its tensions, paradoxes, and ambivalences, and showing how these are resolved into an organic unity. It was an approach well suited to the fifties, the decade when the Red scare in the United States and Britain made political engagement dangerous and academics found it easier not to see social and political meanings in literature.

But Empson's poetry and criticism were more politically engaged than the New Critical approach that derived from him. He was concerned with the "gathering storm" of world crisis (his second volume of poetry was titled *The Gathering Storm,* and Churchill's title copied his, not the other way around).[17] Though he was born into the gentry and had the confidence of his class, he was critical of its smugness and parochialism. He was a friend and admirer of George Orwell and believed that Britain should recognize the Soviet Union. (Yet he did not refuse a knighthood when it was offered him in 1979. He had "ridden the hounds," as Raine says. "He remained in his own secure world of the Yorkshire gentry. . . . Even in Bohemia . . . such as he cannot lose caste.")[18]

When, in the late sixties, Marxist, feminist, and deconstructionist critics reinstated social and political meanings to literary studies, Empson was the New Critic who best survived the critical revolution.

He had a capacious intellect, a restless soul, a tormented and passionate nature. His poetry, published in *The Gathering Storm* (1941) and *Collected Poems* (1949), is often compared to that of John Donne for its complex and condensed imagery, its verbal play and ingenuity, its tortured, knotty, yet colloquial quality, and imagery that combines references to the English classics with allusions to biology, astronomy, entomology. Conflict and pain are his subjects: he once said that his poems were written to get rid of neurosis.[19] "He took to his troubled heart the contradictions of the age," writes his biographer John Haffenden.

"I think what he wanted from me was a quiet life," says Alice, "something he did not get from his marriage. I think I provided him with the right . . . temperature for his life."

William and Alice were both busy people. "I had the survey and interests besides medicine. I had children, then grandchildren to take care of. But in such spare time as I had, it was wonderful to have William. He was a sort of law unto himself.

"At one point it was thought we might get together. Hetta had more or less moved out. But then she came back—with two children, not William's. William was very good about them.

"He was interested in my work. He had done a degree in mathematics and stayed involved with science and was interested in the statistics end of things. He knew a lot about my work. He was there from the beginning of the survey, always interested. He became quite a familiar figure in the office.

"He saw things from a different angle. I had a totally simpleminded view, and he was forever saying, things are more complex. He taught me to read advertising, its subliminal messages. He taught me how the way you see a painting has been influenced by what's been written or said about it. He taught me how to drink a black and tan. He helped me understand George. He discussed Byron with my mother. I can still look at an advertisement, at a landscape, and see it the way he saw it.

"I once needed a speaker. It was our turn to host a meeting of the British Society of Social Medicine. Everyone wanted to hold the annual meeting in Oxford but I knew we'd never get anyone to speak—nobody on the medical faculty would be seen dead with us, and we had no money to pay. So I rounded up three distinguished lecturers who had nothing to do with Social Medicine. There was a professor of biochemistry, Hans Krebs, who'd won a Nobel Prize and was a friend of my family in Sheffield; there was a professor of anatomy at Oxford, a Sir Wilfred Legros Clark—he was a great walker and had come upon my cottage at Fawler while on a walk. There was Empson, who did something from *Seven*

Types of Ambiguity—well, it was quite lost on most members of the British Society of Social Medicine. We dressed up members of my department as waiters because we couldn't afford to hire anyone to wait tables. It was very original, we had a marvelous time.

"Wherever we were it was understood we would walk; it didn't matter where we were, we just wanted to get off together. We'd take these long walks and find a country pub. Every now and then he'd lark off on something and I'd just sort of listen, though I'm sure we talked more about my work than his. We didn't always talk about deep things—we'd talk about our families or about what we were going to have for lunch. Sometimes we'd just sit and not say much of anything."

The two families were intertwined. "We had Christmas and Easter together. Hetta would come down to Fawler sometimes. I think she originally saw me as a sort of convenience but then got a bit jealous because we stayed together so long. She was torn between feeling glad to be free to pursue her interests and not wanting to feel usurped. Their children visited us in Anglesey and it was through me that Mogador got settled as a student in Leeds. Though I think they must have felt a bit conflicted in their loyalties at times.

"I remember once William and I ran into the poet Louis MacNeice. He saw us together and said, 'I don't know how you do it, William, but you always do seem to manage your affairs so well.' I'd never seen myself as a well-managed affair.

"He did offer to marry me—he said we could marry if our arrangement was putting me at a disadvantage. But I said it was better left alone, and I think he thought so too. I suppose I'm not the sort to fly into a tizzy about this sort of thing.

"It's a pity; we two together would have been a much better match than either of us made. But it never happened."

Alice has assembled a file marked "Empson" for Empson's biographer, John Haffenden. It contains his letters to her and some napkins with doodles, mostly mathematical equations. There's a letter to the *Times* on pesticides and tobacco. "He helped me write that—I remember being quite shocked that tobacco could be sprayed with worse chemicals than were allowed on food. The lung is more vulnerable than the stomach; it has fewer mechanisms for defense. I thought, what if lung cancer is from what the tobacco is sprayed with rather than the tobacco itself?"

Reading through Empson's letters to Alice, one is struck by the sense of familial intimacy. His letters seem often to begin in midsentence, as though resuming a conversation. They have a domestic tone and include plans for holidays and vacations, worries and hopes about their children. There are references to their work, to dealings with publishers, to home repairs. There are no passionate avowals of feeling, only the "Dearest Alice" or "My Dear Alice," with which the letters are addressed, and an obvious pleasure in recalling or anticipating their time together.

He reads her articles and encourages her to be more self-assertive: "I thought your piece was very good except you won't blow your own trumpet" (May 10, 1965). "It seems plain to me that you need to write a book; your self-effacing modesty doesn't work out as yielding enough credits in the end. I forgot to leave the balaclava and will return it later. A lovely two days. Love, William" (May 20, 1965). There is reference to a "dark secret."

> After a suitable pause, I shall ask Hetta to make you a leather jerkin exactly like the one she insisted on washing as mine. She was boasting that she still had a lot of the same leather in reserve, so I expect she would do this happily, and free me from a dark secret. The balaclava I will bring too.

And in later years, there are references to his failing health—his eye trouble, deafness, cancer.[20]

When he died, April 16, 1984, Empson was hailed as "one of the most distinguished and widely influential English poets, critics, and university teachers of this century." A reviewer noted that he occupied a position in English letters like that of a "magician, seer or sage." Praise poured in from students and colleagues, for his personal and professional courage and his generosity. A British Academy biographical memoir compared him with Dr. Samuel Johnson as "a great man and a great Englishman."[21] His obituary in the *London Times* quoted "a close fellow poet [who] said that while there could be argument about whether he was a great writer or great head of Department, there could be none about whether he was a great man" (April 22, 1984).

His epitaph for himself was—it was said—"No more bother."

"How could one possibly sum up this myriad-minded man?" asks Christopher Ricks, a friend and fellow critic.

> He knew intimately every dismay and despair, and yet wrote and lived in high-spirited hopes. . . .
>
> He was the least snobbish of men, and he did not disdain a knighthood. . . .

> He was at once the heir to a long line of poet-critics and the first poet-critic of genius to thrive from within the universities. . . . He was profoundly radical, and yet he rejoiced to concur with the common reader: he really did believe that the great writers are those who have commanded an enduring loyalty. He was urgently discriminating, and yet he wrote always from love. He did not find his energies where it is easiest to find them, in repudiation, but in welcome. . . .
>
> For all his intricacies of mind and his perturbations of heart, his central propositions always had a sublime simplicity. He never narrowed his mind or his eyes. . . . English life will miss his incomparable mind, his candid heart, his straight gaze.[22]

Alice certainly missed him. "His death hurt me more than anyone's. I just couldn't get it out of my system, how much it hurt me. Eventually it faded of course and I was able at least to wake up in the morning without the same dreadful feeling. No other death hurt me like William's."

In 1952 the poet and critic William Empson reentered Alice's life. Empson had been the center of the Cambridge literary scene when Alice met him in the late twenties, and went on to become one of the most influential critics of the century. Their relationship lasted until his death in 1983. Photo courtesy of the Hulton Getty Picture Collection, London.

Hetta Empson (Hester Henrietta Crouse) and Alice Stewart at a party in the early sixties in Hamstead Heath, where Empson kept a flat.

Alice at a seminar in Munich, 1960.

Dr. Thomas Mancuso, professor of occupational health at the University of Pittsburgh, headed the Atomic Energy study of nuclear workers that drew Alice into the international controversy. Photo by Robert del Tredici.

Alice, who was called in on the Mancuso study in 1974, struck more than one American observer as resembling Katharine Hepburn.

Dr. George Kneale, Radiation Conference, Kiel, 1992. Kneale came to work for the Oxford Survey as a schoolboy in the early sixties and has been crucial to Alice's work ever since.

Alice and the Oxford Survey of Childhood Cancers, 1983 (twenty-three thousand manila envelopes, each containing the records of a live and a dead child, stored in an out-of-the-way corridor on the edge of the Birmingham campus).
Photo by Robert del Tredici.

Dr. Alice Stewart versus the Department of Energy lawyers, Denver, July 1983. Photo by Robert del Tredici.

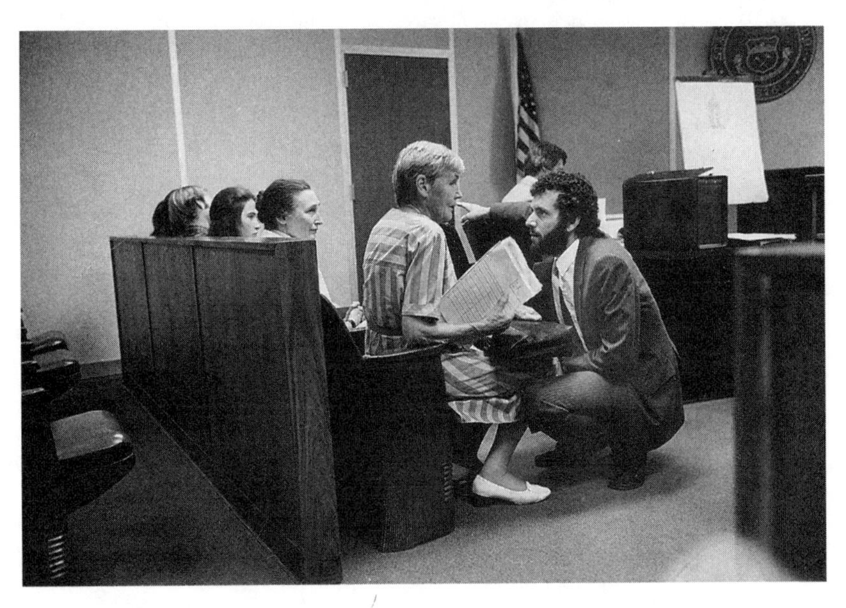

Dr. Alice Stewart and Bruce Deboskey, Denver, 1983. Billingsley Case; Roberta Billingsley (widow) and Marianne (daughter) seated next to Alice. Photo by Robert del Tredici.

In May 1994 Alice Stewart had an unprecedented meeting with a DOE head, Hazel O'Leary, Secretary of Energy in the Clinton adminstration. Photo by Kitty Tucker.

Alice with Rudi Nussbaum, professor of physics, who invited her in 1984 for a sabbatical at Portland State.

While in the northwest, Alice became involved with the Hanford downwinders and began working with HEAL (Hanford Education Action League), Spokane, WA, 1986. Photo by Carole Gallagher.

Alice with friend and fellow scientist Klarissa Nienhuys at Fawler, June 1985. Photo by Elmer Spaargaren.

In September 1998 the STAR foundation and Helen Caldicott organized a conference, at the New York Academy of Medicine, honoring Dr. Karl Morgan and Dr. Alice Stewart.
Photo by Robert del Tredici.

Part 3
Through the Looking Glass, onto the International Nuclear Scene . . .

Chapter 8

Up Against the Department of Energy

"I have no idea what's in it, but I can pretty well guarantee that we'll have our expenses paid."

Leaving Oxford, Alice arrived in Birmingham, "with no plan, really, but to wind up the Oxford survey." But within three months she'd received a phone call from America. Dr. Thomas Mancuso, who had been commissioned by the Atomic Energy Commission (AEC) to study the "biological effects, if any, of low-level ionizing radiation among workers employed in atomic energy facilities,"[1] had turned up something peculiar in his study of Hanford workers.

It was 1974 and Alice was officially retired. She was sixty-eight, rather older than usual for retirement age. Yet she was about to embark on the most dramatic chapter of her career.

Hanford is the oldest and largest nuclear weapons complex in the world. Built in 1943, and built quickly in response to the production demands of the Manhattan Project, it was assigned the task of manufacturing plutonium. It is a huge federal reservation, approximately 620 square miles, located in a sparsely populated area in southeastern Washington. It sprawls along the Columbia River, which is now, thanks to it, the most radioactively polluted river in the world (water was pumped from the river directly over the reactors' cores and discharged back into the river). The plutonium produced at Hanford was transported to Los Alamos, New Mexico, 20 miles from Santa Fe, where, at a top secret scientific laboratory under the direction of Robert Oppenheimer, a team of scientists and army technicians was at work constructing the first atomic bomb.

Within two and a half years, Hanford had produced plutonium for the bombs used in the Trinity test and dropped on Nagasaki. During the Cold War that followed, the facility expanded into a giant complex, including nine plutonium production reactors, several reprocessing plants, and other kinds of chemical processing plants.[2]

Alice, called in to look at Mancuso's Hanford data, stepped through a strange looking glass indeed.

Meeting Mancuso

Alice had met Mancuso several years earlier, under somewhat bizarre circumstances. In 1966, she'd been invited by the Atomic Energy Commission to speak at Hanford, to publicly rebut Dr. Ernest Sternglass, professor of radiological physics at the University of Pittsburgh Medical School. "Sternglass had been tremendously excited about our findings, which confirmed his theories about the vulnerability of the fetus to radioactive fallout, and had cited the Oxford Survey in one of his publications. But he had exaggerated what we'd said, grossly exaggerated, and we commented on this in the *New Scientist*.[3] He'd said that we'd shown that fetal x-rays had doubled the infant mortality rate, when all we'd said was you'd doubled the chance of a child's dying from cancer. Well, the difference is that one is measured in thousands and the other in single figures.

"I'd never even heard of Hanford and now the AEC invites me there, hoping to put me in the stand to refute Ernest. Sternglass was a supporter of our work, but he had got our figures very wrong, and we couldn't have our statistics misused like that. It was an awkward situation—I did in fact have to contradict him, though I steered it away from the sort of direct confrontation the organizers had hoped for. Ernest had had to declare in public that he'd made a mistake, but he bore me no grudge, and this shows his good nature: instead of turning on me, he said, 'are you doing anything on your way home or could you call by Pittsburgh? They'd love to hear you there.' I went to Pittsburgh, more or less as his guest. His wife was very annoyed with me, but not Ernest.

"That was where I first met Mancuso."

Dr. Thomas Mancuso, a professor of occupational health at the University of Pittsburgh, a physician who also holds a graduate degree in public health, had been at work on the AEC study since 1965. It was an extensive project, originally intended to have looked at the employees not only of Hanford but of all the AEC facilities, Oak Ridge, Los Alamos, Savannah River, and Rocky Flats. It was lavishly funded—"the money was astronomical," says Alice—"millions."

"Mancuso embarrassed me by introducing me to everyone as this great lady, making me feel like Madame Curie herself. He got me to talk in his department in Pittsburgh and then invited me to Oak Ridge, where I gave a presentation on the Oxford Survey and found myself on the platform with the esteemed Karl Morgan." (Oak Ridge, the other major facility involved in the Manhattan Project, produced uranium for the Hiroshima bomb; it had been, like Hanford, carved out of a sparsely populated area, eighteen miles from Knoxville, Tennessee. Dr. Karl Morgan was director of health physics there for twenty-nine years.)

"I spoke there remarkably unaware of the fierce undercurrents—I didn't realize there was such controversy about these issues. I was asked to give my opinions, to share my expertise, and I just gave a few everyday principles about how you go about gathering data. I was subsequently told by one of the statisticians, 'you were a breath of fresh air, you had such common sense.' Well I had to have common sense—didn't I?—since I didn't have any special knowledge. I knew very little about radiation, except what you learn in medical school, which is next to nothing.

"Nor had I any idea of the nature of these weapons complexes—I remember the alarm going off at Oak Ridge, which was apparently an everyday occurrence.

"At any rate, four years later, when I returned to the United States for some months with the Bureau of Radiological Health in 1970, I got to know Mancuso. He put me on the steering committee of his study. As a result of this, from then on I got Mancuso's annual reports. They'd come to Oxford and as far as I was concerned, I just opened the envelope and said to myself, 'dear old Mancuso,' and never did anything except find a space on the shelf to put them. I was far too busy with our problems to pay much attention to his."

The Hanford Study

The AEC nuclear worker study had been prompted, in 1965, by public concern over atomic weapons and the expansion of the nuclear industry. Mancuso had been chosen to head it because he had designed and conducted several long-term epidemiological studies of cancer hazards in industries. He had pioneered a way of using Social Security records for tracking the deaths of workers that made it possible to trace deaths that occur much later than the termination of employment and far from the place of work. Since cancer can have a latency period of up to thirty to forty years, and since companies usually destroy worker records after ten years, Mancuso's Social Security method represented a breakthrough in investigating cancer cause and effect. He had used it to link brain tumors to industrial chemicals in a study of rubber workers in Ohio.[4]

Mancuso had been a protégé of Dr. Wilhelm Hueper, a brilliant researcher and a pioneer in occupational cancer whose discovery of links between cancer and industrial pollutants earned him the reputation of a troublemaker. Head of research on chemical carcinogenesis at the National Cancer Institute (NCI), he'd had his funding frozen and his publications censured by the NCI, which did not want to hear the bad news about the toxins industries were producing, since the good will of industry was important to its funding.[5] Mancuso was soon to follow his mentor into similar disfavor.

But at the time Mancuso was put in charge of the AEC study, he was reputed to be one of the foremost occupational epidemiologists in the world, a reputation that would lend the study the authority it needed. No one really thought the study would turn up anything.[6] The official story was that radiation at low dose was negligible. Most health physicists in the new field of radiation biology accepted this: the AEC, which was paying their salaries, was energetically promoting a massive development of nuclear power that depended on public confidence.

The official story came out of the Atomic Bomb Casualty Commission's studies of the Hiroshima and Nagasaki survivors, the most lavishly funded and extensive research on the health effects of radiation ever conducted, which assured that a threshold existed beneath which radiation exposure incurred no risk. This position was upheld by the medical establishment—health physicists, researchers, radiologists—in the United States and United Kingdom. Since Hanford workers were well below the determined limit, it was assumed that Mancuso's study would confirm that nuclear work was safe and worker standards were adequate.

The only study that had turned up evidence of damage at low dose was the Oxford Survey of Childhood Cancer.

When Alice first got wind of Mancuso's study, he and his research team had been at work for more than a decade, collecting data on thirty-five thousand people. He was looking at records going back to 1946, following the workers' health through more than twenty-five years. A thin, shrunken man, he had quit teaching and seemed to have quit eating, and was devoting himself tirelessly to the project. His office on the fifth floor of the University of Pittsburgh Public Health Building was cluttered with charts and computer printouts that took every inch of space and filled the two dozen four-drawer filing cabinets lining the walls of his office.[7] But it was slow work, partly because of the enormity of the task. He'd had to travel to Tennessee, Washington, New Mexico, and elsewhere to track down records of the workers who'd died, the causes of death, job histories, radiation exposures, and then work out a system for collating the data and analyzing it. It was slowed further because he kept encountering roadblocks; as he says, "I discovered that the past data on the workers was being systematically destroyed, and only the current information was being kept. They told me that this was being done to save file space. . . . I got them to stop."[8]

"When I met Mancuso in the early seventies," Alice recalls, "all he'd managed to find was that the workers who were still alive had received more radiation than those who had died. So everyone thought he was on a safe wicket and there was a feeling he wasn't going to turn up anything—in fact, on the basis of what he'd found, it looked like radiation might be good for you. Though I remember George saying, 'well, if there is nothing there, then you shouldn't be finding *any* difference between the live and the dead. Besides,' he added, 'they're *not dead yet.*' "

In the course of the study Mancuso had been urged repeatedly by the AEC, subsequently the Energy Research and Development Administration (ERDA), to publish these negative findings.[9] (During these years, the AEC was being reconfigured as ERDA, which became, in 1977, the Department of Energy, or DOE, and the Nuclear Regulatory Commission, or NRC. In an attempt to divest AEC/ERDA/DOE of its regulatory functions and to prevent the conflict of interest inherent in the agency's dual roles of promoting nuclear technology and regulating it, the NRC was mandated with regulation.)[10] Mancuso refused to publish: he felt it would be premature and might be misleading, given the long latency of cancer.[11]

Then in 1974, a bomb fell in the form of a paper published by a Dr. Samuel Milham in the health department of Washington state. Milham found a higher proportion of cancer deaths for men who had worked at Hanford than for men of comparable age in other industries—about a 25 percent higher death rate.[12] Mancuso himself hadn't known about the Milham report until he heard about it from the AEC. "There was a tremendous flurry of activity," Mancuso recalls; "I was called by the AEC and Hanford and was on the phone by the hour over a period of weeks." His project director, Dr. Sidney Marks, "had prolonged discussions, frequent discussions with me on the telephone relative to Milham's findings and also relative to my potential response to the press."

Marks dictated a press release for Mancuso to issue. He wanted

Mancuso to say that his years of research and the negative results they'd produced contradicted Milham's findings—that there was "no evidence" of cancer or other deaths occurring more frequently among Hanford workers than among their brothers and sisters who did not work at Hanford.[13] "But I told them they couldn't use my study to counteract Milham's," said Mancuso; "My findings were much too premature. For all I knew, Milham might be right."

At this point, he says, he sensed a "chilling" in the agency's dealings with him.[14]

"He was naturally rather upset," says Alice, "and, as a responsible scientist, he refused to bend to pressure. He felt that an insufficient number of years had elapsed to show what was going on. But he felt he needed some advice, some peer review, so he had the research circulated to all members of his steering committee.

"Come November and Mancuso is trying to call me in Oxford, not realizing I had moved to Birmingham. He finally traced me to Birmingham—could he send me his findings? I had no idea why he wanted to, but I said I'd take a look."

Not Dead Yet

"George said immediately, 'there's something very wrong with this study—if you accept this at face value, then Hanford could be a health farm.' So we wrote a critique, saying that though it may look like those who have been exposed to radiation have longer lives, you can't accept these differences between the live and the dead at face value—this could be an overhasty summary of data that may well be pointing in the other direction. Besides, the live people are *not yet dead* so you can't be sure that this situation is going to stay looking favorable.

"This idea of the *not yet dead* was important—there were far more living than dead, something like 85 percent of the people were still alive. So it was obvious you had to take this into account, since if you're looking for the effects of radiation, the real judgment has got to be when everyone has completed their life span. You have no right to make assumptions about what is going to happen to the living; for all you know, they're going to be contributing to the cancer deaths. But the general attitude seemed to be that anyone who was still living was all right.

"We were the only members of the steering committee who gave an informative reply, so I got a telephone call from America asking if I would come over and take a closer look at the data. 'Well, yes,' I said, 'I'm quite free to'—there I was, newly retired and with no commitments—'but the person you really want to see is George.' Mancuso said, 'bring him too.'

"George hadn't been in very good health. He'd never been on a plane trip, let alone to America, and I had no idea if it was a good idea to ask him to come. I remember approaching him tentatively, asking, 'will you come? I have no idea what's in it, but I can pretty well guarantee that we'll have our expenses paid.' And he just said, 'yes!' We set off with so little luggage that we had to buy him clothes.

And that is how a British statistician and epidemiologist ever came to have anything to do with the American nuclear controversy." That is how Alice entered the scene, striking one observer as "tall, white-haired, elegantly efficient in the manner of Katharine Hepburn."[15]

"We arrived in Hanford in the summer of 1975, and George, who is very shy and doesn't talk to anyone, went straight to the computer and started making calculations. I remember how surprised Mancuso was at George's immediate ability to master everything—here he was, a mere junior assistant who knew exactly what to do. George hadn't spoken a word until he surfaced for a moment and said, 'there is something very odd about this industry which you, Dr. Stewart, must find out. I, in the meantime, am going to do a study of people who have completed their life span.' He was going to compare people who died of cancer with people who died of other diseases in much the same way we'd done in the Oxford studies.

"So here was I being asked to figure out what was wrong with the American nuclear industry. Now I must tell you that I hadn't the slightest idea about the kind of radiation these workers were getting—I only knew about the x-rays you use in a machine. The idea that there were these radioactive particles flying about because we had split the atom—it hadn't even dawned on me that we were in this sort of situation.

"All I had was a series of printouts and notes and various things, so there I was nosing around the records and trying to get myself familiar with what the whole thing was about, and asking people a lot of stupid questions and chatting up Mancuso's assistant, when I stumbled on something strange.

"I was looking at a list of workers. There would be the man's name, date of birth, and so on, and then a dose and then a configuration of four letters in a row which read either NNNN or YYYY or any combination of these. 'What is this symbol that always seems to appear on these lists?' I asked Mancuso's assistant. I was told, 'Oh, you needn't bother about that, that just tells about urine testing.' Well I wasn't about to be put off that way. I asked what that meant and was told it stood for *no* or *yes* to whether the worker had ever been urine tested or worn a badge—that is, whether he'd been monitored for internal or external radiation. The first two letters stood for whether he'd been monitored for internal radiation and, if so, whether he'd had a positive or negative reading. The second pair was the same thing, only for external radiation.

"You see, there are two risks to workers making weapons: external radiation, in the air, which works like an x-ray and doesn't stay inside the body; and internal, which gets inside from being ingested or inhaled and goes on emitting radioactivity. Workers were monitored for external contamination by wearing a badge and for internal contamination by urine testing—you can tell from urine testing if there's been internal contamination, since part of the radioactivity will be excreted. Badge monitoring was very extensive—about 90 percent of the workers wore badges—but only about 50 percent of them were monitored by urine testing. Everyone had been looking at the badge monitoring until I spotted these records for the urine testing. This was a side of the story that hadn't been told.

"So I said to George, 'look, do this—put this one in.' We were able to show that there were many more YYs among the living than the dead, which meant that the living had been monitored more closely than the dead—which meant that they'd been put in the more dangerous jobs.

"And from then on it was clear to me that there had been selective recruitment of exceptionally healthy people into the most dangerous jobs.

"You see, those who are employed by any industry tend to be healthier than the general population. You have to be healthy to get these jobs—physical exams are required, regular screening—whereas the population at large has lots of people too ill to work, invalids and elderly. This is why the accepted method in most worker studies, the Standardized Mortality Ratio (SMR)—the ratio of the death rate among workers to the death rate in the general population, adjusted for age—doesn't prove anything. When you compare workers with the population at large, workers almost always come out looking healthier. It's very easy to impress people this way but it doesn't mean a thing.

"High-tech industries employ a large proportion of people with university degrees and technological qualifications, which means that they come from a higher socioeconomic status than the general population. The educated tend to be in better physical condition than others—they have lower rates of smoking and other exposures, they're more resistant to disease, they have a better life expectancy. The nuclear industry employs an even larger proportion of such people than most high-tech industries.

"But there's something else about the nuclear industry. Normally the professional element in an industry is put in the safest jobs, they live in the safest conditions, and it's the workmen who go out and dig the coal or mine the asbestos. Whereas in the nuclear industry, it's the best educated and the healthiest who get placed in the highest risk situation—because, for the riskiest operations, you need someone who understands the technology and you need someone supervising safety.

"So we realized that if you are a member of this industry you will have to be exceptionally healthy and you are likely to be healthier still to be put in the most dangerous sort of work. This was why those who had received the most radiation seemed to be living longer: we were seeing a *healthy worker effect,* and what's more, an *internal* healthy worker effect, that is, a further effect within the industry, where the healthiest are in the riskiest jobs.[16] This was not a representative or homogeneous population we were looking at, and if it looked like the cancer death rate among the workers was the same as or even lower than that of the general population, we could conclude nothing on this basis. But if we found *any* cancer effect at all, we'd better pay attention.

"We also realized that Hanford hadn't been in operation long enough for us to see what was happening. We were dealing with a very young industry, an industry that was literally born in 1944, unlike, say, the coal mining industry, which would be full of people who had been in it all their lives. If you were to continue studying these men long enough, a different pattern might emerge. These workers were alive, but, as George pointed out, 'they're *not dead yet,*' the story's not over. In fact, the story is still unfolding—time hasn't yet told. The industry will have to be older yet before we know the secrets of radiation.

"Then George did a study of the dead and found evidence of a radiation effect. The overall dose for people who'd died of cancer, the total amount of radiation collected by each of these individuals, was significantly higher, according to badge monitoring, than for those who'd not died of cancer. And it was largely the result of radiation received at least ten years before death, and mainly by people who were *over forty* when exposed."

Alice and George were finding that the radiation effect was turning up more in older people than in younger. George was to conclude that by age fifty, a person is as sensitive to radiation as before birth and that he or she is least sensitive at twenty years of age. This put them in major disagreement with the A-bomb studies, which said that young adults are more sensitive to radiation than older.

They were discovering that the cancers were of three kinds, and these findings also put them at odds with the A-bomb studies: lung cancer, pancreatic cancer—which nobody had ever noticed in the A-bomb survivors—and cancers of the blood-forming tissue, which affect tissues of the RES, or reticulo-endothelial system, tissues that are known from animal experiments to be sensitive to radiation. RES neoplasms normally account for between 5 and 10 percent of all cancer deaths, but at Hanford they were accounting for far more. RES neoplasms were turning up in the Hiroshima survivors, but there researchers were finding leukemia, not myeloma; whereas at Hanford Alice and George were finding myeloma. Myeloma would soon—within the next three years—turn up in the survivor population.

Alice and George were finding a definite relationship between low levels of radiation and the development of certain types of cancer, in spite of the fact that these workers had been especially selected for health. It was statistically significant, occurring at a dose that was minute compared to what was allowed, a tenth of the safe dose—and in general, *well below a tenth* of what the International Commission on Radiation Protection (ICRP) said was safe. The doses received by workers at Hanford were very low compared to what workers got in Sellafield or la Hague—the United States has better technology and worker protection than facilities in England or France. There was virtually no one there who'd exceeded the maximum permissible annual dose set by the ICRP, of five rem per annum. This implied that people were far more sensitive to radiation-induced cancer than anyone had thought—as much as twenty times more sensitive.

Alice in Blunderland

Mancuso's funding was cut off, he was ordered not to publish his findings, and he was denied further access to the workers' data—on the grounds that the study was being moved to Oak Ridge and his retirement from Pittsburgh was "imminent." Actually, he was sixty-two and had no intention of retiring and would have been allowed by the university to work until age seventy.

After twelve years and $5.2 million, the Energy Research and Development Administration (formerly the Atomic Energy Commission) removed Mancuso from the study.[17] In the fall of 1977, he was ordered to give up his files or have them seized. An ERDA officer actually did manage to seize the data from Oak Ridge, where part of the study was transferred, but Mancuso had kept a complete file at Pittsburgh. ERDA sent several summons, but Mancuso remained firm—firm but shaken: "I don't know how to handle people like this," he said; "they come at you with official documents and lots of legal talk and try to scare you. They just keep at you and at you. They never let up. And believe me, I get scared."[18]

When he refused to yield the data, ERDA asked the University of Pittsburgh vice chancellor to intervene.

"They tried to get Mancuso's university to cooperate with them," recalls Alice; "but the university fortunately saw that there would be implications for other research and would have nothing to do with this dirty work. Mancuso sent George and me home with a copy of the data, everything he had. He thought the data would be safer in England, and he was right—it was important that we have that copy.

"So the door clanged shut in our faces against the further examination of data. And I used to think Mancuso was being paranoid. Practically everybody who took our side in America lost their funding—Irwin Bross, Rosalie Bertell, John Gofman, Carl Johnson.

"He was wonderfully loyal, Mancuso was, he bravely put his name to our study. He was subjected to character assassination, very badly. They tried to make out that he had got some of his information illegally. And all because of a report for which George and I are largely responsible. We didn't suffer, but Mancuso did, very badly. He had to stay in the country and take the flak, whereas we could go back to England and carry on."

("I thought this sort of thing only happened in Russia," Mancuso said. But he never regretted calling Alice in on the project—he said it was "the most important step that I took on the study.")[19]

"We got copies of the data across the Atlantic, so the work could go on. We worked on it for free, in our spare time. I had a pension from Oxford and could afford to; George worked on it because he was fascinated. Mancuso dug into his personal retirement money." With assistance from the Environmental Policy Institute, a public interest organization headed by Bob Alvarez, Mancuso managed to carry on his research for a few more years.

The MSK team, as it was now called—Mancuso/Stewart/Kneale—had enough evidence to publish, even without the confiscated data. The three scientists went public with their results in a 1977 issue of *Health Physics,* a journal edited by Karl Morgan.[20] "We were finding that there was a 6 or 7 percent increased cancer effect," reports Alice. "It wasn't much of an effect, but the shock was that there was any effect at all since the cancers were occurring at radiation exposure levels well below the official limit of five rads per year. It meant that the current standards for nuclear safety might be as much as twenty times too high."

But even a tenfold reduction of the limit would alter the economics of nuclear operation beyond recognition, as Karl Morgan explains: "Were we to reduce the maximum permissible exposure by a factor of ten, I seriously doubt that many of our present nuclear power plants would find it feasible to continue in operation."[21] Moreover, to admit that standards had been too lax would open a floodgate to compensation claims being made not only by radiation workers but by hundreds of thousands of veterans and downwinders.

There were powerful incentives to keep the researchers away from the data.

Unsafe at Any Dose

"For years and years our prenatal study had been the only evidence that there was anything dangerous with low-dose radiation, but now, at Hanford, we were turning up further evidence. At Hanford we were looking at people being exposed day in and day out over a period of time to doses only a fraction higher than background radiation, and we were finding a cancer effect. We were finding an effect at levels comparable to those absorbed by the general public. This meant there was a serious health hazard not only to workers in the atomic energy industry, but to the general public as well.

"The guidelines say if you control the dose and stay beneath the limits, which are laboriously spelled out in lots of publications, there will be no risk, or a negligible risk, a risk too small to be measured—therefore you can forget about it. The official view is that if you split the dose and reduce the dose rate by spreading it over time, risk becomes negligible. The idea is that dose fractionation, as it's called, mutes radiation's dangers by giving time for the cells to repair.

"That would be correct if radiation worked like a sleeping pill," Alice explains; "five pills on five nights don't have the same overwhelming effect as five on one night. But radiation doesn't work that way. Any dose of radiation, however small, is going to have an effect in the body—*any* dose has the potential of damaging a cell. Dose fractionation produces *more* effects over longer periods of time, which makes it *more* dangerous, since each effect increases the chance of a mutation. One reason for this is that low dose may injure cells rather than kill them. A single high dose of radiation has a greater chance of killing the cell, and the body can lose cells without injury; but a surviving mutated cell can cause cancer or a birth defect.[22]

We were challenging the official story that says, if the dose rate is reduced and given over time, you'll get less cancer—and if you lower it enough, you'll get no effect. This principle (the term is Dose Rate Effect of Radiation, DREF), attached to linear extrapolation from high dose, is the basis of the standards adopted by all government committees. Now we were saying, lower doses received over time might actually produce *more* cancer per unit of exposure than a single large dose. Dose fractionation *increases* the cancer risk."[23]

Karl Morgan explains, "We had, all of us, a serious misconception in that we adhered universally to the so-called threshold hypothesis, meaning that if a dose were low enough, cell repair would take place as fast as the damage would accrue. In other words, we believed there was a safe level of radiation." Now along comes Alice who says that "damage per unit dose is greater at . . . low levels."[24] The Oxford Survey, and now the Hanford studies were challenging the linear extrapolation from high to low dose and were saying that there's no threshold beneath which radiation is safe.

Morgan soon came to agree with her. He announced in the September 1978 issue of the *Bulletin of Atomic Scientists* that he thought the current radiation risk had been underestimated by a factor of ten.[25] Since he was chairman of the International Commission on Radiation Protection, the commission that set the standards, and since he had, as he says, "for a quarter of a century supported rapid expansion of the nuclear industry," this caused quite a stir.[26] "When we first began to assert this, many people thought we were crazy."

Alice herself was shaken by her findings. "Not even I had thought the effect of such a small dose would be as great as this. The Oxford Survey had upset the medical profession, but I now saw that these findings would upset a more formidable set of authorities.

"It came at me all of a sudden that what we were going to say would shock the world."

Aftershocks

"I can remember the airplane skidding into the runway in Heathrow and George was beside me and he of course never speaks, and I said, 'Mission accomplished.' And I said, 'by the way, George, I hear that they're going to have an inquiry about Windscale, and I think as good citizens we ought to be prepared to speak out about this situation.' And he agreed.

"At this point I was sure that we'd be contacted by the industry to find out what we knew. I assumed that the sensible thing was for them to get hold of us as soon as they could, find out what all the fuss was about, and try to convince us to see things their way. No such thing—we never heard from the industry at all, though we got innumerable calls from the other side. And it was from the opposition that we found out what the industry was saying about us.

"It seems that our work was being sent all over the world to collect opinions, mainly adverse, and these opinions were being circulated and sent to every nuclear installation in the world. Leaflets were being sent to atomic plants around Britain, saying that it had all been a mistake, that there was no need to pay any attention to us. Suddenly everyone knew about us. People whom I'd never heard of were writing me and asking me questions, asking me for data."

In the United Kingdom the National Radiological Protection Board, the government's powerful regulatory adviser, produced a special report rebutting Mancuso/Stewart/Kneale (MSK). In the United States, scores of published papers, most by the DOE-sponsored epidemiologists, challenged MSK's methods and conclusions. "There was very harsh criticism of our 1977 paper which was orchestrated by ERDA-DOE and followed by what was described as a 'federally sponsored' 'reanalysis of Hanford data.' "[27]

ERDA was circulating its critiques without making them available to MSK for response. Mancuso only managed to get hold of them through the Freedom of Information Act.[28] After obtaining them, the researchers prepared responses and sent copies to each of the government officials and their consultants, but ERDA continued to circulate the critiques—without the authors' responses.

"We were eager to get on with the research, but we didn't have a nickel to do it with," Mancuso said. Nevertheless, MSK managed to publish five more papers within the next several years, based on more refined analytical methods and an enlarged database.[29] Their findings continued to be published in refereed journals—they passed the test of independent peer review, despite the formidable efforts of the government to discredit them.

"Because of the strong criticism of our findings, we updated them with a larger sample and instituted more careful checks for control procedures, making sure we weren't confusing the effects of radiation with other features of the workers' experience, and the checks have shown that we certainly had not exaggerated the risks the first time round," Alice told the 1978 congressional investigation.[30] "We were confirming the original Milham finding."[31]

ERDA continued to critique Mancuso/Stewart/Kneale for its disagreement with the A-bomb findings and to circulate its critiques. "They said our methodology was wrong. Their own studies were based on the standard method, the SMR, Standardized Mortality Ratio analysis, which compared workers with the population at large. We pressed the point that it wasn't fair to compare the industry with the population at large because workers in the nuclear industry are healthier than average.[32]

"George developed a subtle way of measuring dose that followed a man getting radiation exposure every year of his life as his age changed, taking him through time and adjusting for variables such as the year of exposure (since ways of measuring exposure depended on the year). His method revealed that sensitivity to dose increased with age. Our critics invoked the A-bomb data to assure us that it did not.

"They said our methods were no good since they'd never been used by anyone before. We were doing something for the first time—well, that's hardly a reason for not doing it, because it hasn't been done before. The experience of workers in places like Hanford, hundreds of thousands of them, functioning in an environment of chronic, low-level radiation exposures—this was unprecedented too. A situation where you've got exact records of when workers came into the industry, what jobs they held, with doses recorded on a badge and often measured in a urine test—this gave us many more variables than anyone had before. Naturally we had to develop new methods; older methods of analysis did not apply.[33]

"They said our population was too small and therefore our database is weak compared with the strength of the size of the A-bomb survivors. We said, 'well, then, let us expand our population. *Give us the records of the nuclear workers.*' "

It seemed a simple enough request, but it was to take the U.S. government nearly two decades and several Freedom of Information Act lawsuits to comply.

Chapter 9

Taking on the International Nuclear Regulatory System

"The A-bomb studies have set standards that are patently false."

"It was simply unacceptable," says Alice. "We were turning up effects ten to twenty times higher than those claimed by the A-bomb studies." Naturally, she found this troubling, for these are the studies that set radiation safety standards for the entire world. They determine international guidelines for licensing new nuclear facilities, for deciding risks and benefits, and for settling the claims of veterans, workers, and downwinders. Who were Alice Stewart and George Kneale to go up against this Goliath?

It pushed her, as she says, "to think about the situation. Because you can see, they weren't fitting with what we were finding, they weren't fitting to an extraordinary degree. We said we really must take a closer look at the A-bomb studies, there's something very wrong."

Damage Control

"With a violent flash that ripped the sky apart and a thunderous sound that shook the earth to its foundation, Hiroshima was pounded to the ground."

"In an instant, the city was reduced to ash and rubble. A fireball of several million degrees centigrade formed, and from the incinerated buildings rose a huge cloud of dust and ashes, casting the city in darkness."

"Then from where a city once was, a huge column of fire bounded straight up toward heaven. A dense cloud of smoke rose and spread out, covering and darkening the whole sky. . . . Fires broke out all over and soon merged into a huge conflagration. . . . Out of the fierce whirlwind, half-naked and stark naked bodies darkly soiled and covered with blood, began moving. . . . When their hands hung down, the blood accumulated in the fingertips and caused throbbing pain, so they held their hands up and forward; burned so badly that their skin peeled and hung loosely, their raw hands and arms oozed and dripped blood."

Human beings evaporated into dust. "No one knew what had happened, only that people turned to charcoal here and there . . . or were tossed through the air . . . or crushed under falling buildings. Limbless or headless bodies rolled about or piled up like logs on the ground, and around and between them writhed the still living, their flesh torn and tattered."

"In corpses near ground zero the eyeballs were blown outside their heads. The skin was a black-tinged yellowish brown and very dry; it was clear that these persons had died in agony."[1] Those who lasted longer suffered longer agonies, dying of the failure of vital organs, lungs, kidneys, liver, their skin so burned that it peeled off to the touch, their stomachs so badly destroyed that they vomited chunks of them along with blood. Those who could threw themselves into the river to quell the burns. Many lay charred and writhing, maggots festering in open wounds.[2]

The essential life-support systems of the city ground to a halt. Eighteen hospitals and thirty-two first-aid centers were destroyed or rendered useless; most of the doctors, nurses, and hospital staff were killed or injured.[3] A few weeks after the bombing, a typhoon swept through Hiroshima, killing untold numbers of convalescents, and the winter that followed, 1945–46, was unusually harsh. In both Hiroshima and Nagasaki, survivors were without adequate food, shelter, and health services for several years. Water was contaminated, there were severe shortages of drugs and antibiotics, infections were rampant.

Many who had seemed to come through the blasts unhurt began—within hours, days, weeks—to manifest alarming, mysterious symptoms. They were astonished when their hair began to fall out by the handful, when black spots appeared on their skin, when muscles contracted, leaving their limbs and hands deformed. Only gradually did they begin to understand that these were no ordinary bombs.[4]

Of all issues surrounding the new weapon, the question of radiation poisoning was the most sensitive: it seemed comparable to the issue of poison gas, which warring nations had stockpiled but rarely used. When Tokyo Radio announced that people who entered the cities after the explosion were dying of mysterious causes, American officials dismissed the allegations as propaganda intended to imply that the United States had used an inhumane weapon. General Leslie Groves, who had headed the Manhattan Project, ordered a team of Manhattan Project doctors and technicians in, with the mission of proving "there was no radioactivity from the bomb." The first scientists and doctors allowed into the cities, in late 1945, were with the U.S. Armed Forces. Most journalists also entered the cities under U.S. military escort and similarly concluded that tales of radiation poisoning were groundless. A *New York Times* headline proclaimed, "No Radioactivity in Hiroshima Ruin."[5]

Wilfred Burchett, the first independent journalist to enter Hiroshima without an army escort, told a different story, warning that a month after the bomb fell, people were dying "mysteriously and horribly" from "an unknown something which I can only describe as the *atomic plague*." "In these hospitals I found people who . . . suffered absolutely no injuries, but now are dying from the uncanny after-effect. . . . Their hair fell out. Bluish spots appeared on their bodies. And the bleeding began from the ears, nose, and mouth. They have been dying at the rate of 100 a day."

Burchett managed to get his story through the censorship office, probably because it was written for a British publication. (Not a single article appeared in U.S. publications describing conditions in hospitals.) When he got back to Tokyo, he was confronted with a press briefing given by General MacArthur's staff to discredit his sensationalist reports and discovered that his camera containing film shot in Hiroshima was missing. MacArthur ordered him out of Japan.[6]

The A-bomb was not "an inhumane weapon," General Groves insisted, and Occupation authorities carefully suppressed material that might indicate that it was.[7] Japanese doctors and scientists who had been on the scene and were eager to publish their observations about this bomb that produced symptoms for which no treatment was known, were ordered to hand over their reports and found Japanese medical journals heavily censored. Many became suspicious that data were being manipulated in the interests of denying future compensation claims from victims.[8]

In late 1945, U.S. Army surgeons issued a statement that all people expected to die from the radiation effects of the bomb had already died. No further physiological effects due to radiation were expected in subsequent years.[9]

It was in this highly politicized situation that the Atomic Bomb Casualty Commission came into being.

Official Stories

Two years after the bombs were dropped, in 1947, the Atomic Bomb Casualty Commission (ABCC) was established by the U.S. National Academy of Sciences to study their health effects on the survivors. In 1975, funding was partially shifted to the government of Japan and the ABCC was given a new name, the Radiation Effects Research Foundation. (It was agreed by both Japanese and Americans that the words "atomic bomb" should be removed from the name.)[10]

The actual studies did not begin until 1950, five years after the bombs were dropped, when Japan took a national census. Citizens were asked where they had been on the fateful days of August 6, 1945, and August 9, 1945, and a survivor population of 195,000 was identified. Over the next ten years, about half of the survivors were interviewed and asked where they were standing in relation to the blasts and whether they'd experienced symptoms associated with radiation exposure—whether they'd suffered bleeding, coughing up blood, blood in the stool, purpura spots (vivid, purple spots in the skin that are evidence of bleeding), patches of spontaneous bruising, acute lesions of lips and tongue, hair loss. On the basis of the symptoms they described and their distance from the explosions, the dose they'd received was estimated. Dose estimates ranged from low to over four hundred rads, with every possible gradation in between, though estimates were approximate, taken on trust and memory, and dose reconstruction was crude. The radiation the bombs gave off was calculated on the basis of tests conducted in the Nevada desert.[11]

The blasts had created a kind of human laboratory for studying the effects of radiation. The survivors became the largest population of humans exposed to radiation for whom estimates of doses are available. "Ah, but the *Americans*—they are wonderful," exclaimed Japan's radiation expert, Masao Tsuzuki, who lamented that he'd had to confine his experiments to rabbits; "It has remained for them to conduct the *human* experiment!"[12]

No medical study ever has had such resources lavished on it and so many scientists involved. It had—and still has—staffs of hundreds, scientists from all over the world; it has state of the art equipment, computers, and data-gathering and analyzing facilities. Since the enterprise is so large and lavishly funded, and since in epidemiology, the larger the sample, the greater the statistical accuracy—or so it is believed—there has been a tendency to accept these studies without question. The ABCC/RERF has produced "pages and pages, volumes and volumes of official reports, unofficial reports and what have you," as Alice says.

There are many powerful committees concerned with radiation protection, and they all accept the calculations of the Radiation Effects Research Foundation (RERF). The International Commission on Radiation Protection (ICRP), the United Nations Scientific Committee on the Effects of Atomic Radiation (UNSCEAR), the International Atomic Energy Agency (IAEA)—all agree that RERF estimates are applicable to all situations involving radiation risks, including nuclear work and medical x-rays. In addition to the international committees are the national committees—the National Radiation Protection Board (NRPB) in England and the Biological Effects of Ionizing Radiation Committee, a special committee of the National Academy of Sciences (NAS), in the United States. The national committees are free to modify the recommendations of the international committees, but in practice they do not—they take the word of the International Commission on Radiation Protection and the ICRP takes the word of the Radiation Effects Research Foundation. The ICRP's 1990 report, ICRP 60, is based on RERF data.

The U.S. committee on Biological Effects of Ionizing Radiation (BEIR) is the most prestigious of the national committees. It has produced several reports, numbered in sequence, BEIR I, BEIR II, and so on. The 1990 report, BEIR V, is commonly regarded as the gold standard for radiogenic risk estimates and the most comprehensive overview of the health effects of radiation. It too is based on RERF data.

The RERF assumes a linearity of dose-response relationships: you can move down a line from high to low dose and can figure out, according to a principle of linear extrapolation, radiation risk at low dose. BEIR V and ICRP 60 proclaim that if the dose is small and is accumulated over time (as it is in the case of nuclear workers), cancer risk will be *less* than that predicted by linear extrapolation. It has been assumed that if the dose is low enough, risk will disappear entirely; this implies there is a safe level of radiation, a threshold beneath which radiation presents no danger (though this position is being modified, as we'll see).

When the ABCC began its study in 1950, it concluded that the population had returned to normal; the RERF concurred. According to their calculations, the death rate from all causes except cancer had returned to normal.

"A normal death rate means a normal population, they decided. They admitted that the blasts may have had short-range health effects that might be missed by a study starting five years after the disaster, but they have stuck with this position that within five years the population had returned to normal.[13] There is nowhere in RERF studies the slightest doubt expressed that they are dealing with a normal population—and therefore with a representative population," Alice explains.[14] "The only delayed reaction they acknowledge is cancer and birth defects—very few birth defects, at that.

"It's nonsense! It's rubbish! It would have been impossible to undergo the worst holocaust in recorded history and pop back to normal in five years—it couldn't have happened! My every medical instinct said it couldn't possibly have happened.

"Common sense tells you that the death rate dropped because you've already killed off the weaklings. Thousands of victims must have died within months of the bombings and thousands more in the years to come. During the winter of 1945–46, when both Hiroshima and Nagasaki were battling hunger and cold, when people were fighting to stay alive without adequate medical services or protection from the weather, anyone with a weak constitution, anyone who was infection-sensitive, would have died off. But these don't have a part in the story.

"Besides, those who'd survived the blasts were damaged—they were physically changed by the high doses of radiation, psychologically changed by the trauma they'd lived through. My experience as a doctor has shown me there are many types of trauma from which you never recover. You cannot recover. I saw from the London air raids that the people who went through those were never going to be the same—certainly not after five years. And there was no radiation in that story. I once read about a flood in Bristol—one of the few disasters that's ever been studied: after the flood, the death rate from *every* cause went up. There was one death from drowning, but that was the least of it—there was this sort of generalized disaster effect, from shock, stress, infection, bad water. And you get this from a tiny disaster, without the added horror of radiation. *Imagine the case in Hiroshima.*

"If you get severe trauma, if you get destruction of tissue for any reason, you may cope with it, but you will not be the same person you were—you are changed. Given an accident on this scale, there will be irreparable damage, even without the effects of the radiation."

Colonial Science

It is extraordinary, the blind trust in the A-bomb studies, considering the guesswork that went into them.

The Americans encountered a variety of problems doing a study of survivors in an Occupied country halfway around the world, a country they had little understanding or sympathy for. From the beginning, the Atomic Bomb Casualty Commission set out to study rather than treat the victims, an objective that did not endear it to the population. It was said that the ABCC "does nothing but experiment on patients and has as its ultimate goal an autopsy."[15] The bad feelings toward the project were exacerbated by many insensitivities: patients were summoned to the ABCC clinic for examination during working hours, which might cost them a day's wage, for which they were offered the consolation that they were contributing to scientific knowledge. There was also the segregation of the facilities—American and Japanese doctors ate in separate dining rooms.[16]

There were many difficulties gathering data. The identifying of the survivors was based almost entirely on voluntary self-identification, dependent on the survivors coming forth with their experiences. Many survivors had fled the cities, and there was difficulty recruiting subjects and a constant attrition of subjects.[17] Many were reluctant to discuss intimate details with American researchers; a woman interrogated about her abortions, spontaneous or induced, might be understandably reticent. Much of the day-to-day work of the genetics project was carried out by Japanese midwives who acted as links between the subjects and the ABCC; early on, it was clear that they were holding back reports of malformations, stillbirths, or neonatal deaths, knowing that such reports would attract the attention of the ABCC and involve the family in bureaucracy and social stigma. There was shame attached to being a victim of the bomb's radiation, based partly on the aversion to contamination strong in Japanese culture, and partly on fears that the survivors would be unable to produce healthy children.[18]

Problems of data gathering were compounded by the fact that the systems for accounting for people—record-keeping offices, hospitals—were not functioning in the months after the bombings. Assessments of how many people were killed remain to this day inconclusive. There were difficulties estimating the dosage. Levels of exposure depended on atmospheric humidity, on the exact position of the body at the moment of detonation, on whether it was shielded in any way from the explosion and if so, on the properties of the shielding material. They depended also on the subject's activities immediately following the explosion and whether he or she had ingested radioactive materials.[19] Such information was difficult to come by, especially five years after the event.

Added to this was censorship on the part of the U.S. military that minimized the effects of fallout and made it impossible for Japanese scientists to publish their findings. The project was assigned to the National Research Council, the research arm of the National Academy of Sciences, which was the only body with sufficiently high scientific status to undertake it. But ABCC funding depended entirely on the Atomic Energy Commission. AEC members sat on its committees and members of the ABCC Advisory Committee served on AEC committees—indeed, nearly every radiation researcher in the United States in the fifties was engaged in the atomic bomb program.[20] In 1950, the ABCC took almost one quarter of the AEC's medical research budget for the year. It is not surprising that the AEC "was interested in the operations of the ABCC to a degree bordering on intrusion."[21]

The expensive new laboratory complex on Hijayama Hill, overlooking the broad river delta of Hiroshima, was completed in December 1950. "It had nineteen departments, structured to resemble a university medical school, including pathology, bacteriology, biochemistry, hematology, parasitology, radiology, serology, biometrics, vital statistics, employee health, and public relations." The structure, of seven two-story-high concrete half cylinders, was said to resemble a fish cake. It inspired resentment and derision.[22]

But the studies produced here have remained the last word on the subject of radiation risk.

Survival of the Fittest

"*Imagine the situation in Hiroshima.* Imagine the physicians—they hadn't a clue about what they were seeing. On-the-spot observers, even those with medical training, were witnessing a situation without precedent. They didn't know what to look for; they had no names for what they were seeing. Most had never heard of radiation sickness. What was happening to these survivors had never been observed before. Later, the data gatherers came along and asked questions based on what had come to be known about radiation damage—'did you have hair loss, bleeding, bruising?' and so forth."

Most on-the-spot observers were in any case Japanese, and the ABCC was suspicious of their data and discounted or suppressed much of it. "But the ABCC data gatherers who came along later were not physicians who had had direct experience of human illness," Alice points out. "They were health physicists and nuclear physicists, radiobiologists, people with M.A.s or Ph.D.s in biology or physics, biostatisticians—number crunchers. They were not people who had dealt directly with human illness, who had a feeling for the body—they got their knowledge of the body through textbooks." So there was a situation where those who were on the scene with medical experience had little or no information about the biological effects of radiation, and those who came later to collect the data had little or no medical training.

Alice's experience as a physician told her that the survivors represented a highly select population. "They are *by definition* survivors, exceptionally tough, because only they would have survived the awful mess of the first five years." She describes a *healthy survivor effect* something like the *healthy worker effect* among Hanford workers, a selection on the basis of strength.[23] But how to explain this curious fact that the death rate had returned to normal? If the survivor population consists of those strong enough to survive the debacle, should it not have a *lower* death rate than normal?

Alice was able to account for this, again, drawing on her experience as a physician: "Yes, they're exceptionally tough, but they've also been damaged by high doses of radiation. The death rate looks normal because there's this other effect canceling out the healthy survivor effect, and therefore, what would have been a lower than normal death rate is brought back within the range of normal."

In what way, exactly, are they damaged? "Only those with superior immune competence would have survived, and it would be their immune systems that would be impaired.

"Look at what happens in a health emergency: when you have an attack of pneumonia, white blood cells rush to the defense and engulf the bacteria. The white cell count goes up to fantastically high levels at the height of recovery or attempted recovery periods. With the very high dose of radiation that came from the bomb, the white cells came rushing to the rescue, but when they did, they were hit by the radiation—which created just the right circumstance for causing cancer. This puts leukemia at the top of the list of cancers to come, since leukemia is a mutation of the white blood cells."

Leukemia is a cancer of the reticulo-endothelial system (RES), the blood-forming tissues and bone marrow that, together with the lymph nodes, spleen, and thymus, are the basis of the immune system. Alice finds it useful to think of RES neoplasms etiologically, as a family, and to think of them as cancers of the immune system. There are several kinds of RES cancers, she explains, diffuse and localized. The diffuse forms are the leukemias, cancers of the white blood cells. The localized RES cancers are tumors in the lymphatic glands, the lymphomas—Hodgkin's and non-Hodgkin's lymphoma—and myeloma, which invades bone marrow and appears as a tumor of the spine. (Diffuse forms of a disease are always more dangerous than localized forms—the leukemias are more lethal than the lymphomas.) Tumor forms take longer to manifest, which explains why in the mid-seventies, when Alice and George were at work on the Hanford data, myeloma had not yet appeared among the A-bomb survivors—though within the next three years, it would become prominent.

Damage to the white cells produces leukemia; damage to the red blood cells produces aplastic anemia, a type of anemia that is particularly recalcitrant to treatment. "You try to treat it as you do other forms of anemia, with transfusion and iron, and find that it won't respond because the source of the red cells is damaged, there's some primary fault in the marrow. The marrow refuses to function, to produce hemoglobin and other constituents."

A high dose of radiation like that delivered by the atomic blasts would severely damage the bone marrow, which is the source of both red and white blood cells and is, of all tissues, the most susceptible to radiation. Bone marrow has remarkable powers of recuperation: it can be almost totally destroyed and recover, or seem to recover, in the sense of return to normal cell count. But animal studies show that bone marrow damage leaves the organism more infection sensitive; the tissue appears to recover, but under stress it can no longer produce the cells that are needed to fight infections. "You may look normal, but when things go bad, you have no tolerance, you can't respond to crisis; you're more prone to death from all causes, especially infection.

"So the survivors now have this compromised immune system. Their systems might be adequate to the day-to-day job of keeping them alive, but when they meet with infection they get wiped out because they have no resistance. Leukemia and lymphoma, cancers of the immune system, also leave you more prone to infection, as the Oxford Survey demonstrated.[24] So what gets recorded as a death from infection or unknown cause might well be a death from a cancer, or from a condition that would have developed into cancer. Many who died before the government study even began must have died on account of infection sensitivity because they were incubating cancers.

"Imagine the physicians and data gatherers. The question no one asked was, *'did you get damage to your immune system?'* And even if they had asked it, there would have been no answer because there would have been no outward and visible sign. It would have been the commonest effect, but it would have been invisible. Anything even resembling the *epidemic of acute bone marrow depression* that must have occurred as a result of high blasts of radiation was unheard of, and unlooked for.[25]

"There are so many things going on that the cancer story gets masked, but we can be sure that the cancer effect has been grossly underestimated and that cancer wasn't the only late effect of A-bomb radiation. George and I reckon that fully ten times more people died of radiation-related effects than the official studies estimate."[26]

Silent Forces

"It's two effects of the blast canceling one another out, neutralizing one another and producing what looks like normalcy. It's blinded everybody.

"In 1982 I published a paper expounding this theory. I called it the 'Theory of Silent Forces.' There are two forces operating in opposite directions and creating a false appearance of normality. There is, on the one hand, selective survival of exceptionally fit individuals; there is, on the other hand, a loss of immunological competence from bone marrow damage in the survivor population.[27]

"The RERF had no use for my theory. They had found things were back to normal, and that was that. Why have two theories where one would suffice? According to the principle of Occam's razor, the simplest thing is the most likely, and my theory was unnecessarily untidy.

"I was criticized for being overly subtle. It was said I was searching around for reasons—well, yes, I admit to it freely, I am a scientist, and I am searching for reasons. But I say they've accepted a gross oversimplification of what is medically speaking, a very complex story. It's remarkable, given this complex situation, that they have been satisfied with this simple assumption that the population has returned to normal.

"Of course they knew on some level that they were dealing with a survivor population—it was short of babies and old people, which told you straightaway that something had happened. But they never did anything with this. I spoke to a geneticist who'd been in the A-bomb studies from the early days. He said, 'oh yes, you're not saying anything new; from the very beginning we all assumed that there would be a strong selection effect, and we had all sorts of meetings about how we were going to handle it. But we found nothing. So we assumed it had disappeared.'

"It spontaneously disappeared by 1950? There had been an abnormal situation that lasted for five years, but by 1950 everything was back to normal? There is no biological way that could happen. Yet analysts of the RERF data have continued to find the normal non-cancer death rate reassuring rather than puzzling.

"They have accepted too much at face value."

Something else Alice feels has been taken at face value: the A-bomb studies conclude that radiation affects the young and healthy more than the old, since they found more cancer in the young. "But this would make radiation unlike almost any other known cause of death. Anyone coming at this from a first-hand experience of medical problems would be perplexed by this conclusion. Older people are less able to rebound from onslaught of any sort—infection, sickness, trauma.

"How did they come up with this? The young adults were the ones most likely to live long enough for the cancer to manifest—so it looks like they are more vulnerable; the old had already, most of them, been bumped off. Who knows how many of them might have developed cancer, if they hadn't died of other causes, but these cancers never became part of the story. So when we began finding an age effect at Hanford, we were naturally assumed to be wrong. We found that sensitivity to radiation increases with age, such that the addition of eight years age more than doubled the risks. And this makes intuitive sense—the older we get, the more vulnerable we become."[28]

Alice concludes that the survivor studies are not a reliable indicator of low-dose radiation risk because the survivors are not a normal or representative population. "You cannot learn about low-dose radiation from a population exposed to high dose. Once you've exposed people to high dose, you've altered the picture. If you want to study the effects of low-dose radiation, you need to start with a healthy population, expose them periodically to a specific agent and keep track of their dose so that you know the exact amount and type of exposure they got—which is exactly the situation we have with the Hanford workers.

"You cannot use the A-bomb studies to estimate radiation health effects for the general population. The nuclear workers provide a much more reliable standard.[29] Since they are working in exposures near background radiation, receiving radiation in small dollops over time rather than in one dose, they are also the best population for understanding the effects of background radiation, which is what concerns the public most.

"Talk about building a house on sand—this is like building a house on a wave! The errors and errors that have been perpetuated!"

RERF errors have skewed all subsequent calculations about the cancer effect of low dose radiation. They have also confused the question of genetic defects. The official story is that the survivors and their descendants provide a suitable population for estimating the genetic effects of radiation and that few genetic defects have turned up. "But again, you've been left with the healthiest—the weak didn't survive to become part of the story! Records indicate there was a high incidence of infant mortality following the bombings, and a high rate of abortions. Few people were able to have children during the first years after the blasts; probably there were many spontaneous abortions, damaged fetuses that didn't survive long enough to become part of the statistics.

"How can you possibly say what genetic effects there would have been, on the basis of this population?"

Excluding the Noise

"Back in the seventies, when I heard that the *in utero* exposure to A-bomb radiation had not produced any signs of leukemia, I wrote a letter or two and was invited to speak to the BEIR Committee. And I said, 'don't you think there would have been a disaster effect that would have made it likely that *in utero* cancers had been killed off and make it impossible to see properly what was happening?' and I cited the flood in Bristol. Well, they didn't want to hear my theories about the flood in Bristol. They consulted and got back to me, saying, 'even conceding that Dr. Stewart might be right about a disaster effect, it would have had a very short duration.' So the BEIR II report has a paragraph saying they've considered the issue and decided there was no effect. And that was that. And I think I am right in saying that the next time anybody protested was when I wrote again in the 1980s and this time argued, 'couldn't we possibly be taking two diametrically opposite effects of the radiation and confusing them with no effect? Let's put this theory on the table and leave it for analysts to pick up and work out.' Well, nobody picked it up. They said it was impossible to test.

"When awkward findings come up, the RERF sweeps them under the carpet—they ignore whatever doesn't fit their interpretations. They don't do what a scientist must do and say, 'this seems impossible, but if we can't figure it out at this time, let's at least not forget about it.' Too many things get forgotten.

"For example, they have not looked at the non-cancer deaths with anything like the same attention that they've given the cancer deaths—they've simply decided they don't have anything to do with radiation.

They've got any number of different groupings for the cancers and pages and pages of details about cancer, but they've taken all the non-cancer deaths and put them into six groups: two cardiovascular, which account for approximately half; tuberculosis and digestive diseases, both of which are fairly large; blood diseases; and miscellaneous. And blow me down if the *miscellaneous* isn't the biggest category of the six. It includes deaths from every type of infection not mentioned, all respiratory and renal infections, and a very large group of ill-defined symptoms where you die for no obvious reason—'a precise diagnosis was not made.'

"Why aren't they looking at these groups? Tuberculosis and blood disease are right up to the sky. But it had been decided that there was no correlation between these and radiation, so they weren't worth considering. Yet tuberculosis and miscellaneous infections might well be linked to radiation, since radiation makes a person more infection sensitive; and blood diseases have an obvious relation to bone marrow damage.

"There are other things that get ignored. One of the complaints the survivors constantly make is that they feel tired all the time. They say every time they get an infection it sticks to them—they can't get rid of it. According to the official line, they're being neurotic, they're making it up. But it's far too consistent, and it is exactly what you'd expect to find with latent bone marrow damage. I'd no idea that people were actually complaining about these things, and when I first read these stories, I said that's exactly what I'd imagined.

"Then, too, there's the question of aplastic anemia. The rate of aplastic anemia has always been higher in the survivors than the Japanese national average, but the ABCC/RERF analysts decided that what the doctors had diagnosed as aplastic anemia was actually leukemia. Why? The whole of their story hangs on the notion that the only late effect of radiation is cancer, so they have to turn any late effect into cancer, or they'd have to acknowledge that there was another effect.

"If you bring all these together, all these bits and pieces that have been pushed under the carpet, you see that they amount to quite a lot. But nobody ever adds up the things that don't fit. Still the great engine grinds on saying there's no delayed health effect to radiation except cancer.

"Their rule is, however many exceptions there are, just ignore them, and by all means, keep Stewart and Kneale away from the data. The data is so fiercely guarded that it might as well be locked up. The RERF has been as protective of the Hiroshima/Nagasaki data as the DOE has been of the Hanford data. Whenever I've gone near, it's *keep off the grass:* 'we have to get permission from so-and-so and permission for such and such.' George and I have had to make our calculations on the basis of what's appeared in publications.

"I did ask the people in Washington for access and they said, 'oh, you have no idea how fussy the Japanese government is; they won't let us have these data.' Of course they've got a sealed copy in their vault, but forget about that. The Americans have paid enormous sums of money for it—*blood money*—and effectively, they run it. They say it's a joint venture between them and the Japanese, but if you ask them for it, they say it's the Japanese's responsibility, and if you talk to the Japanese they say it's the Americans'. It's the usual runaround."

Flat Earth Policy

Alice has been dismayed at the dogmatism she's encountered from the RERF. "Most scientists don't say, 'I'm absolutely sure of this finding, this conclusion is indisputable.' No, they say, 'here are the data, here is my proposed explanation, but if someone else comes along and explains the data differently, I listen.' But for mainstream radiation experts to say absolutely that anyone coming after me must find the same thing—this is dogma, not science.

"In other areas, people are allowed to change their minds and let their imaginations range. But the radiation field is so politicized, there's so much invested, ideologically and economically, that these scientists cannot change their minds. They simply cannot say, 'we may have been wrong.' RERF dogma has become almost a monstrous cancer itself, dominating the scene in such a way that all sorts of critical faculties have been put to sleep."

Politicized the field has been, and politicized it remains. As Sue Roff suggests, the researchers did not expect to find residual effects from radiation, and so they found none.[30] The Atomic Energy Commission was, in effect, paying their salaries, and the AEC had strong reasons for downplaying the effects of radiation and maintaining a safe threshold theory since it was, in the same years the Hiroshima data were being interpreted, defending above-ground testing of nuclear weapons and promoting the development of nuclear energy.[31]

"The data continue to be used to argue that low doses of radiation are negligible," exclaims Alice; "They are still regarded as the gold standard for setting safe exposure levels. They hang on like grim death." They hang on despite several reanalyses of the original estimates of how much radiation the bombs gave off, reanalyses that call into question all further calculations.[32]

"The RERF and the international standard-setters have underestimated the number of cancers produced by the bombs. They have underestimated radiation's other effects—genetic damage, immune system damage, lowered resistance to disease, infection, heart disease. These are serious misrepresentations because they suggest that it's safe to increase levels of background radiation.

"The approval of standards based on this skewed data is irresponsible and actually wicked."

The Standard-Setters

The International Commission on Radiation Protection (ICRP) came into existence in the late twenties, to review what was known about radiation and draw up guidelines by which workers and the public would be protected. It set permissible radiation doses of up to 75 rem a year for doctors and dentists and those working with radiation. By 1936, ICRP had reduced its recommended permissible doses for workers to 50 rem, then to 25 in 1948, to 15 in 1954—and to 5 in 1958. There the level has remained.[33]

But "there is something disturbing about the repeated assurances, 'this time folks, we have got it right,' " comments Dr. Morris Greenberg, a senior public health official for the Health and Safety Commission in Britain, "when on each occasion, a previous understatement of hazard is revealed."[34]

The ICRP has set a standard of five rem per annum exposure according to a risk-benefit principle that allows a one in five thousand chance of contracting cancer. The standard was proposed with full knowledge that there will be health injury and death to some, but it was decided that the overall societal benefits of atomic energy would outweigh damage to the few.

The five rem standard has been "plucked out of a hat," says Alice.

Who or what is the ICRP, that it has the power to make these decisions?

"I'm not sure it's an organization I would trust with my life," says Karl Morgan, one of its founders and a member for more than a quarter of a century. After the war, it came to consist mainly of physicists who worked on the Manhattan Project. Its members have associations with the military and with the medical radiological societies, all of which have vested interests in promoting the use of radiation and downplaying its risks.[35] Morgan suggests that the bias in favor of these special interest groups, the nuclear industry and medical radiology, invites conflict of interest.

"It's a self-perpetuating committee right out of the military," says Rosalie Bertell, author of *No Immediate Danger.* Its membership is "highly selective and controlled." "It is in every sense of the term, a closed club and not a body of independent scientific experts." In order to belong, you have to be proposed by a present member and accepted by the executive committee. There has been one woman on it, but never an epidemiologist. It has never taken a stand in favor of public health on any of the controversial issues such as nuclear testing, exposure of uranium miners, medical uses of radiation, or radiation experiments on humans.[36]

Alice is not quite so strong in her condemnation. "Membership is pretty incestuous," she acknowledges; "It's like the representatives of large corporations; you find the same type of people on all the boards and often the same people. But I think they have the best intentions." Asked whether most members are employed by the nuclear industry, she assents, "That is an unfortunate fact. It is very difficult to find experts who are not, who haven't been employed in one or other branch of the nuclear industry all over the world, and it is very heavily weighted in that direction. But I maintain that the biggest single weakness is that they have not had represented what I call basic epidemiology. There are very few biologists, geneticists or specialists in occupational health, for that matter. They are heavily represented by health physics and radiobiology."

Yet, as Alice has found, "Anyone associated with these committees has no trouble getting access to the Hiroshima data. Whereas I arrive and I get turned away without a how-do-you-do."

The A-bomb studies continue to be invoked as the standard by which anomalous findings are discounted.[37] Nevertheless, Alice Stewart's criticisms have had their effect. In 1976, around the time the Oxford Survey findings were finally altering medical practice, the ICRP acknowledges that *all* exposures carry a degree of risk. Each successive BEIR report has increased the estimated cancer risks of low-dose radiation,[38] and in 1990, BEIR V acknowledged that there was probably no threshold beneath which there was no risk. In 1995, Britain's National Radiation Protection Board issued a report acknowledging that we should "assume a progressive increase in risk with increasing dose, with no safe threshold."[39] This acceptance of a no-threshold hypothesis was at least partly in response to the Oxford Survey of Childhood Cancer and the Mancuso/

Stewart/Kneale study of Hanford workers. However, the ICRP and BEIR and other standard-setting committees continue to adhere to a linear dose-effect model and continue to maintain—as Alice does not—that there is a proportional relationship between dose and effect,[40] claiming, as well, that their estimations err on the side of caution.

As Patrick Green, of Friends of the Earth, summarizes: "What has happened over time is that science is moving in Alice's direction. Science is catching up with her—she has shifted the terms of the debate. She was very stubborn, and she was right."[41]

But "there's a lot of double-talk in the air," says Alice; "they give with one hand what they take with the other. They agree *in principle* that there is no safe dose, no threshold, but they assume *in practice* that there is a level below which radiation is negligible;[42] and they continue to set standards on the basis of a risk-benefit principle. They hedge their bets." Dr. John Gofman, former head of a major division at the Lawrence Radiation Laboratory, agrees: "all these committees like the ICRP say, 'yes, we consider there's no threshold, yes, we operate on a no-threshold hypothesis,' but let a Three Mile Island happen, let a Chernobyl happen, and all these fine statements about thresholds go out the window; they keep backing into the safe threshold idea, when all the evidence is pointing in exactly the opposite direction, showing a supralinear model, with damage at low dose."[43]

The science is shifting in Alice's direction, but it still has a long way to go.

Multiple Injury Data

Alice did manage, in 1995, to get access to a part of the A-bomb data that had never been studied, and she used it to test her theory that the survivors had been selected in favor of resistance to radiation damage. She heard, quite by chance, that ABCC researchers had collated data on the survivors with and without acute radiation injuries, but that they had decided these data were of no use. Alice hypothesized that those with the highest level of multiple acute injuries might be considered the closest approximation to those who died before the study population had been assembled, the non-survivors, and that studying them might reveal something about the survivor population.

She got hold of the data when Hazel O'Leary became DOE head, under President Clinton, and appointed Alice's old friend Bob Alvarez to a high-level position within the department. Sure enough, she and George found a strong association between the survivors with multiple injuries and cancer deaths—more of them got cancer, with a given level of dose, than did those without multiple injuries.[44] "This shows that the survivors who don't have any injuries are a highly select group, quite different from the people who died, selected in favor of resistance to radiation. It also suggests that what was assumed to be a single homogeneous population—though somewhat short of people under ten and over fifty—actually consists of two populations, those with and without multiple injuries. Those with multiple injuries are nothing like those without them, and it's wrong to lump them together and conclude that this is a homogeneous population—or a representative population." Alice believes that this is strong proof for her theory that there were opposing forces canceling one another out and creating the appearance of normality—that there was selection of the fittest, on the one hand, and marrow damage, on the other hand.

"If we make this critique heard, we will have delivered a body blow to the stories that form the basis of international standards on low-level radiation." But they are having difficulty making it heard. Their paper has appeared in the proceedings of the conference at Ben Gurion University, Israel (November 1996), where Alice first presented the findings, but it is not having an easy time finding a publisher.[45]

Chapter 10

Rogue Scientists

"It smells, doesn't it? when those who are assessing the danger of the industry are in the pay of the industry. It's like the fox guarding the henhouse."

After the Mancuso affair the Department of Energy split the study of the workers' records into groups, assigning them to government contractors in Hanford, Oak Ridge, and Los Alamos. From now on, research on nuclear health was to be performed solely by labs under contract to the DOE.

"Divide and conquer—in this case, literally," says Alice. "The government determined never again to let the records fall into the hands of one person. And it determined not to allow any researchers from outside the DOE to have access—now only their own people could come anywhere near them. From now on, the research would be strictly in-house."

Mancuso's data on the Hanford workers were given over to Battelle-Pacific Northwest Laboratories, a major contractor at Hanford, located on the Hanford site—which meant that Hanford was now in the position of evaluating the health of its own employees. The study was put under the direction of Sidney Marks, formerly of the Energy Research and Development Administration (ERDA, formerly AEC, soon to be DOE)—the same Sidney Marks who had urged Mancuso to issue the press release contradicting Milham's findings and who by this time had left the federal agency to take a position with Battelle. (Battelle had been awarded $80 million in government business by ERDA.) It happened that in transferring Mancuso's work to Battelle, federal officials violated the usual protocol: they failed to put the contract out for competitive bidding, to submit the contractor's past work or present capabilities to peer review, or to ask Battelle to submit a research design. It also happened that Battelle had never done a human epidemiological study before.[1]

Mancuso's Oak Ridge data were assigned to Dr. Clarence Lushbaugh at Oak Ridge Associated Universities. Lushbaugh had no one on staff with the expertise to handle the project; this assignment was made, again, without proper protocol.[2]

The data on the nuclear workers became the virtual monopoly of a small group of government-sponsored scientists and were made unavailable to the larger scientific community. Research on the nuclear workers now went on entirely without reference to the Mancuso/Stewart/Kneale findings, without reference to their caution about the healthy worker effect or the sensitive methods George Kneale was developing to measure exposure over time. "It's as though the MSK study never happened—except for the rebuttals that were flying about the world," comments Alice.

Behind the walls of secrecy, ERDA, then DOE carried on research without independent oversight, shielded from the open processes of scientific inquiry or public accountability. The agency had established a principle of state ownership of science. "Talk about the fox guarding the chicken coop," says Alice: "here you have the fox reporting on morbidity and mortality in the chicken coop."

Not surprisingly, this research turned up no cancer hazards for low-level radiation.[3]

State-Controlled Science

But the Mancuso affair had opened a Pandora's box, and try as the government might to put the lid back on, the scandal was attracting the attention of scientists throughout the world.

Dr. Joseph Rotblat, who would win the Nobel Peace Prize in 1995, wrote in the *Bulletin of Atomic Scientists* in 1979 that the Mancuso incident highlights the "contradictory objectives" of the AEC/ERDA/DOE, to promote nuclear energy and "prevent its misuse," and has ramifications "beyond the U.S. administrative arrangements, because similar situations (whereby the same body is given the task of protecting against the harmful effects of products it promotes itself) can be found all over the world."[4] Dr. Karl Morgan sent a scathing letter to Secretary of Energy James Schlesinger: "One can only suppose that the new Oak Ridge team must get the *right* answer, i.e., prove there is *no* radiation risk to Hanford workers, if it cares to have a continuation of funding."[5]

The scandal raised an outcry from concerned citizens groups, workers, and unions demanding that Mancuso be allowed to continue his study. In a joint letter to the Secretary of Energy in November 1977, representatives of the Oil, Chemical and Atomic Workers Union and seven environmental organizations charged that the termination of Mancuso's contract "reflects a well defined pattern of harassment and intimidation of scientists who do not agree with promoters of radiation technology." The United Steelworkers Union, the International Association of Machinists and Aerospace Workers, and several public interest groups issued a statement to Joseph Califano, Secretary of the Department of Health, Education and Welfare, demanding that Mancuso's funding be restored and research on the health effects of radiation be removed from the DOE.[6]

"The suppression of scientists is something that supposedly only happens in countries like the Soviet Union," said Bob Alvarez, then of the Environmental Policy Institute; "no one seems to realize that it's happening right here."[7]

So began a struggle to wrest radiation research away from the DOE and put it under the auspices of a more disinterested agency, a movement that would succeed—though it would take decades—in transferring the research to the Department of Health and Human Services.

From Silkwood to Stewart

It was the lobbying efforts of Bob Alvarez that persuaded Congressman Paul Rogers (Democrat, Florida) to hold a congressional subcommittee to look into the Mancuso affair. Alvarez had been working for the Environmental Policy Institute, a nonprofit public interest group in Washington, DC, and had become interested in the effects of radiation on public health.

Alvarez had been drawn into the Mancuso affair by a somewhat serendipitous route: his wife, Kitty Tucker, had been instrumental in bringing a case against Kerr-McGee, the plutonium processing plant at Crescent, Oklahoma, on behalf of the Karen Silkwood family and had enlisted Mancuso as a witness on the family's side. Silkwood was a feisty young technician at Kerr-McGee who, alarmed by conditions at the plant, had approached AEC officials in Washington with testimony that Kerr-McGee was violating health and safety regulations. She promised to produce documentation; some weeks later, she was found to be contaminated with plutonium—inexplicably, the source of the contamination was located in her apartment rather than the plant. Several weeks after that, on November 13, 1974, her car went off the road as she was driving to meet with a labor union official and a *New York Times* reporter with the intention of presenting evidence supporting her allegations. The Oklahoma highway patrol ruled that she'd fallen asleep at the wheel, though later investigations by a private detective found signs that the car had been pushed from behind. The documents were not in her car; they were never found.

The incident had been investigated, briefly, by the FBI and a congressional subcommittee but had been dropped until two years later, when Alvarez's wife, Kitty Tucker, became interested in the case and, together with Sara Nelson of the National Organization for Women, formed a pressure group to lobby the Justice Department and Congress to get a proper investigation. Tucker and Nelson were idealistic and charismatic young women determined to get to the truth of the Silkwood case. Thanks to their efforts, Silkwood's parents filed a lawsuit.

With uncanny timing, the case came to trial just as the film *The China Syndrome*—a tense thriller starring Jane Fonda and Jack Lemmon and dramatizing a nuclear-power-plant emergency—opened to packed theaters nationwide. Two weeks after this, Three Mile Island came to the brink of a major nuclear disaster. The federal jury found against Kerr-McGee and awarded damages of $10.5 million, almost the full claim, to the Silkwood estate—a precedent-setting decision, the first time a jury awarded damages to a victim of radiation contamination from a nuclear facility.[8]

In preparing her case against Kerr-McGee, Tucker called Thomas Mancuso as star witness, and this was how Alvarez got wind of the Mancuso story. He initiated a congressional investigation into it, then another—and another. "The hearing on Mancuso was part of a series of hearings that began to look into the legacy of the federal nuclear program on atomic veterans and workers," he explains; "together they drew attention not only to the censoring of government scientists, but the exposure of 250,000 or more military personnel used as guinea pigs during the atomic bomb tests." Alvarez went on to become a widely known critic of nuclear interests and coauthor of *Killing Our Own: The Disaster of America's Experience with Atomic Radiation.* When Hazel O'Leary became Secretary of Energy under Clinton in 1992, she appointed him deputy assistant secretary for national security and environmental policy at the Department of Energy.

Through Mancuso, Alvarez met Alice and became, as he says, her "chief U.S. promoter and publicist."[9]

Congressional Rebukes

The congressional investigation into the Mancuso affair was held in January and February 1978. Alice was pleased that the matter was finally getting a public airing and traveled to Washington in midwinter to testify.

Mancuso protested that he had been subject to "governmental interference;"[10] and indeed, the hearings turned up a whole host of contradictions around the termination of his contract, revelations that were the source of some embarrassment to ERDA. Sidney Marks defended the termination on the basis of a review ERDA initiated that (Marks claimed) had turned up evidence of "poor performance." (Mancuso had not been informed this review was going on, so he'd had no chance to defend himself.) But Congressman Rogers had before him a summary of the reviewers' remarks written by one of the participants indicating that their "consensus" had been that the Mancuso study was a good one and that the "University of Pittsburgh should continue as the prime contractor." Of the six reviews, only two were unfavorable; only one of those had recommended termination—and even that one had gone no further than to suggest that Mancuso be transferred to another school of public health.

In the questioning that followed, Marks had to admit that the peer reviews he'd described as critical had actually been "quite strongly favorable."[11]

ERDA also claimed to have been dissatisfied with Mancuso because he'd refused to publish his findings. Mancuso pointed out that he *had* in fact published "the equivalent of a monograph of 180 or more pages in the proceedings of the Health Physics Society . . . in 1971, which includes precisely the methodology, plan of study and some preliminary results." He said he'd refused to publish his results prematurely because they might lead to false negative findings and be misleading. He pointed out that there was never any question of the "quality" of his work as long as he was coming up with negative findings: on the contrary, as long as his findings were negative, he was being urged to publish.[12]

(Later, documents obtained under the Freedom of Information Act from the DOE revealed that federal officials had plans to use Mancuso's negative findings to fend off workers' compensation claims and for public relations purposes.)[13]

Alice was asked whether in her opinion, Mancuso's study had been deficient. "No," she replied, "I would like to say that I don't think one minute of time had been wasted." She explained that the A-bomb studies had raised false expectations about the nuclear industry, lulling the world into assuming there would be *no* cancer effect from low-level exposures, and that this assumption, together with the negative findings Mancuso first turned up, had built "a climate of opinion that we were dealing with a safe industry." The shock was not that we'd discovered a great cancer effect, she said, but that we'd discovered any effect at all.[14]

The investigation found that Mancuso had been terminated without justification. Congressmen charged that his firing was "insupportable," that the decision to transfer his study to Oak Ridge was "highly questionable at best" and possibly illegal, and that the reasons given for transferring the study were "bunk." Rogers referred to an "attempted government cover-up" and concluded, "we need some policy changes. I am not sure but what I question the whole concept of allowing the Department of Energy to conduct research work." He noted that since Battelle had never done an epidemiological study, it would have made more sense to have turned the study over to the Centers for Disease Control than to the DOE's "friends."[15]

But in spite of the congressional rebuke, in spite of a congressional recommendation that the Justice Department undertake a thorough investigation, and in spite of a recommendation on the part of an interagency task force set up by President Carter that the studies be transferred to the Department of Health and Human Services—the DOE stood firm. It refused to release the records and retained its control of its $60 million a year budget for research into the health and environmental effects of radiation. The press largely ignored the incident.

The nuclear industry continued to claim that it was a "clean industry," that nuclear operations were safe, that there had been no deaths in the industry and no serious accidents. It remained firm in its insistence that existing health and safety precautions were protecting the workforce and the public. It promoted its claims with the costly and extensive public relations campaign it had cranked up in the early days of Atoms for Peace, using full-page newspaper ads, brochures, booklets, and television time. Always it made its assertions with total certainty, arguing—whether in Hanford, Oak Ridge, or Sellafield—that nuclear power was essential to economic growth and prosperity, that the industry brought jobs.

But try as industry and government might to silence the voices of opposition, they were getting louder.

The Anti-Nuclear Movement Revives

As the Mancuso scandal was coming to a head, a powerful international anti-nuclear movement was making itself heard. After the Partial Test-Ban Treaty between the United States and the Soviet Union was agreed upon in 1963, the bomb vanished underground and the anti-nuclear movement receded; but in the seventies, as governments and the utilities industries began to move forward with their plans to build reactors, opposition revived and the isolated protesters of the sixties were brought together by national groups such as Friends of the Earth, the Union of Concerned Scientists, and consumer advocate Ralph Nader. Suspicions about the industry were fueled as U.S. citizens came to realize, from Vietnam and Watergate, that their government was not above lying to them. By the late seventies, there was widespread uneasiness about nuclear power.

The Mancuso affair became a cause célèbre,[16] and Alice found herself at the center of an international controversy.

On both sides of the Atlantic, protests erupted, first at local, then at national levels, as groups of concerned citizens and scientists organized to oppose construction of nuclear plants. Citizens' groups informed themselves about the science, economics, and dangers of the new technology and learned to use public hearings and the licensing system to delay construction.

In 1966, when Long Island Lighting Company announced plans for building a nuclear plant in Shoreham, in the town of Brookhaven (site of Brookhaven National Laboratory), a coalition of environmental and civic groups formed to oppose the facility and demand a hearing. The Shoreham inquiry was one of the earliest public inquiries; it went on from 1970 to 1972. In 1970, Alice's Oxford Survey of Childhood Cancer was the only proof positive that low-dose radiation had a cancer effect, and activists called on her as an expert witness. In her testimony, Alice described her findings, emphasizing that they challenged the claims of the industry and the AEC that there was a "threshold dose" below which radiation would not cause cancer. Dr. James Watson, winner of the Nobel Prize for codiscovering the DNA molecule, warned, "you have to pay a great deal of attention to the prevention of cancer" because we may not find a cure for it: the mechanisms that cause it appear to be "integral components" of the body's system that can become activated by agents including radiation.[17]

Plans for the plant were nevertheless approved and construction began in 1972.

In the mid- to late sixties, local organizations opposing particular plants joined together into regional and national coalitions. Demonstrations escalated into casts of thousands that occupied plants, blocked access, and committed acts of civil disobedience. In February 1975, the first large-scale anti-nuclear demonstration occurred in Whyl, in the Rhineland's wine country. What began as a small, local event accelerated to thirty thousand demonstrators when police brutality and media attention brought supporters from all over Germany and France. Demonstrators succeeded in occupying the construction site for a year and actually halted construction of the plant. Other mass actions followed as plans were announced to build nuclear power stations in locations all over Germany.[18]

Whyl provided the model for the Clamshell Alliance, a coalition of environmentalists, local fishermen, and clam diggers protesting the Seabrook site on the New Hampshire coast. The occupation of Seabrook by more than 30,000 in April 1977, followed by the arrest of 1,414, established nuclear energy as a major political issue. A Clamshell rally in June 1978—barely a year before the Three Mile Island accident in 1979—drew some 20,000 protesters, making it the largest U.S. antireactor action ever and creating a proliferation of other demonstrations throughout the country, including the Abalone Alliance movement against the Diablo Canyon reactor in California. Alice was—again—called in as expert witness.

In February 1976, the West German government announced plans for a reprocessing and waste disposal plant at Gorleben, a sparsely populated site near the East German border. The government assumed that the remote location would allow the project to proceed unnoticed, but within weeks, more than 15,000 people had turned out to demonstrate. In May 1979, hearings were held in Hanover on the proposed plant, and 100,000 people turned out to protest. The hearings coincided with the Three Mile Island accident; Alice, called in as expert witness, recalls, "the accident was reported while we were there. The town was out in banners, there were demonstrations in the street." The government voted against the nuclear reprocessing plant later in 1979, making Gorleben one of the great successes of the anti-nuclear movement in Europe.[19]

As the international anti-nuclear movement gathered momentum and the Mancuso affair attracted more attention, Alice began making frequent trips to the United States, testifying in congressional hearings and compensation cases, speaking to community groups and public inquiries, and giving papers at scientific conferences. She felt it was her responsibility to speak out, for, as she says, "the number of people with expertise in this area are few, and of these few, most are dependent on government for their research grants."

Rogue Scientists

The anti-nuclear movement has always been highly dependent on the support of scientific experts because nuclear technology is complex and radiation is invisible: its detection and understanding require special knowledge, expertise, training, equipment.[20] The nuclear establishment has used this complexity to mystify issues and has, further, invoked national security to shroud the subject in secrecy. Since most scientists are directly employed or indirectly funded by industry or government, those who have been willing to challenge industry or government are few and far between.

Most who have spoken out have paid a price.

There were from the beginning scientists who dissented. Joseph Rotblat walked off the Manhattan Project when he realized that Germany had abandoned its bomb project. Attempts were made to prove that Rotblat was a spy, and on the train from Washington, DC to New York, his papers and research notes mysteriously vanished, never to be found. (Rotblat later helped found the anti-nuclear group Pugwash, an organization of scientists from around the world that helped bring about the Partial Test-Ban Treaty.)[21]

Other scientists who had been eager to be part of the war effort were to recoil in horror from subsequent events—Leo Szilard, Robert Oppenheimer.[22] As the military took control, scientists found themselves with less and less say about the use of nuclear physics. Within a year of the Hiroshima and Nagasaki blasts, the United States was exploding bombs in the Marshall Islands and shortly thereafter, the United Kingdom began its own tests in the Christmas Islands. In 1951, the United States brought the tests home to the Nevada desert and between then and 1963 exploded more than one hundred bombs in the atmosphere. Eventually, the eleven thousand scientists Pauling got to sign his petition helped bring about the 1963 Partial Test-Ban Treaty. In 1944, concerned scientists formed the Federation of Atomic Scientists, the organization that published the *Bulletin of Atomic Scientists,* the powerful voice for an end to the arms race. In 1961, the Physicians for Social Responsibility came into existence and published a series of landmark reports on the medical consequences of nuclear war; it defined the prevention of thermonuclear war as a whole "new area of preventive medicine."[23]

In the late sixties, as Alice was expanding the Oxford Survey of Childhood Cancer and critiquing the AEC's claims about a safe threshold, there emerged another set of scientific challenges to government reassurances.

Dr. Ernest Sternglass, born in Berlin in 1923 to parents who were both physicians, had worked for Westinghouse as a researcher but soon found that his concern with radioactive fallout conflicted with his job. In

1967 he joined the faculty at the University of Pittsburgh to head the newly created lab for radiological physics.

"And then I ran across the work of Dr. Alice Stewart," he said. "When I saw her figures . . . I realized that a nuclear war could destroy the next generation. It was madness. . . . But the military keeps assuring us, 'yes, there's evidence of a safe threshold of radiation below which nothing will happen. And therefore you can have a nuclear war.' "[24]

Alice liked Sternglass, though she first met him in an adversarial stance, when she was brought to Hanford by the AEC to correct his use of her work on x-rays and childhood cancer. She found him a genial, generous man of enormous energy who responded to her rebuke by inviting her to Pittsburgh as his guest. Sternglass's 1963 article in *Science* may have got Alice's figures about fetal x-rays wrong, but it did turn up something important: it suggested that when nuclear testing began, the rate of infant mortality stopped declining and that each rise in infant mortality could be correlated with peaks in the testing. In 1969 he published a powerful article in *Esquire,* "The Death of All Children," aimed at Congress, which was then deliberating a proposed Anti-Ballistic Missile system. Sternglass contended that by the mid-sixties, some 375,000 American infants had died "in excess of normal expectations" as a result of testing—"Each nuclear test meant a loss of thousands of babies."[25]

Sternglass's findings were vigorously contested by the AEC. In 1969, one of his studies came across the desk of Dr. Arthur Tamplin at the AEC Lawrence Livermore Laboratory in northern California. Tamplin had been asked to refute it, and he did find that Sternglass had overstated the case—his own figures indicated that fallout had probably killed 4,000 infants, not the 375,000 Sternglass had estimated.

The AEC did not want to hear that fallout had killed *any* babies. The agency approached Tamplin's boss, Dr. John Gofman, director of radiobiological studies at Livermore, urging him to get Tamplin to publish these figures in an obscure, specialist journal, so as not to draw attention to them. Gofman told them to "go to hell, essentially," and Tamplin published his paper in the *Bulletin of Atomic Scientists.*

Gofman was a brilliant nuclear chemist who had discovered a way of separating plutonium from uranium that had provided the Manhattan Project with plutonium for its bombs. After the plutonium work, he completed medical school. With a Ph.D. in nuclear/physical chemistry and a medical degree, he taught molecular cell biology at the University of California at Berkeley. In the sixties, he accepted a position as head of a biology and medicine lab at Livermore: "the AEC was on the hot seat because of the 1960s testing. They thought that maybe if we had a biology group working with the weaponeers at Livermore, it would take the heat off." But since he didn't trust the AEC, he arranged to be allowed to return to his teaching position if he chose. He then set to work with a $3 million a year budget.

By 1969, Gofman and Tamplin were coming up with findings that corroborated Alice's. They went public with these in October 1969. They concluded that there was no basis for the AEC's claim that there was a so-called safe threshold of radiation and estimated that the cancer risk from radiation was roughly twenty times worse than previously thought, and that the hazard to future generations in the form of genetic damage had been underestimated even more seriously. Their staff and budget were slashed, their work censored, and they became known as "the enemy within." Gofman resigned in 1972, calling Livermore a "scientific whorehouse," and resumed his professorship at Berkeley. Tamplin resigned in 1974.[26]

Gofman and Tamplin began appearing at public debates, hearings, and anti-nuclear rallies. Even strong nuclear proponents had to recognize them as highly reputable scientists, and their criticisms gave the anti-nuclear position considerable credibility. Following the publication of their findings in 1969, criticism of allowable standards mounted. Other eminent scientists joined in, including Nobel laureates James Watson, Harold Urey, chemist at the University of California, and George Wald, biologist at Harvard. Dr. Henry Kendall, a MIT physicist and former consultant to the Defense Department, founded the Union of Concerned Scientists in 1969. Kendall drew in consumer advocate Ralph Nader, who spearheaded the anti-nuclear campaign in the United States.

Gofman, a sprightly, gnomelike man with lively, compassionate eyes, is now retired. He lives in San Francisco and heads an independent research institute, the Committee for Nuclear Responsibility. He has worked indefatigably to publicize the dangers of radiation and has published several books and a regular newsletter. He speaks feelingly of the wrong that has been done—by scientists like himself. He wrote, in 1981:

> At Nuremberg we said those who participate in human experimentation are committing a crime. Scientists like myself who said in 1957, "maybe Linus Pauling is right about radiation causing cancer, but we don't really know, and therefore we shouldn't stop progress," were saying in essence that it's all right to experiment. Since we don't know, let's go ahead. So we were experimenting on humans, weren't we?

But now that we *know* that nuclear plants "release radioactivity and kill a certain number of people," Gofman concludes, "scientists who support these nuclear plants—*knowing* the effects of radiation—don't deserve trials for experimentation; they deserve trials for murder."[27]

Defrocked, Defunded, Debunked

"Everyone in America who took our side in the years subsequent to the Mancuso incident lost their funding," says Alice. "They don't burn you at the stake anymore, but they do the equivalent, in terms of cutting you off from your means to work, your livelihood."

Mancuso never worked again. Tamplin and Gofman were cut off from all further government grants. Sternglass was a tenured professor with patents of his own and so he was financially beyond the grasp of industry or government, though not out of range of the character assassination that he and those associated with him continue to suffer. He has been retired for some years and continues to publish, working with Jay Gould, a statistician.

Karl Morgan became a pariah to most of his former colleagues for his criticism of nuclear policies and his outspoken support of Alice Stewart. In 1971, scheduled to speak at a nuclear symposium in Nuremberg, about to give a presentation that critiqued the AEC's new reactors, he had his first experience of government censorship: Oak Ridge ordered the Germans to destroy the two hundreds copies of his paper already printed and forced him to read a revised version. In 1980, he was dropped from his post at Georgia Institute of Technology. A former colleague describes him as having "gone round the bend."

Morgan, now retired, a tall, gentle man with a slow southern accent, shrugs off the criticisms and says, "I don't have to please the boss. I don't have to leave out things in my research to get the next raise, or to hold my job, or to be on the list of candidates for a prize."[28]

Those without academic salaries or pensions have had a harder time of it.

Dr. Rosalie Bertell had been working on the Tri-State Leukemia Survey, at Roswell Park Memorial Research Institute in Buffalo, with Dr. Irwin Bross. When in 1973 she spoke out against building a nuclear power plant in an area next to farms that grew produce used in Gerber baby food, she "experienced a smear campaign in the newspapers and reprisals and censoring at the cancer institute where I worked." Bertell is a mild-mannered, unassuming woman, a nun and a mathematician who had done research on radiation. "Given my fairly innocent speech, I was taken aback at this reaction," she said. "I was very naive. I had no idea what I was wandering into."[29]

The Tri-State Leukemia Survey, inspired by the work of Alice Stewart, was designed to study the increase in leukemia that had taken place in the fifties. It drew on tumor registries in New York, Maryland, and Minnesota and followed sixteen million people from 1959 to 1962. By the early seventies, it was showing a cancer effect of x-rays.[30] When in 1977, Bross coauthored an article blaming doctors for the misuse of x-rays, within weeks, the National Cancer Institute terminated funding for the Tri-State Survey.[31]

His research team of nine had to disperse and find other jobs.

Bertell found herself without a job or means of support. She began writing *No Immediate Danger,* researching this compendious work and writing it without funding, in such time as she could spare from making a living. It is a remarkable achievement that brings together a vast collection of historical, political, and scientific information into a powerful warning in the tradition of Rachel Carson's *Silent Spring. No Immediate Danger*—the title refers to the "mindless assurance which automatically follows every nuclear accident or radiation spill, namely that there is no immediate danger"—was published by the Women's Press in England. Sales were increased by the Chernobyl accident, which occurred within a few months of its publication and gave her warnings added urgency. "Our present path is headed toward species death," she cautions.[32] Of radiation, she writes, "the bullets are invisible, the dying long and painful, and the wounds are carried by the children and grandchildren."[33]

In that same year, 1986, Bertell won, along with Alice Stewart, the Right Livelihood Award, the so-called Alternative Nobel. Unlike Alice, however, she has had difficulty continuing original research, since early on she lost her academic footing; and unlike Alice, she has never had a steady salary or a particularly robust constitution. The mid-1990s found her heading an independent nonprofit organization, the International Institute of Concern for Public Health in Toronto, earning what money she can by speaking, consulting, and advising—"living on the edge," as she says; in her sixties, in precarious health without a pension, she faces an uncertain future.[34] She has had to take occasional retreats into nunneries to be able to get on with work that is "pretty exhausting" and can be "pretty ugly."

Bertell decided to move to Canada after she was nearly killed in a mysterious accident. In 1979, four years after the death of Karen Silkwood, she was driving home from giving an anti-nuclear talk in Rochester, New York—where, the last time she'd lectured, she'd received threats. It was rush hour and she was in the middle lane of a highway three lanes wide each way when suddenly, "I could feel a car on my left getting too close. So I pulled back, and when I did, the driver shot forward into my lane directly in front of me. I could see his hand come out of the window and drop an object in the road in front of my tire, a very heavy sharp object—metal, I think—out of the car, in line with my front tire. I saw it coming, but I couldn't move out of the way. I tried to straddle it, but it caught the inside of the tire and totally blew it. . . . It cut a three-inch hole in my brand-new steel belted radial tire. It was heavy enough to bend my tire rim. I think if I had hit it head on, it could have turned the car over because I was in a small Toyota."

When she got out to assess the damage, a car marked "Sheriff" pulled over. "There were two people in it. I didn't see the driver. The passenger did not have a uniform on. He asked what happened and when I told him, he wanted to know if I had either the license number or the piece of metal itself. I said 'No' to both questions. Then he told me that this wasn't their jurisdiction, but they had radioed the local Rochester police who would be there any minute." Bertell stayed by the side of the road for more than an hour, and no police came. "Later, when I contacted the sheriff, his office verified that the car had not been a sheriff's car. . . . Their office combed all the written reports that had been broadcast on the police radios for the local cities, towns, county and state police. Nothing made reference to the incident."[35]

"These people play for keeps," said Sternglass in an interview April 30, 1994.

Many others whose findings challenged the party line on radiation lost their jobs and livelihoods. In 1974, Dr. Carl Johnson, the county health director in Jefferson County, Colorado, where Rocky Flats is located, sixteen miles north of Denver, was asked to approve a routine expansion of the residential zoning adjacent to Rocky Flats. His analysis of the soil convinced him that the nuclear facility was a health hazard, and he began his own investigations. His investigations turned up plutonium concentrations more than three thousand times higher than normal in the soil of neighborhoods around the facility, and he found 491 excess cancer cases where the DOE said there would be less than one. Rocky Flats, which produced plutonium triggers for all the nuclear bombs and missile warheads made in the United States, had a record of accidents, fires, and such carelessness with radioactive waste that the seepage into groundwater is suspected to threaten Denver.

Johnson was no rebel: he was a stolid, conservative man, an officer with the Army Reserve who had a top secret clearance; but his findings helped fuel a movement to close the plant. Local real-estate interests applied pressure to have him fired from his job as Jefferson County health director. In May 1981, they succeeded.[36]

He died of a heart attack in his fifties. "He was a real hero, a courageous, outspoken man," says Bruce DeBoskey, the firebrand young lawyer who was later to call Alice as star witness on compensation cases for Rocky Flats workers. "He'd been publicly ridiculed and condemned, but he fought on tirelessly, using his own money to do these studies, travel around, testify in cases. After his death it became apparent that what he'd been saying was right all along."[37]

There are many such stories. There is a steady stream of reports about DOE and contractor lab scientists who are demoted, defunded, ostracized for reporting positive findings or questioning DOE methods or assurances.[38] Dr. Irwin Bross, testifying at the 1978 congressional hearings on behalf of Mancuso and Stewart, described the attempted suppression of Mancuso/Stewart/Kneale and his own Tri-State Leukemia Study as part of "a sordid story extending back to the furor over fallout from weapons testing. It is a story where researchers were rewarded for not finding any hazards and punished if they failed to support the official AEC line that these low levels of radiation are harmless."[39]

Alice was to meet these people again and again in hearings, trials, and inquiries.

"So now I've become associated with the activists," she says, "which is not always a good thing. I'm very sympathetic with the activist point of view, but I'm determined that this sympathy not influence my interpretation of the data." She is a scientist who insists first and foremost on fidelity to the data: "You must see what the numbers say."

But she feels that she has a responsibility to speak out because she is in a position to do so. "Often I've been the only expert that activists could get to put in an appearance because I wasn't afraid of losing my job—I hadn't got one. I have no department that anybody's dependent on for work. Maybe they can bump me off the road—that might be all right. It would save me getting a chronic illness and dying that way. I speak out because there are not a lot of people in such a good position. I have nothing to lose. A lot of people do.

"This area of research can be shut down. I've watched it happen."[40]

Chapter 11

Alice in Blunderland: Back in Britain

"It never occurred to me that the British nuclear industry wouldn't care what we'd found. I imagined they'd want to know the truth, but they had no interest in the truth. No, from the word go *we were Public Enemy Number One."*

"When we returned to England with the Hanford data, there was an inquiry about whether they should be expanding the Windscale facilities. I remember telling George, 'we must be prepared now to let the world know what we've found.' " Alice assumed the nuclear industry would be in touch immediately, eager to know what the Mancuso/Stewart/Kneale team had turned up about the Hanford workers, but she was wrong. "They were sending out refutations of MSK behind our backs, but never once did they consult us directly.[1]

"They hire someone to investigate the safety of their product, we come up with findings they don't like, and so off with the head of the messenger. I can't think this is operating sensibly—you never hear the truth if you chop off the heads of the people who bring it to you; you live in a fantasy world.

"The anti-nuclear side did want to know what we had to say, however, and was in touch immediately. I was contacted by anti-nuclear groups everywhere, in the United States, Canada, Scotland, Australia. In fact that was how we found out what was going on behind our backs. But never once by the industry."

Windscale, 1977

The first public inquiry she and George were invited to was the Windscale inquiry, 1977. Windscale, the oldest, largest, and dirtiest nuclear facility in the United Kingdom, was already site to several reprocessing facilities when British Nuclear Fuels (BNFL) announced plans in the mid-seventies to build a new thermal oxide reprocessing plant (THORP) that would significantly expand its reprocessing and waste management activities.

Windscale—whose name was changed to Sellafield in the 1980s as part of an effort to clear up its tarnished image—has a dismal safety record. A three-day fire in 1957 that sent a cloud of radioactive fallout across Britain and into Europe was, before Chernobyl, the worst nuclear disaster on record, releasing nearly a thousand times more radioactivity than the Three Mile Island accident would (though not until 1983 did the British public learn the full implications of the fire). Since the early fifties, the facility has been discharging nuclear waste through a pipeline into the Irish Sea. What doesn't go into the sea gets vented into the air through smokestacks, on the assumption that it will dilute and disperse.[2] (In 1958, John Dunster, then Windscale's chief health physicist, later chairman of the National Radiological Protection Board—the agency responsible for recommending safety standards for the British public and workers—argued that "the sea had always been regarded by coastal and seafaring people as the ideal place for dumping their waste and this is, of course, a very reasonable and proper attitude. Almost anything put in the sea is either diluted or broken down . . . or stored harmlessly on the seabed . . . not the least of the attractions of the sea as a dumping ground has been the lack of administrative controls.")[3]

Windscale/Sellafield is located on a scenic strip of coastline bordering the Lake District, one of England's vacation paradises. No one is cautioned against vacationing or living in the area or eating the fish, though the Irish Sea is one of the most radioactively polluted seas in the world. There is an underwater lake of wastes including one-quarter ton of plutonium not far off the coast, and plutonium levels in the mud of river estuaries near the facility are approximately twenty-seven thousand times greater than in other parts of Britain. Local residents have long been concerned about the disappearance of entire species of fish and birds and mutations in insects and flowers.[4]

Now in the mid-seventies, British Nuclear Fuels declared its intentions to expand Windscale's reprocessing and waste management activities. The announcement attracted attention because reprocessing, which takes spent nuclear fuel and extracts plutonium and uranium, enormously increases the volume of waste—so much so that it was banned in the United States in 1977; but it was the preferred method in Britain because it provided plutonium for bombs. The expansion drew further criticism because it involved plans to receive radioactive waste from Europe and Japan. Many were alarmed at the idea of making the United Kingdom "the world's nuclear dustbin," as *Daily Mirror* headlines put it.[5] A wide range of groups, local and national, joined together to oppose the facility and demand a public inquiry. These included the Scottish Campaign to Resist the Atomic Menace (SCRAM), Friends of the Earth (FOE), and the Town and Country Association, which invited George and Alice. (Notably absent was the local authority, Copeland Borough Council, which was in favor of the expansion, since BNFL is the largest source of employment in West Cumbria.)

The Windscale inquiry went on for one hundred days, between June and November 1977. Alice and George testified September 27, 1977, and were cross-examined at length. Alice argued that even if Mancuso/Stewart/Kneale criticisms were not accepted, they nevertheless indicated heated disagreement among the experts that made expansion of the facility unwise.[6] "I began to see at the Windscale inquiry that they were dead set against us," she recalls. "There was a deliberate attempt to trip us up, very nasty indeed. They refused to allow George and me to testify together, on the grounds that two witnesses weren't allowed in an English court of law. I pointed out that this wasn't a court of law, it was an inquiry. They directed all the statistical questions to me and all the medical questions to George—they set out to make us look foolish, a very unpleasant business. They paid no attention to our written reply. I was rather shocked."

In his report, Justice Michael Parker, who conducted the inquiry, acknowledged that there was considerable controversy over Alice's data and that "if her conclusion is valid it would seriously affect the whole picture." But, he asserted, the proper place for this debate is "the forum of science." He then went on to declare that her figures are "clearly wrong" and the answer she gave under cross-examination was "untenable."

Since he did not, by his own admission, have the Hanford data at hand, he relied on the industry's critique of it—written by Ethel Gilbert, of Battelle. Alice had argued that Gilbert's criticisms did not apply to the main body of the Mancuso/Stewart/Kneale study, but Parker insisted that "the manner in which Dr. Stewart had dealt with Dr. Gilbert's criticism gave me no confidence." His final report completely ignored Alice's objections.[7]

Nicholas Hildyard summarizes his sense of the proceedings: "whenever uncertainty over data arose, Parker consistently found in favor of

British Nuclear Fuels." Hildyard recounts that when Edward Radford, one-time chair of the National Academy of Sciences committee on the Biological Effects of Ionizing Radiation, cited studies indicating that safety standards are too lax, Parker discounted his testimony because Radford didn't have the studies with him. He describes how, when Friends of the Earth argued that projections of energy demand had been overestimated in the past and that Britain's energy needs could be met by alternatives such as oil or coal and renewable sources of energy like the sun, wind, and wave power, Parker dismissed arguments about alternative energy sources because they were "of necessity uncertain"—though he accepted, without question, BNFL's claim that the technical problems of waste disposal would be overcome.[8]

As Alice pointed out, "if BNFL had nothing to hide it would have taken opponents' arguments more seriously, but this utter stonewalling of the opposition indicated an underlying anxiety that what we were saying would lead to the economic impracticality or failure of THORP." Reprocessing was assumed to be essential to Britain's nuclear program, and the nuclear program was viewed as a basic imperative by industry and government: the profits anticipated from reprocessing foreign fuels were assumed to justify the risks. Parliament approved THORP by an overwhelming majority in May 1978.[9]

The proceedings made it clear that public inquiries in Britain were manipulative strategies, not genuine airings of problems.[10]

On November 1, 1983, a Yorkshire Television documentary, "Windscale—the Nuclear Laundry," drew attention to a shocking rate of leukemia cases and deaths in children living near the plant. In the area around Windscale, the incidence in children under ten was reported to be ten times the national average. Within twenty-four hours, the government responded by setting up a committee chaired by Sir Douglas Black. (Alice, an acknowledged authority in the field, was not invited to be part of this study, nor of any of the other U.K. cancer cluster studies.) The official *Black Report,* 1984, confirmed that there was indeed a higher rate of leukemia in these areas but denied any connection between this and nuclear discharges.[11]

Sizewell B: The Longest Inquiry on Record

As the Black Commission was clearing Sellafield of responsibility for the leukemia in the area, the local press in East Anglia began turning up high leukemia rates in the vicinity of the Sizewell A nuclear facility, in operation on the Suffolk coast since 1966.[12] Margaret Thatcher, who had become prime minister in May 1979, chose this moment to move forward with her ambitious nuclear expansion program. The first reactor in the new plan was to be built at Sizewell, where one facility was already causing consternation to the local population. It was similar in design to that at Three Mile Island, a design that U.S. nuclear utilities had abandoned after the accident.

Opposition sprang up at once. Thirteen local authorities, from Cornwall to Cleveland—though not including Suffolk—joined together with Friends of the Earth, the Campaign for Nuclear Disarmament, and established conservation groups to demand a hearing.

Alice again presented evidence that overreliance on the A-bomb data had created a false sense of complacency, and that it was a mistake to extrapolate linearly down from the results of high-dose radiation to determine low dose, as the A-bomb studies do. George Bartlett, who presented the final submission of the Central Electricity Generating Board, said, "as far as the inquiry is concerned, the essence of Dr. Stewart's evidence is founded on an unsupported hypothesis." Of her written submission, he commented, "The length was inversely related to the complexity and obscurity of the thesis it contained."[13]

The Sizewell inquiry dragged on from 1982 to 1985, for more than two years. It was the longest-running and most expensive public inquiry on record, generating sixteen million words—twenty-four times as long as *War and Peace*—from two hundred witnesses. Practically anyone with any possible objection to the nuclear program was allowed to present evidence, but it made no difference: the event was largely ignored by the public and press. The inquiry was brilliantly used to drown the issue in paper and boredom.

As the report was being prepared in the spring of 1986, the Chernobyl reactor blew, and opponents called for a reopening of the inquiry. But the report was written as though Chernobyl hadn't happened. Even the siting of U.S. cruise missiles at Greenham Common and the vigorous protest by Greenham Common women, which took place during the years of the Sizewell inquiry, didn't succeed in stirring up interest. Sir Walter Marshall, Central Electricity Generating Board chairman, announced, "We have ended up with a clean bill of health on the nuclear safety issue and we have sustained our economic argument."[14]

The project was endorsed, permission was granted, and construction began.

Greenham Common Women

On January 1, 1983, television audiences throughout the world saw women dancing on a missile silo at Greenham. Support poured in, in the form of money, letters, and food. Even the local people, who were not inclined to be sympathetic since many worked for the weapons facility, were impressed.

The Greenham Common peace camp began when a Welsh housewife led a group of fifty women, children, and men from Cardiff on August 27, 1981, to march the 125 miles to the USAF/RAF Greenham Common near Newbury, in protest of NATO's decision to site U.S. cruise missiles there. When they arrived ten days later and asked for a televised debate on the issue, the request was denied. They set up a peace camp to get wider publicity and were joined by dozens, then hundreds, of supporters, mainly women, from all over the world.

Nobody knew how to respond to these women: their style was unlike anything that had ever been seen in British politics. They were deploying performance and theatrical gestures traditionally associated with women's activities—keening, weaving, dancing around the missile silos, decorating the fences with photos of loved ones—to make a statement about the death-dealing missiles.

They camped out winter after winter and succeeded in drawing worldwide attention to the base. Forbidden to live in campers or to build permanent structures, they made domelike dwellings from plastic sheets draped over trees. They blockaded entrances and roadways and disassembled fences, and one spring day they put on teddy bear suits and picnicked on the runway. At Christmas, they climbed the fence dressed as Santa Clauses and delivered presents to the soldiers.

There were hundreds of arrests—for trespassing on the common, for disturbing the peace, for creating a public disturbance.[15] "They endured repeated harassment from the authorities—some were physically removed," says Alice. "They were living under the most appalling conditions, camping out in bitter weather, for years and years; some were jailed again and again. They became an international symbol."

On New Year's Day, 1983, the forty-four women who had danced on the missile silo were arrested and charged with breach of peace. Expert witnesses were called in, among them, Alice Stewart and Rosalie Bertell.

"The women packed the courtroom at Newbury. I can still see them, coming into the law courts, taking the opportunity to get a good wash.

"They argued that they had a right to disrupt the peace because the missile base was threatening their lives, creating a genetic hazard—they were dead right. Rosalie Bertell and I defended them. We helped substantiate their cause and we got one or two of them off.

"It was marvelous, simply super," recalls Alice; "they really made their presence felt. They succeeded in closing down the air base and made it very difficult for the government to continue taking over commons or open spaces. They got good media coverage and the tide of opinion was on their side."

Dumping the Waste—and the Wastebin

The halting of the Billingham waste project was another resounding success for the activists.[16]

After the efforts of Greenpeace and the National Union of Seaman succeeded in calling a halt to sea dumping, the British government was forced to look for land sites for nuclear waste. In October 1983, it announced Billingham, Cleveland, near Newcastle, as a proposed site.

"Billingham is a company town, built around an old mine run by Imperial Chemical Industry (ICI)," Alice explains. "It had been an anhydrite mine [a mineral used in fertilizer] but was now defunct, so Billingham was on the skids. Then NIREX came along and said this is a perfect place to store atomic waste." (NIREX, the Nuclear Industry Radioactive Waste Executive, was established in 1982 by BNFL and the U.K. Atomic Energy Authority to supervise waste disposal.)

Opposition erupted immediately. "The town was up in arms—bad enough to be losing their jobs, but to be made a wastebin! They had a big meeting, tremendously attended, in a theater, absolutely packed—the whole city turned out." Alice was invited by Friends of the Earth and found herself, November 1, 1983, on the platform with David Bellamy, the well-known botanist and TV personality, who was chairing the event. The crowd listened attentively to Alice, Bellamy, and Peter Taylor, the third anti-nuclear speaker, but shouted down NIREX officials as they tried to reassure the community that the waste dump posed no danger.

NIREX persisted in its efforts to persuade local residents, with press releases, handouts, a newspaper entitled *Plaintalk,* and more public meetings; but at these meetings, industry representatives continued to be heckled until ICI finally realized that its interests would not be served by the plan and withdrew support for the NIREX proposal. The Billingham project was formally abandoned in January 1985.

"We won—but the Billingham victory is only partial," Alice cautions; "it still leaves the question of where to bury the waste." NIREX has since turned its sites to areas already accustomed to dumping, such as Windscale and Dounreay, but that doesn't solve the problem. "They should never have gone ahead with this technology until they could demonstrate that they knew what to do with the waste. They thought they could solve the storage problem by letting the radiation out so slowly you wouldn't feel it, which would be true if reducing the dose actually reduced the effect. Now we come along and say no, even if you reduce the dose, you will not eliminate the risk. The slow leakage of nuclear waste could do more harm to the human gene pool than all the bombs that have been set off."

Alice is fond of quoting John Dunster on nuclear waste. Dunster, director of the National Radiological Protection Board of Great Britain, member of the ICRP, and proponent of Windscale's use of the sea for dumping, met Alice's friend Bob Alvarez at a meeting of the Health Physics Society. Alvarez asked, "What are you fellows going to do with your nuclear waste in England?" Mistaking Alvarez for a government representative, Dunster replied, "Between you and me, we are going to wait for the next generation to deal with it."[17]

Dounreay: A Grueling Inquiry

On May 31, 1985, BNFL and the United Kingdom Atomic Energy Authority applied jointly for permission for the construction and operation of the world's largest reprocessing plant, estimated at £300 million, to be built at Dounreay. Dounreay, on the remote north coast of Scotland, had been the site of two reactors and a reprocessing plant since the mid-fifties. "It stands right on the edge of the sea, juts out, surrounded by wild countryside," says Alice. "Its towers and pale green sphere dominate the craggy coastline, giving it an eerie feeling."

The expansion was to be run by a European consortium. It was to service a series of three of four collaborative power stations to be built in France, West Germany, Italy, and Holland. British scientists were eager to get the planning permission through quickly because they saw no prospects of building such a plant in England.[18]

Opponents were outraged. It was proposed that the new facility would receive some sixty shipments of fuel per year by sea and thence by rail, more than one shipment per week. SCRAM calculated that this transport schedule meant—based on the U.S. accident rate for trucks—that "we could expect an accident involving nuclear waste every five weeks."[19] Critics argued that such trade would move Britain toward a "plutonium economy," making plutonium an ordinary item of commerce and increasing the risk that terrorist groups could get hold of it.[20]

Opponents were further outraged when they learned that discussion at the hearings would be restricted to local planning and environmental issues. This effectively muzzled the debate on broader issues of nuclear waste and proliferation, as well as the question whether Scotland should do the work of reprocessing other countries' spent fuel. Several national anti-nuclear groups boycotted, describing the inquiry as "a sham, a farce and an absolute disgrace."

Moreover, the facility had the support of many in the local community. The area had prospered with the Dounreay facility.[21] "The area was dying, but Dounreay transformed it," said a local resident. Critics pointed out that now it was the children who were dying.

The news of a Dounreay cancer cluster broke in October 1986 in a letter to the *Lancet* from Dr. Michael Heasman. Statisticians working for the Scottish Health Service reported that in the five years preceding 1984 there were six cases of leukemia among those under the age of twenty-four who lived within twenty-five kilometers of Dounreay, where less than one case would have been expected.[22] The government researchers who investigated concluded that this did indeed constitute a cluster, but more data would be necessary to implicate the nuclear facility.[23]

Alice testified in mid-October, invited by the Scottish Conservation Society. She had just returned from a few weeks' speaking tour in the United States, where she had been awarded the $2 million grant to study U.S. nuclear workers. She had, in September, delivered a paper in Pisa suggesting that background radiation is responsible for most childhood cancers, a paper that should have been of considerable interest to anyone wanting to know about cancer clusters around nuclear installations. She would later that year travel to Stockholm to receive the Right Livelihood Award.

She had, ten days before the Dounreay inquiry, turned eighty.

On the morning of October 14, she took the long trip to Dounreay and was immediately put on the stand. "They had me on the stand for two days," she recalls. Roger Milne, covering the event for the *New Scientist,* wrote, "She rarely faltered. I watched her field five hours of often rigorous cross-examination. . . . And she positively enjoyed slugging it out with the counsel appearing for the nuclear industry. Her footwork was neat and her counterpunching effective."[24]

Questioners were particularly interested in whether nuclear installations can cause cancer clusters. Informed that there were no cases of leukemia within a 25-kilometer radius of Dounreay between 1968 and 1978 but that between 1978 and 1984, five cases had been found within an area less than 12.5 kilometers from the facility, Alice responded, "that would perhaps cause one to raise one's eyebrows. That does seem rather suspicious of something new in the environment. . . . I would not leave it alone; no, not for a minute."

Her study of background radiation had yielded complex results, and her testimony was accordingly complex. The study had shown that cancer clusters may occur in rural areas for reasons not related to nuclear installations: rural areas have fewer epidemics of infections that may kill children off before cancer has a chance to manifest. "Since nuclear installations also tend to be in rural areas, you have to be wary of jumping to conclusions." Her findings had nonetheless convinced her that "any increase in background radiation is going to do harm," and "nuclear facilities are bound to increase background radiation, since they cannot operate without routine discharges of radioactive material both to the sea and air." This means that even the routine operations of nuclear installations increase the risk of cancer—"Therefore I believe we ought to be putting the brakes on. I am afraid one is forced to the conclusion that perhaps this is not the moment to go forward fast because you may accidentally do something that is irreparable."

Alice was particularly critical of the idea that radioactive wastes would dilute and disperse. She explained that if you get a larger effect from radiation in small doses than is currently assumed, you cannot accept the current division of waste "into high level and low level" and you can by no means relax standards for low-level waste.

The investigators were taken aback by her unyielding opposition to the official position.

Q: They [the A-bomb studies and authorities that accept them] are saying precisely the opposite of what you are saying.
A: Precisely the opposite.
Q: They are saying, in fact, that a linear extrapolation is a conservative assumption, whereas you say, far from being conservative, it is an under-estimate?
A: Yes.
. . . . . .
Q: . . . if you are right, then there is an under-estimate [by ICRP and NRPB] on the order of something between 10 and 15?

A: That is exactly the way it stands. . . .

Q: The only trouble with that, Dr. Stewart, is that you are not able to persuade at least a majority of the Radiological Protection fraternity about whether your view is correct, is that right?

"The argument continues," Alice replied.[25]

The argument did indeed continue, with investigators firing question after question at her until at one point a member felt compelled to cut in.

Dr. Delbray: Mr. Bell, could we possibly have a short adjournment of 10 minutes?

Reporter: Why?

Dr. Delbray: Dr. Stewart has traveled a long way today and has hardly had time to even have a cup of tea. Perhaps it would be kind of you to consider that she had just celebrated her 80th birthday too.

Reporter: Would you find it convenient to have a short adjournment, Dr. Stewart?

A: I am willing to fall in with whatever anybody wants.

The inquiry went on for ninety-five days, the longest public hearing ever held in Scotland. In the end, the reprocessing plant was given the go-ahead.

In 1994, however, when utilities were privatized in the United Kingdom and the private sector divested itself of nuclear installations, the fast breeder reactor at Dounreay was closed down and on April 1, 1994, Dounreay became a demolition site, with a £500 million program to decommission its three reactors. The people in the area need not have feared for their jobs: the dismantling will provide employment for the foreseeable future.[26]

Hinkley Point: White Elephants Parade

Hinkley Point, in a remote farming area on the Somerset coast not far from Bristol, was already site to two nuclear power stations when in 1983 the Central Electricity Generating Board (CEGB) announced plans for a third, a virtual replica of the reactor being approved at Sizewell.

Opposition sprang up, instantaneous and spirited, led by a Consortium of Local Authorities (COLA) representing twenty-three county and district councils in southwest England and Wales and headed—surprisingly—by the local county council in Somerset. COLA had not only the cooperation of FOE, CND (Campaign for Nuclear Disarmament), and the Town and Country Planning Association; it also had the lively participation of an inventive local grassroots group, Stop Hinkley Expansion (SHE).

As Lord Silsoe, the planning lawyer, outlined his argument in favor of the nuclear power station, a herd of demonstrators wearing papier-mâché white elephant masks, led by a fifteen-foot-high inflated balloon white elephant, stomped through the hall. Demonstrators offered to sell share certificates in the "White Elephant Electric Company," and they kept the pressure on, sitting in nearly every day, questioning every nuclear industry and government witness, sporting placards proclaiming Hinkley was "as safe as the Titanic." When the inquiry ended in November 1989, they held a burial ceremony with a somber procession in which the white elephant was ceremoniously laid to rest.[27]

A study carried out by Somerset health authority's public health department, published after the hearing began, showed 19 cases of cancer in a population of children and young adults living within a 12.5 kilometer radius of the power station in the ten years after nuclear operations began at Hinkley Point in 1965, where only 10.4 would have been expected.[28] This report failed, however, to establish a causal link between the nuclear installation and the leukemia.

Once again, Alice and George met with incredulity for their disagreement with the experts, Sir Richard Doll, Ethel Gilbert, Sarah Darby, each of whom were invoked to discount them.

Q: It is fair to say, isn't it, that you disagree with the best estimates and conclusions of virtually every other study that has been conducted?

A: For a good reason!

Q: What are you urging? Are you saying that [the inspector] should reject all the other studies and all the other indicators and rely on your estimate of 15 times ICRP?[29]

The inquiry went on for fourteen months, ending in November 1989. It cost £10 million and generated an official transcript thirteen times as long as *War and Peace*. The debate covered the usual subjects—economics, safety, health, impact on local environment, what to do with the waste, the cost of decommissioning, the relevance of Chernobyl.[30]

But meanwhile, processes were going on that would make all these arguments irrelevant, as the government moved to privatize the electricity industry. As the figures began coming in, as the industry was scrutinized by city economists in preparation for privatization, the private sector realized that the actual costs of operation, the costs of decommissioning and of waste disposal, made nuclear power a horrendous investment. The government took over the nuclear power program, deciding that no more nuclear power stations would be built until a review had taken place in 1994. This meant an effective moratorium on nuclear power development in Britain.

Even so, the CEGB pushed ahead with its application to establish an option to build Hinkley Point—in case the 1994 review turned out to be favorable. When the report finally came out, eight months later, it gave the go-ahead to Hinkley, though the final decision whether it would be built was deferred to the government review of the entire nuclear enterprise in 1994. The 1994 review concluded that nuclear power stations were economically unviable, once government subsidies to the industry were removed by privatization.[31] Only in 1997 were plans for Hinkley dropped.

"What was the outcome of the Hinkley hearing?" I asked Alice. "Nil, as usual. Up in smoke. They blur, one into the other. I never felt I had much impact on the decision. But perhaps there was a cumulative effect, and these were all little battles in the war."

The inquiries did have some effect: they brought anti-nuclear activists together and drew attention to scientists' disagreements with the regulatory bodies, NRPB and ICRP. They raised public consciousness, caused delays, increased costs, and helped halt the expansion of the industry. But they were also brilliantly used by the British government to give the illusion that issues were being aired, while matters relating to nuclear weapons and energy continued to be settled without reference to democratic processes.[32]

A Prophet in Her Country

Alice may be right that her participation in the inquiries, and the inquiries themselves, had little political effect, but her being there made a difference to many individuals.

"Alice has been a great inspiration to me. From talking to Alice, I made the decision to go on and try to get to the truth," says Janine Allis-Smith of Cumbrians Opposed to a Radioactive Environment (CORE), who lived fourteen miles downwind from Sellafield and believes her son got leukemia from the facility.

Allis-Smith and her husband bought a farmhouse near the coast. "I came to the Lake District because it is very beautiful. . . . I thought life could not be more idyllic." Then her son Lee became ill at age twelve and tests confirmed that he had leukemia and gave him a 40 to 50 percent chance of surviving. "Lee was diagnosed in 1984, just after the report came out about the connection with radiation and Sellafield and childhood leukemias." She recalls, "in the 1960s and 1970s, we spent a lot of time on the beach. . . . When Lee was born we had a camper and so we used to travel up and down the coast. . . . We used to play in the streams that come from the dump; we used to put Lee in the pools to play."

To her surprise, she found herself up against industry public relations people who assured her there was nothing to fear. "I thought there would be a keen interest in . . . finding out why the leukemias happened," but instead, "they talk down to you, try to placate you. . . . Some of the things they say, calling people hysterical and ignorant and saying that the fear of leukemia is something only in the lower social classes and women. I think anybody who has seen what the children go through couldn't possibly say something like that."

Allis-Smith's questions led her to CORE and to "some articles by Dr. Alice Stewart" which helped her make connections. Other people have also been inspired by Alice. "A family in Preston, Jo and Stella McMasters, now in their seventies, lost three daughters to leukemia. He worked at the plant; they've had a battle to get BNFL to acknowledge that he ever had a radiation dose. Alice told them to keep going. The inspiration she's given to a lot of people has been really important."[33]

In 1985, the Severnside Campaign against Radiation (SCAR) invited Alice to address their conference on Low-Level Radiation and Health. There are two power stations on the River Severn, and residents had long suspected that the cancer clusters in their area had something to do with them. But, as Barbara French of SCAR explains, "it was very very difficult to get information. After about a year of struggling on our own, we set up an international conference on the health effects of low-level radiation. I thought it was really a long shot to invite Alice—I didn't think she'd be at all interested in a little group like ours, which was obviously very amateur. But she straightaway wrote back and said she'd be thrilled to come, which was a huge boost to us. We held it at Gloucester Cathedral. It was run on a shoestring. She was our main speaker and was greatly applauded. She gave the conference a huge lift."[34]

CORE and SCAR decided to turn this conference into an annual event, hosted in different parts of the country each year, to bring together medical experts, academics, local authorities, environmental health officers, trade union members, members of the public and the nuclear industry. Alice took part in these for the next ten years, from 1985 to 1995, addressing audiences in conferences at Glasgow, Carlisle, Bristol, Edinburgh.

"It was surprising to me that although we were just laypeople, it mattered to her that we had read her work and valued it," says Allis-Smith. Says Dr. Jill Sutcliffe, independent environmental scientist, "She was more than happy to give her support to something started by ordinary members of the public and put her science into a language that ordinary people could understand. She didn't just come to the conference to speak—she joined in all the social activities, she was there late into the night, sitting up with everyone, drinking and chatting and going on till midnight." Sutcliffe describes a moment at the Carlisle conference.

> Alice stands up to give her lecture and the person operating it has got the slides in the wrong way, and slide number ten is in the number one position. Alice continues to lecture, absolutely clear, using what slides came up. She was not at all fazed. She just carried on, adjusting her talk to whatever slide turned up, while giving instructions to the guy operating the slides at the same time. Not at all cross with the guy, either. She was eighty-eight.[35]

At the Glasgow conference in 1994, SCAR dedicated the opening lecture to her, officially naming it the Alice Stewart Lecture. "We wanted to say that we think she's done a brilliant job and we'd like to thank you."

And so Alice became a hero to the activists. She found herself in this odd position, of being nonexistent in official circles and in demand by citizens throughout the world. "It was as though—suddenly I woke up one day and realized that people all over the world, people everywhere, knew what we'd been doing. It was the activists, the activists who kept us alive."

Chapter 12

Fallout

"They say we might imperil national security—such feeble excuses!"

Meanwhile, back in the United States, the Mancuso affair had opened a can of worms that continued to spill out in embarrassing places.

The 1978 congressional investigations had blown the cover off the secrecy that had shielded DOE operations for decades. An accident in 1979, near Harrisburg, Pennsylvania, followed by revelations in the mid-1980s of scandalous conditions at DOE facilities, blew other covers.

Three Mile Island

On March 28, 1979, a combination of mechanical failure and operator error brought the Three Mile Island (TMI) facility to within a half hour of meltdown. It was—as Alice says—"the sort of accident that was supposed never to have happened."

It took several days before the accident was contained, during which time official sources attempted damage control, issuing press releases that compared maximum exposures to a single x-ray. A citizens' health survey of three hundred people living to the north and west of the plant, the area toward which the winds were blowing at the time of the accident, found that many residents experienced what seemed like sunburns, some of which were sufficient to blister their skins; many reported hair and weight loss, respiratory and heart disorders, and deaths of pets and livestock.[1] Seven infants died in the Harrisburg hospital in the three months following the accident, as compared to one in the three months preceding, and thirty-one infants died within a ten-mile radius around the facility in the six months following, as compared to fourteen in the year 1978.[2] Yet official sources steadfastly denied that there was any connection between these health effects and the accident.

The Kemeny Commission that investigated the accident criticized the personnel operating the plant, Metropolitan Edison, for its training of the personnel, and the Nuclear Regulatory Commission for its low standards of worker performance and inadequate response to the accident.[3] It concluded that far from being unlikely, as the industry had claimed, an accident like TMI had been all too likely.

The Three Mile Island accident marked a turning point for the nuclear industry, making it impossible for it to continue its boast that nuclear plants presented no danger. Protest movements in the United States and Europe grew in size. Picketers carried signs, "Today, TMI—Tomorrow . . . ," "Hell No, We Won't Glow," "No Nukes Is Good Nukes." On May 6, five weeks after the accident, more than 100,000 protesters gathered in Washington, DC; on September 23, more than 200,000 gathered in Manhattan; and in July 1982, a New York City peace rally attracted 1.3 million people.[4]

Alice heard about Three Mile Island while she was in Hanover, Germany, testifying at hearings on the Gorleben facility. It was to change her life in ways she could not have imagined.

In 1981, citizens' groups won a class-action suit against the Three Mile Island facility, an out-of-court settlement of $25 million, with a provision that part of that money be used to study the health effects of low-level radiation. From this settlement came the Three Mile Island Public Health Fund. When the Fund set out to assess damage to the population around Three Mile Island, it became clear that the local population would not be sufficient for a meaningful investigation: not only was it too small but there had been no measurement of individual exposures. What also became clear was that the data collected on workers in the weapons industry would provide ideal material for such a study.

"There was a sort of world advertisement for people to come forward with good ideas on how to study radiation effects," Alice explains, "and George and I were selected. I said I'd take part only if Mancuso was brought in, but every time his name was put up, it was knocked down. The DOE had done such a good job of character assassination that anyone who didn't know the inside story had doubts about him. It was all wicked nonsense. They said, 'you can have him as an assistant, but you can't name him anywhere in the application, and what's more, you must go and tell him this.'

Poor Mancuso was very hurt and thought we were in on it against him. What he had been through doesn't improve a person's temper, and he was getting touchier and touchier. We really had ruined his career—he could have turned on us. But we're on good terms now."

In 1982, Alice was awarded a $2 million grant from the Three Mile Island Fund. She had been coming to see herself as "working for the activists," but at this point she began working for them in a very real sense. "We were within a month of running out of funding—if we hadn't got their grant, we'd have to have stopped."

It was to take four years of legal hassle before she could begin to collect the money, and a decade-long struggle before she could get the workers' records.

A Spreading Stain

In April 1986, the disaster at Chernobyl set off further alarms about the safety and viability of nuclear reactors. But accidents, as it happened, were not the only problem.

In 1985, shortly after he took over as Energy Secretary, John Herrington, responding to public pressure, ordered safety inspections at sixteen nuclear sites nationwide. Such was the secrecy surrounding DOE operations that Herrington himself had difficulty getting information about conditions at the plants. When he asked the National Academy of Sciences to assess the installations, the study revealed a trail of accidents, violations of safety procedures, and inadequate DOE oversight of the private contractors who ran the facilities. That report led to further investigations by journalists and congressional committees, which led to further revelations of safety and environmental violations and widespread contamination of the sites and surrounding populations.

At every major DOE facility—Hanford, Fernald, Rocky Flats, Savannah River—there were disclosures of flagrant disregard of the environment and health of workers. Nuclear facilities, it turned out, had been overseen by no one—not by the Environmental Protection Agency, nor by the Occupational Safety and Health Administration, nor by any other federal agency. The DOE had been exempted from federal environmental and occupational health regulations. Behind the wall of secrecy that shielded its operations, the contractors had felt free to dispose of waste as they pleased and had simply dumped it out back, burying it in the ground or pouring it into the nearest pit, ditch, pond, creek, or river. This meant that radioactive contamination had been seeping in unknown quantities, over long periods of time, into the drinking water, soil, and air of nearby communities.[5]

In October 1988, a series of powerful articles in the *New York Times* drew worldwide attention to the national disgrace.[6] A 1988 congressional investigation, led by Ohio senator John Glenn, turned up festering conditions at Fernald and other sites. Glenn summarized, "we are poisoning our people in the name of national security." Congressmen referred to "the 35-year secret chemical warfare which the DOE has waged against the workers and nearby residents of the 17 nuclear plants around the country."[7]

By 1988, the four largest weapons plants were shut down: Hanford, Savannah River, Rocky Flats, and Fernald. Eleven of the seventeen plants were declared Superfund sites. Estimates to clean up the complex were in the hundreds of billions of dollars.

The government steadfastly denied that workers and residents had suffered as a result of DOE negligence, but citizens suspected otherwise. Citizens' groups like the Hanford Education Action League began to organize and gather data on radiation releases, trying to identify exposed populations and determine their level of exposure. Researchers soon realized what poor records the DOE had kept, how few studies it had done, and how difficult it was going to be to do such studies, given its inadequate monitoring and poor cooperation. Many health professionals felt that they were facing a public health emergency; some called it a "creeping Chernobyl."[8]

The Physicians for Social Responsibility (PSR), founded twenty years before, created a Physician's Task Force on the Health Risk of Nuclear Weapons Production. The PSR and other public interest groups—the Sierra Club, the Environmental Policy Institute (EPI, Bob Alvarez's organization)—launched campaigns to investigate health conditions in affected communities. The Childhood Cancer Research Institute (CCRI), founded in 1988 by David Kleeman, an old friend of Alice who generously endowed the organization so it could carry on her work when Oxford Survey funding was running out, went to communities to raise public awareness and encourage citizen action.[9]

The campaigns of these public interest groups came to focus on a clearly defined goal: to open radiation health research to public scrutiny and to wrest epidemiological studies away from the DOE and transfer them to the Department of Health and Human Services so that independent researchers, unfettered by conflict of interest, could take them over. "This data has been kept secret," said Dr. Jack Geiger of Physicians for Social Responsibility and professor of community medicine at the City University of New York Medical School, "in the name of national security. That's an absurdity. As if anyone could tell from the incidences of leukemia or other cancers in nuclear plant workers how to make our latest version of the bomb. It's simply outrageous that, in the name of national security, a child at Fernald should suffer the risk of leukemia or someone in Denver walk around with plutonium deposited in their bones—as if national security didn't include the health of the population! These risks have been assumed without people's knowledge or consent. There is no national security issue involved here except the security of the DOE from outside scrutiny."[10]

Central to this campaign to open the DOE data to scrutiny was Dr. Alice Stewart.

Science to the People

Alice was awarded the Three Mile Island grant of $2 million in 1982, but it wasn't until 1986 that she received the first installment, of $1.4 million. In 1986, she turned eighty—"we celebrated the funding at my eightieth birthday. The long delay was because at every step of the way there had to be negotiations between the utilities and the lawyers who fought the case, and who in any case got most of it."

There followed the long wait for the nuclear worker records.

In the meantime, Alice and George were engaged in vital research on the Hanford data and the A-bomb studies. Alice was using some newly released Hiroshima data to demonstrate that the survivors were not a representative population just because the death rate had returned to normal, that opposing forces were at work in the population that accounted for the appearance of normality.[11] She and George were turning up evidence from the Oxford Survey that inoculations against infectious diseases might act as cancer inhibitors. Still bothered by the question of why children get only lymphatic, not myeloid, leukemia and why lymphatic leukemia peaks at ages two to four, Alice was mining the Oxford data and developing a theory of Sudden Infant Death Syndrome as the missing leukemia.[12] The TMI funding also made possible a study of background radiation she'd long wished to do: the National Radiological Protection Board had been taking measurements of natural background radiation throughout Britain, measuring dose rates for every ten-kilometer square mapped out on a grid, and she began matching this information with Oxford Survey data to discover whether there was a correlation between levels of background radiation and patterns of childhood cancer deaths. (We'll hear more about this in chapters 13 and 15.)

The Three Mile Island Fund lawyers persevered in their efforts to get the workers' records away from the DOE, submitting request after request under the Freedom of Information Act. Bob Alvarez and the Environmental Policy Institute had also been submitting requests, but they were similarly denied. By the end of the decade, Alice still had not received the nuclear worker records.

The DOE objected that the Fund's request was in violation of the Federal Privacy Act, then claimed that it would violate certain agreements between the contractors and the states to reveal the causes of workers' deaths. The department argued that the requests constituted interference with its "deliberative processes," that they'd interfere with the rights of the DOE researchers, that it would be unfair to ask these researchers to share their work with others. Affidavits referred to the data files as the "intellectual property" of the investigators who prepared them and claimed that their release could lead to "spurious . . . results that would confuse the public and generate irrational criticism that could . . . undermine public confidence in DOE." Dr. Shirley Fry of Oak Ridge Associated Universities said she was concerned that the data might be too complex for outside researchers to understand.[13]

In the last years of the 1980s, amidst breaking scandals throughout the nuclear weapons complex and the shutdown of DOE facilities, amidst the growing furor of citizens' groups and contaminated communities, Alice took her science to the people. Though she was primarily engaged in research during this time, publishing about ten papers a year, she began making regular visits to the United States to meet with the Three Mile Island Fund lawyers and give papers at conferences; and while there, she responded to requests from activists and public interest groups. Between 1988 and 1990, she made twenty-five to thirty appearances, giving lectures and workshops to citizens' groups and health professionals. Her efforts to get the workers' records became part of a larger struggle of public interest groups such as the PSR and CCRI to get communities to put pressure on Congress to break the DOE's stranglehold on radiation health research.

Alice in the late eighties is in her early eighties. Her hair is nearly white, cropped short, curly. She dresses efficiently, in suits or pants suits set off by the token scarf or brooch; she wears sensible shoes built for covering ground. She has a brisk, no-nonsense efficiency about her—more than one person notes the Katherine Hepburn likeness. But she also has a disarming informality about her, mischievous, even impish. She describes herself as "an old woman standing up against the DOE."

She lectures on east and west coasts, in the northeast, the northwest, the south, the midwest. She speaks to gatherings at hospitals, churches, city clubs, veterans' groups, medical schools; she gives lectures and seminars at universities from Yale to Berkeley. She addresses a wide variety of audiences, from grassroots activists to physicians and researchers; she speaks in auditoriums, living rooms, church halls, high school gyms, and recreation rooms. In a given audience there may be radiologists, radiobiologists, epidemiologists, nuclear physicists, mothers worried about the health of their children, workers wanting to know where they got their cancer.

She is sought out by groups at all the major DOE facilities—Fernald, Savannah River, Rocky Flats, Hanford. She is invited by organizations concerned with the contamination of their communities, by groups resisting the siting of waste dumps in their neighborhoods. She combines speaking engagements with legal consulting and courtroom appearances. She prefers to stay in people's homes but sometimes gets lodged in a local motel or school dormitory.

At the Feed Materials Production Center at Fernald, Ohio, congressional investigations revealed that the government had known for two decades that thousands of tons of uranium waste were being released secretly, exposing untold thousands of workers and residents to radiation. The plant had not been upgraded in the seventies because it was assumed it would soon close, but with the Reagan military buildup, production had actually increased. Nearby residents and workers had long suffered health problems but had been reassured that they had nothing to fear.

Alice testified in a $300 million class-action suit against the Fernald facility, representing fourteen thousand workers and residents who claimed they had suffered damage to health and property values. People living near Fernald "have suffered worse than those at Three Mile Island," she told an audience at the University of Cincinnati. "At Three Mile Island, it was only one big accident, whereas here the contamination has been going on for a much longer time."[14] (The DOE eventually settled out of court for $73 million, claiming the money was compensation for emotional distress and diminution of property values, rather than for any association with illness.)[15]

"She spoke to those people with such a genuine sense of caring," recalls Dianne Quigley, the Childhood Cancer Research Institute representative who organized Alice's tours. "I remember at one Fernald gathering, she sat down next to a farmer and put her hand on his. She was really there with them, with genuine understanding of what a doctor is—that a doctor is there to heal, not to preen their own ego or protect data. Audiences loved her, I never saw a disappointed audience. She was very charismatic. Often audiences wanted her to go on talking, to take more time for her presentation, which was very unusual. She was equally at ease with all types of people and would give incredibly ambitious presentations.

"These communities had been so intimidated—for forty years they'd been told there was nothing wrong, the government assuring them, 'there's no problem, there's no problem,' and all the while feeling in their gut there *is* something wrong, that their health is very wrong. They knew they couldn't trust public officials because of special interests and conflicts of interest. To have, suddenly, a scientist come along with all these credentials and say, 'you're right, there's a problem here, we don't know what radiation does, new research has to be done.'

"She took her science to the people and emboldened them to stand up and protect themselves. Doctors were inspired by a woman in her eighties, still producing research. Her humor could disarm any uptight person. Her folksy expressions and tales, her obvious compassion, were irresistible."[16]

"People are smarter than experts credit them with being," she told one group of activists. "They know when they're being conned. I always say, 'they may not know physics, but they can recognize a con man when they see one—he's the one who showed up at my back door selling brushes last week, and off he went with my money.' People are instinctively frightened of nuclear contamination, and they're right to be—we don't know what effect it is having. The experts thought they understood it, but they were wrong, they were too arrogant. People smelt them out."

"Especially in a debate situation," Quigley recalls, "she'd come in with some human, genuine response, where the other side would get up and be so stiff and formal and uncaring. She always won."

At Hanford, where nearly nine thousand curies of radioactive contamination were released in the postwar years (Three Mile Island released twenty curies); where over 200 billion gallons of radioactive toxic liquids have been dumped into the ground—enough, as Bob Alvarez notes, "to cover Manhattan to a depth of 40 feet"[17]—residents have suffered so many thyroid cancers and disorders that local pharmacists order thyroid medicine in bulk. Ethel Gilbert of Batelle, in a public forum with Alice, claimed that she'd been unable to detect any significant ill effects from radioactive exposure at the site. Alice pointed out that there could be a thyroid cancer epidemic going on and no one would notice because there's no tumor registry in the state.[18]

"It was never much of a debate," recalls Larry Shook of Hanford's citizens' group HEAL (Hanford Education Action League); "there was never anyone who could get in the ring with her. She both wowed people and charmed them—it's a formidable combination. She's an extremely articulate speaker, with a great deal of precision, yet very humble and humorous, with a wonderful capacity for translating medical issues into lay language."[19]

During those years, Alice made frequent stops in Denver to consult with lawyers and testify in compensation cases for Rocky Flats workers. Addressing an audience in a church near the Rocky Flats facility, where Dr. Carl Johnson had linked high amounts of plutonium concentration in the soil to an excess of cancers, she cautioned "whatever the experts tell you about risk, multiply that figure by 10 or 20, and that's what we've found."[20]

"Most of the stored radioactive waste hasn't had time to affect the population yet, so don't be complacent about it," she told an audience at the University of South Carolina in November 1989, referring to the 32 million gallons of liquid waste stored at the Savannah River site and thousands more cubic yards of low-level waste buried in solid form. Though the facility had been shut down for some months, "there will be an effect, even if you are unable to detect it," she told a press conference. "Damage probably won't become apparent for many years."[21]

To an audience near Idaho National Engineering Laboratory, she said, "so far we are only midway in the story. There eventually will be a great impact because you'll never get rid of the radioactive waste and we are only beginning to see what it will do. Nor can Idaho simply export its waste to another state because eventually it will get into the environment and affect everyone."[22]

"These appearances really had an effect reaching people, helping them to know how to think about these issues and organize around them," recalls Quigley. "It's terribly important that her findings are there in the literature, so that people can point to them—it makes their case respectable. Alice has made a huge contribution to the work of our organization, the Childhood Cancer Research Institute, in our efforts to teach communities how to empower themselves, to know how to ask questions and assess risk. And she's still available to us."

An Old Woman Stands up against the DOE

In 1989, Alice had been waiting seven years for the DOE to release the nuclear worker records when she was invited by the Three Mile Island lawyers, as part of their effort to get the data, to testify at congressional investigations of the weapons facilities. In February and again in August 1989, she testified that it was indefensible for the DOE to lock up the medical records of its 600,000 weapons workers, since they were, in effect, the best source of information on the effects of low-level radiation, and that DOE management of health studies was obstructing freedom of scientific inquiry.

Secretary of Energy James Watkins, who had become head of the department in January, admitted that the weapons complex had been "cloaked in secrecy and imbued with a dedication to the production of nuclear weapons without a real sensitivity for protecting the environment."[23] He agreed that the worker data should be made available to qualified researchers but argued that the Department of Energy should keep control of radiation health research. "There is no other agency within the Federal Government that is more appropriately positioned to conduct epidemiological research in the energy field," he insisted; "As an employer, DOE has a moral and ethical responsibility to monitor the health of its workers."

Watkins proposed that the department should set up a repository of epidemiology data that would be open to qualified researchers but remain in control of the DOE. The time to complete this repository was projected at six years, at a cost of $36 million. It would then be left to the National Academy of Sciences to decide what types of researchers would be allowed access and in what form the data would be made accessible.[24]

Alice was dismayed. She could imagine the years and years that would pass before a researcher like herself would be allowed access. She protested at the DOE's lack of sense of urgency and argued that Watkins's proposal would "only lead to further delay and interference with . . . access."[25] She was backed by Senator Timothy Wirth of Colorado, whose constituency included the Rocky Flats nuclear weapons facility: "How in the world can we ask the people who are working in those plants, who are living around those plants, to believe the health research of an agency whose job it is to make nuclear weapons?" To address this conflict of interest, he introduced legislation that would transfer epidemiological studies of the effects of radiation from the Department of Energy to the Department of Health and Human Services and create an independent—not a DOE-sponsored—advisory panel to assist in developing reliable data.[26]

"It does require independent assessment," agreed Alice, who argued that the DOE's standards for assessing radiation risk at weapons plants were badly flawed. "It's high time they let in some fresh air. There are a lot of questions to be answered about the weapons complex facilities."[27]

Her position was supported by *Dead Reckoning,* a 1992 review of 124 DOE epidemiological studies of nuclear workers, conducted by the Physicians for Social Responsibility. It found that the department had looked at only a fraction of the 600,000-person workforce and there had been major problems with the design and methodology of the few studies it had done. It found also that most DOE studies had neglected the healthy worker effect, which made it possible to come in with more cheerful risk assessments than warranted.

"And these are the epidemiological studies that have been the basis of DOE's endless assurances over the years that there's no evidence that anybody has been harmed, either in the workforce or in surrounding populations, by DOE weapons-producing activities," said Dr. Jack Geiger, one of the authors of the study. "All of this was occurring behind a virtually impenetrable wall of secrecy—none of it was subject to external scientific review." Geiger refers to "an ugly combination of monopoly and secrecy—secrecy with regard to access to the most useful kinds of basic data on the exposure of human populations to low-dose radiation and monopoly in which so many of the people in the field were funded by the agencies that were producing the hazards. Secrecy corrupts and absolute secrecy corrupts absolutely."[28]

Dead Reckoning found that the DOE had discounted findings that might be alarming to the workers and the public and that the secrecy protecting its operations had allowed it to censor, intimidate, demote, and dismiss scientists whose discoveries were seen as harmful to DOE interests. It cited Mancuso as the primary, though by no means the only, instance.[29] It saw this pattern of denial and manipulation as part of the DOE's effort to minimize its liability for compensation and recommended that the DOE get out of the health research business.

Alice Gets Her Data

Finally, in 1990, under pressure of a Freedom of Information Act lawsuit and the threat of congressional legislation that would transfer epidemiological research to the Department of Health and Human Services, the DOE consented to transfer the workers' records to the Three Mile Island Public Health Fund.

The attempts at legislation had gone nowhere. Senator Timothy Wirth and Senator John Glenn had introduced bills in 1984, then again in

1988 and 1989, to transfer studies of the health effects of radiation to the Department of Health and Human Services (HHS). The bills had been defeated, but the threat of legislation did bring pressure that made the DOE yield. Finally, DOE head Watkins agreed to a "Memorandum of Understanding" that gave HHS major responsibility for the $17 million annual program for studying the effects of low-level radiation on the nuclear workers. The National Center for Environmental Health at the Centers for Disease Control was mandated with conducting research on people who live near weapons facilities. The National Institute for Occupational Safety and Health was assigned the epidemiological research on worker populations. Another Memorandum of Understanding was negotiated between the DOE and the Agency for Toxic Substances and Disease Registry, founded in 1980 for evaluating the health effects of radiation on contaminated sites.[30]

The DOE agreed to these administrative shifts in order to forestall legislation, but critics would have preferred legislation. "The Memorandum of Understanding does not have the force of law. It can be altered or canceled by private agreement between the two current Secretaries—or future Secretaries. It fails to decisively remove DOE involvement in measuring, monitoring, analyzing and reporting on the human health consequences of its own weapons production activities. Overall, some change has occurred—but much more remains the same."[31]

Still the records were withheld.

"Even after Secretary Watkins promised all possible cooperation," said Geiger, "I can tell you, as a participant in the negotiations, that the department was slow and grudging. It took several years of the most intense kind of litigation and negotiation." The DOE kept finding problems that needed resolving before the data could be released, most of them simple procedural matters. "You name the excuse, and we heard it through the years," grumbled Senator Glenn, who had been involved in efforts to secure the records since 1980. "First they couldn't find them. Then they said they couldn't release them. Then our request wasn't proper—they seemed bound and determined to keep that data inside the DOE's system."

"Even after the DOE agreed to give us the data, they stopped us from getting it for a long time," said Alice; "they put every conceivable obstacle in our way. It took ages of time to get it, years of negotiating."

The DOE then stipulated that one of its own officials should work with the TMI researchers and that analyses be shown to the department before publication. The TMI Fund rejected these terms. Release of the data was further postponed at the last minute because "the DOE forgot to tell us a consent form was required from us," says Daniel Berger of Berger Law Firm, attorney for the Fund. "That took another few weeks. Then information that could identify individuals, such as Social Security numbers, had to be removed." Berger and the DOE lawyers finally signed an agreement that met Alice's requirement that the information be handed over in a computerized form that she could use immediately.[32]

Finally, in 1990, Berger announced, "All preconditions have been swept away and safeguards, including an arbitrator, have been included to prevent the department from sabotaging the agreement by yielding data in a form that cannot be used. We've established the scientific principle that there should be independent assessment of health risks of workers for any danger. Now any reputable research department can have a look at DOE data. There should never be any secrets. It's the end of a story about science and government that is not pretty at all." Geiger called it "a victory in a war that should never have been fought." Senator Glenn summarized, "It's been a long push. But the release of all this data will finally allow researchers outside the DOE to start assessing whether workers and the public were hurt by radiation hazards."

In the summer of 1990, the *Observer* carried the story: "A ten year battle to gain access to the world's largest data bank on the health of nuclear workers will end tomorrow when U.S. officials hand over a series of computer tapes to veteran British scientist, Dr. Alice Stewart of Birmingham," tapes that include the "histories of 200,000 workers employed in nuclear plants over the past 40 years."[33] The records were to be released in four stages, beginning with those from Hanford, followed by data from Rocky Flats, Los Alamos National Laboratory, the Mound facility in Ohio, and the Pantex plant in Texas. Within ninety days of that, the DOE was to turn over records from Savannah River and most of the units at Oak Ridge; within the year, it was to turn over data from the Feed Materials Production Center in Fernald, the Mallincrodt Chemical Company in Missouri, and the K-25 plant at Oak Ridge.[34]

Alice and George were to take possession of approximately one-third of all the health records of workers employed in the U.S. nuclear industry since it began in 1942.

It was a historic occasion, described by the *New York Times* as a landmark victory against the DOE. It was hailed by many scientists as a blow for scientific freedom, though others expressed doubt that the independent researchers would turn up anything new. Dr. Ethel Gilbert was quoted by the *Times* as saying that "if Dr. Stewart evaluated the data according to methods accepted in scientific circles, it was unlikely that increases in cancer deaths caused by radiation would be found among the Hanford workers." But she "was concerned that Dr. Stewart, an avowed opponent of the nuclear weapons industry, would evaluate the data in such a way that increases might appear falsely."[35] Alice bristled at that: she does not see herself as "an avowed opponent of the nuclear industry," nor do the scientists who have paid close attention to her work so consider her.[36]

Finally, Alice and George could take possession of enough of the data to continue their investigations. The records arrived in Birmingham on a large disk. There were many gaps in the data and many inconsistencies, since regulations controlling data gathering had been different in the various communities. "Some are very messy; none is quite complete. Hanford is complete up to 1986, but we're still missing four years. Oak Ridge is complete only up to 1984."

The records continued to dribble in through 1992.

"With Fernald, there were 13 people who were still working five or more years after they were dead, according to the record. . . . The data should never have been sent out in that shape! Workers at Fernald were employed for four years before there was any dose recorded." The Rocky Flats records were very incomplete: "the data are not worthy of epidemiology," she said.[37]

"But for two key places we've got sufficient data, of sufficient quality control, to surmise the condition of the whole industry. These records are uniquely important. We can finally complete our work. We can finally settle the question whether the A-bomb data are of value in assessing risk to the workers and public.

Some very hard work has gone into this struggle," says Alice; "The government has wasted ten years of everyone's time and they don't even blush. I only wish I were twenty years younger so that I could see the end of the story."

Follow-Ups

By early 1993, Alice and George had produced "A Reanalysis of the Hanford Data" that confirmed the initial Mancuso/Stewart/Kneale claims. The study was announced on the front page of the *New York Times* three months before it appeared: "the first independent study of the health records of 35,000 workers at a Government bomb plant in Washington State presents a new, more sinister picture of the risks of small doses of radiation."[38]

The study was carefully devised to ascertain whether, when everything else is equal, differing radiation doses produce different cancer results. It considered a wide range of factors that might have been operating as confounding variables, such as age, sex, date of hire, duration of employment, type of work performed, and time between end of work and death. Of the workers studied, 7,342 had died by 1986, and 1,732 of those deaths were from cancer. Of these, about 3 to 5 percent resulted from workplace radiation exposure. "By the time all the people under study have died, there will be 200 'excess' cancer deaths," Alice explained in an interview. She called this "a very conservative estimate," adding that its significance lies not so much in the number but in the fact that the risk was largely a result of exposures after forty years of age and in the divergence of these findings from the A-bomb studies—according to which there should have been no extra cases, and if there were, they should have been between ages twenty and thirty.[39]

"The nuclear establishment will be very upset: our analysis has made it difficult to deny the link between low-level radiation and risk of cancer and suggests that the official BEIR report [Biological Effects of Ionising Radiation] has understated the risk by four to eight times."

Alice was pleased by the front page *New York Times* coverage. "I think it takes twenty years for an unpopular discovery to become accepted. It took that long for our findings about prenatal x-rays to be accepted, and it's nearly that since we started with Hanford."

But she has been dismayed to encounter further roadblocks.

"When we followed this with a second paper adding Oak Ridge and sent it to the same journal, *American Journal of Industrial Medicine,* the editor sent it to a reviewer who claimed he couldn't understand our methodology. This reviewer happens to have close ties with the A-bomb studies, and he insisted on seeing the whole of our data to make sure we'd analyzed it correctly—well, nobody ever does that! But we sent it. Then they asked us to rewrite, which we did. We didn't want to withdraw the whole thing and start somewhere else, so we waited it out, and it was rejected—after two years—so we published it in *Occupational and Environmental Medicine.* What an uphill struggle it is! The papers are held up on technical grounds that you feel, in a cooperative atmosphere, wouldn't be there. There's this curious resistance.

"We're still up against a very strong feeling that nobody wants to hear the truth about low-level radiation."

"Actually, nothing much has happened since then as far as I'm concerned," said Alice in 1995. "I've been rather upset that we've been bogged down with unnecessary interference. I suppose there are a lot of people hoping we'll go away, or disappear, by shortage of money or by old age."

A further follow-up, a 1995 Hanford paper, once more corroborates the age effect.[40] "We're doing our best to convince the world about this important question—it's very crucial because whoever wins this battle will determine whether the A-bomb data are of value in assessing risk to workers and the public." She is gratified to see others validating her findings—including "some of the DOE's own people." Steve Wing, a University of North Carolina epidemiologist working under DOE contract, found a cancer effect in the Oak Ridge data and published his findings in March 1991; in a subsequent study, he corroborated the age effect—as we'll see in chapter 14.

Epidemiological studies need time, as Alice is fond of saying. They also need patience, and it helps, if you are an epidemiologist, to be long lived. But the Three Mile Island Fund that's supported her and George ran out in June 1996. "I may be a bit gloomy, but I'm very old, and time is running out."

Chapter 13

The Invisibilizing of Alice

"The best way not to find something is not to look for it."

There are a lot of people wishing that Alice would go away, since her findings imply that there are risks to nuclear workers and residents around facilities, risks that are, at best, embarrassing to the industry and, at worst, financially ruinous.

The reason the nuclear industry continues to get away with its claims that it is a clean, safe industry is that there is, as yet, no way of proving beyond a shadow of a doubt that illnesses associated with radiation have actually been caused by working with radiation or living near nuclear installations. Cancers don't come with clear tags on them, their latency period is long and variable, and there is the further complication of the background radiation that bombards us continually. Also, by some curious logic, it is left up to those who have been contaminated—those who are the *least* likely to have resources—to produce evidence of harm, rather than to the multibillion dollar industry that is responsible for the contaminant.

Alice and George have been developing methods that would be useful in establishing proof—if anyone would use them.

Do Nuclear Installations Make Cancer Clusters?

Cancer clusters had been detected around all the installations at question in the U.K. public inquiries—Sellafield, Sizewell, Dounreay, Hinkley Point. All these facilities had a history of fires, accidents, seepage. Official investigations confirmed that there had been contamination of the environment, acknowledged that the population was suffering cancer in excess of what would be expected, but denied that the contamination caused the illnesses. Such studies made it possible for the industry to go on claiming that its activities are safe.

Their conclusions were based on standards of low dose radiation risk set by the National Radiological Protection Board in Britain and the International Commission on Radiation Protection, according to which the contamination detected around the sites wouldn't have been sufficient to have produced the cancer cases. The standards are derived from the Hiroshima studies and supported by reference to background radiation: if low-dose radiation were that harmful—the argument goes—areas where the population is exposed to more background radiation, like the mountain states in the United States, would have a higher incidence of cancer; but studies show that they do not.[1]

Now along come Alice and George with a study showing that there *are* more cases of childhood cancer in areas where background radiation is higher. They presented their findings to the International Radiobiological Society at Pisa in 1986, a few weeks before the Dounreay hearing.[2] They got a muted reception, says Alice.

"What happened was this: the National Radiological Protection Board had taken measurements of natural background radiation levels throughout Britain—of the very same areas we'd been collecting information on for thirty-odd years with the Oxford Survey. When we learned that this information was going to be available, we realized the potential of linking up our data with it. We knew the place of birth and death of every child who'd died of cancer in Britain from the year 1953 to the year 1979. We knew whether they'd been x-rayed or not and quite a lot else besides—about the mother's age and health, the father's profession, the child's position in relation to siblings; we had figures on children who had died of other causes. We had information on a total of 22,351 childhood cancer deaths and matched controls. Up until this time we hadn't concerned ourselves with the geographical distribution of childhood cancers, but we saw now that with the NRPB data we'd be in a position to investigate whether there was a correlation between patterns of childhood cancer and background levels of radiation.

"The initial results turned up a rather odd, inverse relation. In the southeast, where there is less background radiation, there was a higher rate of childhood cancer, whereas in Scotland, Wales, and the north of England, where there is more background radiation, there was a lower rate. But when you control for more variables, you see that something else is going on.

"The southeast has less background radiation, but it is also the affluent corner of England—it is wealthier and therefore healthier. If you live in the southeast, your child has a much better chance of living long enough for you to recognize that it's got leukemia than if you live in Wales or Scotland. In poorer regions, where medical care is worse, where there's less access to vaccines and antibiotics, the child simply doesn't live long enough to manifest leukemia, since children developing leukemia are so much more infection sensitive than normal children—you've got more competing causes that take away the cancer cases. And so, paradoxically, the healthier and more prosperous a population, the more leukemia it seems to have, and the poorer it is, the less leukemia, since children are more likely to die of something else during the preleukemic stages.[3]

"Climate is also a factor, since we think the biggest single competing cause is respiratory infection—and the south has a better climate as well as better medical care; hence it seems to have more cancer."

Alice and George found that childhood cancers tend to cluster in rural areas, which is where nuclear facilities tend to be built,[4] but they see this as related to the greater likelihood of a leukemic child dying of infection in urban areas.

Asked at the Dounreay inquiry to estimate whether the Dounreay cluster was due to radiation, Alice pointed out that this natural clustering of childhood cancers in rural areas makes it difficult to say—"You have to factor in competing causes of death, you have to control for socioeconomic factors, you have to look at health care, crowding, climate, all sorts of variables, some going one way and some going another. You have to have systematic data collecting; you have to know the doses, the exposures, when they occurred—which is extremely difficult, given the secrecy surrounding the enterprise." (Secrecy has been even a greater problem in the United Kingdom than in the United States, where the Official Secrets Act makes a government employee liable for prosecution for information divulged and holds recipients of the information, such as journalists, liable for publishing it. The Official Secrets Act serves as a kind of anti–Freedom of Information Act that makes whistleblowing virtually impossible.)[5]

"When you find a high concentration of cases around Sellafield and Dounreay, you can't automatically assume that this has to do with the reactor. You cannot make a simple correlation. In order to demonstrate a genuine cluster, you'd have to begin from our findings about naturally clustered distributions in rural areas, take this larger picture, and measure the particulars against it. You'd have to look at a whole lot of rural areas and a whole lot of areas with nuclear installations and show that the areas with the installations are different from the others. I think these little studies of individual places are useless: the populations are too small, and you'd have to prove there's an effect occurring over and above the natural effect of rural areas. But this has never been done."

Alice and George attempted to do it. They had data on children covering the whole of Britain for three decades. They had information on prenatal x-rays, inoculations, family, and social class. When they combined their data with regional health statistics on density of population, climate, infant mortality, and stillbirth rates, they found that there was a "statistically significant association" of cancer deaths with background radiation for most areas. Alice estimates that "the proportion of early cancer deaths caused by background radiation lies between 50 and 80 percent. This, plus another 5 percent of deaths attributable to x-rays, means that radiation is the cause of a clear majority of all childhood cancer deaths."[6]

Alice also maintains that the cancer effect of background radiation is stronger than that of prenatal x-rays, because it can affect the fetus within the first days of conception. It was "lucky for the human race," as she says, that obstetric x-rays were done in the last trimester of pregnancy, since the Oxford Survey showed that the earlier the exposure, the greater the risk. First trimester x-rays—which were fortunately rare—created a greater cancer risk than third trimester exposures, which were more usual.[7]

The conclusion Alice draws is that "you must on no account put up the level of background radiation. . . . It is no longer possible to defend nuclear power by saying, 'Well, we're only adding a fraction to background radiation.' You must not add *anything!*"

The Oxford Study of Background Radiation was widely ignored. Alice and George were not called in on any of the official investigations of U.K. cancer clusters and their findings about background radiation have been left out of discussion even at the academic level.[8] Alice speculates that this was because "the official epidemiology opinion in this country agrees with the Hiroshima studies that low-dose effect is negligible. We are constantly reminded that there were no early deaths from leukemia among 1,630 children who had *in utero* exposures to the A-bomb radiation—this is still given as reason for doubting the validity of the Oxford Survey." Alice and George have pointed out that the Hiroshima findings may be explained by the many miscarriages and stillbirths occurring in the years after the blast and the many infection deaths that would have eliminated the preleukemic child before leukemia had a chance to show. But nobody seems to hear them.

Friend and fellow scientist Klarissa Nienhuys suggests that the study was ignored because it was not useful to either side of the debate: neither anti- nor pro-nuclears could seize on it as proof of their position, it was "too ambiguous"—do nuclear installations cause cancer, or do they not? Alice and George did find cancer clusters, but they did not find a simple correlation between them and nuclear facilities, and the methods they recommend for further study—factoring in a variety of competing causes and looking at many areas in relation to one another—do not lend themselves to easy implementation.[9]

It is Alice's guess that there are clusters around some of the U.K. installations, but she insists on making a clear distinction between her opinion and what her research has shown. "I reckon Sellafield is the only place you've got sufficiently strong evidence, where you can really say there is a cluster. There are one or two other places in England that are very severe—one of them is Aldermaston, which is not easy to study because it's heavily populated and its population is mobile—and the other is Dounreay.[10] I think they've got something. But this is purely my opinion, and I don't like my opinion mixed up with what my research has shown. There needs to be more research."

Martin Gardner versus BNFL: Compensation Denied

Not only has Alice not been called in on any of the official studies of cancer clusters; she has not been invited to testify in compensation cases *against* the industry in the United Kingdom. This is the more remarkable since she has had success in such cases in the United States. Despite the judiciousness of her approach, plaintiffs seem not to want to enlist anyone who has associations with the "radical fringe"—as environmental groups are stigmatized in Britain—though they are ready enough to use her arguments in these cases and even call on her for advice.

The "Sellafield cancers case," as it was dubbed by the media, was one of the longest suits in the history of British civil law. The issue was whether the cancers suffered by Vivien Hope and Dorothy Reay—the daughters of lifelong employees of British Nuclear Fuels—could be blamed on radiation damage to their father's sperm. For the first time, a British court was asked to rule on personal injury claims based on alleged genetic damage from radiation.[11]

The debate came down to a duel between two opposing pieces of research: the A-bomb studies, which found no excess of cancers or genetic defects in the children of survivors who had been exposed to radiation, versus a 1991 study by Martin Gardner, a professor of medical statistics at the University of Southampton. Gardner had been a member of the Black Committee that cleared Sellafield of the leukemia clusters around the installation, but he had gone on to launch his own investigation. By culling birth and medical records, he and his colleagues identified 74 cases of leukemia and non-Hodgkin's lymphoma, diagnosed between 1950 and 1985, in the county of West Cumbria, and matched them with 1,001 controls. They then investigated possible causes: viral illness in the mother, prenatal x-rays, parental occupation and exposure to radiation, a shellfish diet, and other factors that would enhance radiation exposure. The factor that stood out strongest was the father's employment at the plant and especially his radiation dose before the child's conception, as ascertained from film badges worn at the plant. The researchers found that children whose fathers were exposed to the highest levels of external radiation were six to eight times more likely to develop leukemia than were the controls.[12]

This was the first study ever to demonstrate a correlation between paternal exposure to radiation and childhood cancer. Gardner speculated that radiation genetically alters sperm in a way that gives the child a congenital predisposition to leukemia. (Alice was quick to point out that this confirmed Oxford Survey findings that the earlier the exposure, the more dangerous. It was also consistent with a much later finding of the Oxford Survey, regarding paternal smoking—as we'll see.)

The study attracted worldwide attention, and considerable criticism. It was acknowledged to be statistically sound, but it drew fire for the same reasons that Alice's work has: the radiation exposures of the men had been well within the British occupational limit. Gardner was claiming that low-dose radiation was more dangerous than the A-bomb studies acknowledged, arguing—like Alice—that the slow and steady damage incurred by occupational exposures would have a different effect from the single, intense exposures of the bomb blasts.

Gardner died and left his work unfinished. "Follow-up studies failed to support him. But it was the usual story," says Alice; "everyone rushed in trying to tear him apart, invoking the A-bomb studies to refute his findings. I reckon he was right, or on the right track," she says, adding that she admired his study. Sir Richard Doll and colleagues, however, published a paper rebutting him, in which they invoked the A-bomb studies.[13] In a 1994 interview with the *New York Times* headlined "Nuclear Plant Cleared in Leukemia Cluster," Doll said that "Gardner was jumping to a conclusion" and that he, Doll, favored the theory that the spread of leukemia around the Sellafield plant is probably due to a virus. This theory, which is that of another Oxford University epidemiologist, Dr. Leo J. Kinlen, proposes that Sellafield is a potential breeding ground for the leukemia virus because it has "migrants from diverse origins" who are working in close proximity to one another—"9,000 workers spend the day packed into a mile-square, fenced-in area. Doll said this theory was still speculative, 'but it's the best we have got at the moment.' "[14]

Doll's dismissal of Gardner was a powerful tool in the hands of British Nuclear Fuels.[15] As described by Chris Busby, who has been studying for years the causes of the high rates of cancer in Scotland and Wales, "[Doll] is one of the most distinguished of contemporary epidemiologists and is widely considered to be *sans peur* and *sans rapproche*. His name has appeared on an increasing number of reassuring studies, perhaps acting as a kind of guarantee of authenticity."[16] Plaintiffs in the Sellafield cancer case made many of the same criticisms of the Hiroshima data that Alice had been making for years: the failure to take into account the high rate of miscarriages and abortions, the underreporting of birth defects, leukemia misdiagnosed as infection death. Lawyers flew in expert witnesses from around the world, and for ninety days, both sides thrashed it out.

On October 8, 1993, the plaintiffs lost. So, too, in May of the following year, did eight families who were using Gardner's findings to sue British Nuclear Fuels also lose; the judge ruled that "on the balance of probabilities," paternal preconception irradiation was not to blame.[17]

Again, Alice was not consulted and compensation was denied.

"They might as well have called me in, for all the success they had," she comments.

Christmas Island: Compensation Denied

Compensation was also being denied to veterans of the British nuclear tests in the South Pacific. Alice and Tom Sorahan, her colleague in the Department of Public Health and Epidemiology at Birmingham, were called in on this issue—by the veterans themselves.

"After the war, when America stopped sharing nuclear secrets with Britain, Britain had to go on its own and show that it could produce a bomb—which it promptly did and proceeded to test it on these remote Pacific islands, of which the most famous was Christmas Island." Twenty-one tests were carried out between 1952 and 1958 in South Australia and at Christmas Island, the British equivalent of America's Bikini.

In the seventies, a group of British veterans formed an association to fight for compensation. Veterans of the tests had been reporting high rates of leukemia and other radiation-related health problems. The Ministry of Defence denied that anyone had suffered illness as a result of the tests and insisted that safety rules had been stringently observed, that strict radiation dose limits were enforced, that protective clothing had been worn.

In 1982, a news program called *Nationwide* aired on the BBC, investigating safety and protection at the tests. Following this, the BBC and the *Observer* were inundated by calls and letters from servicemen and the relatives of servicemen.

"There were letters from servicemen saying they had various diseases and had always wondered if they might be related to the tests; from relatives of servicemen saying 'my beloved brother suffered such and such, my son died,' and so forth," summarizes Alice. "The official side told how tidy it all had been but the veterans told a different story: 'none of us had the foggiest notion what we were doing, we wore no special clothing, the blasts blew us over, knocked us about'—the usual things. One says they munched sandwiches during a flight. Another tells how he was instructed to sit with his back to the bomb and cover his eyes with his hands, and he was still able to see the bones in his hands when the bomb went off."

John Hall, who died of leukemia, described flying through the bomb cloud, tank louvres open, to collect the radioactive dust. "The louvres were closed when the plane returned to base and the tanks were flown to scientists at Harwell for analysis. The men wore no protective clothing, not even film badges to show if they were exposed to excessive radiation." When they finished, they showered four or five times to wash away the radiation. Ground staff were drenched with water as they washed down planes fresh from their flights through the mushroom cloud.[18]

Alice's colleague Tom Sorahan, a tall, soft-spoken, sandy-haired man, describes how the Birmingham department got called in. "The TV program took the letters and responses of the servicemen to the National Radiological Protection Board (NRPB) and they said nothing could be done with this information; 'but whatever you do, *don't take it to Birmingham University.*' So their next stop was to bring it to us at Birmingham University."[19]

Alice and Tom, together with their colleague George Knox, worked out, on the basis of national mortality rates, how many cancer deaths would be expected in this group by the end of 1982. They established that

27 men from a sample of 330 veterans of atomic tests on Christmas Island had died of RES neoplasms, cancer of the blood-forming organs, where statistically, 17 deaths would have been expected. They reported also a high incidence of skin trouble and eye cataracts. They published a letter in the *Lancet,* April 9, 1983.[20] In January 1983, Alice went on BBC's *Nationwide* with these findings.

"It's horrible to think how few safety precautions were taken," she said. "Few people wore protective clothing. Frigates were sent into contaminated waters and subsequently sold abroad to unsuspecting nations. Clothes on dummies that were standing inside the danger zone were worn and then sold."

But in a follow-up study a few months later, using an augmented list of respondents and refined mortality data, the researchers found less significant results—though they did find an excess of such neoplasms in the younger age group.[21] "The numbers were suspicious because it was a young population and there should be no such cases," Tom explained, and it was clear that a larger study was needed. "Quite honestly," he adds, "we would have liked to have done this study, but there's no way that they were going to assign it to us—we who had caused them the aggravation of making them do the investigation in the first place."

And so the National Radiological Protection Board was brought in. The Defence Minister who commissioned the NRPB investigation, Geoffrey Pattie, made it clear when he announced the study that he did not consider that anyone had suffered ill health as a result of the tests.[22]

Alice and Tom and the veterans were unhappy with this choice: the NRPB is "the wrong captain of the wrong boat," Alice said; "it is directly funded by the government. What is needed is an independent study." What's more, the NRPB was that same year under heavy attack for its handling of the 1957 Windscale nuclear fire. Omissions in the original report were forcing it, in September 1983, to produce a second report, which almost doubled estimates of deaths resulting from the Windscale blaze.[23]

The NRPB handed the Christmas Island studies over to the Imperial Cancer Research Fund, Cancer Epidemiology and Clinical Trials Unit at Oxford, headed by Sir Richard Doll, who was also consultant to the NRPB. It went from there to Dr. Sarah Darby of the Cancer Epidemiological Unit at Oxford, who has worked closely with Doll on several projects. Darby and Doll and colleagues published a study in 1988, finding twenty-two leukemia deaths in the participants and only six in the controls; for multiple myeloma, they found six in the participants and zero in the controls. They conceded that it would be difficult to ascribe these findings to chance, but they nevertheless concluded that "participation in the nuclear weapons test program did not have a detectable effect on the participants' expectation of life or on their total risks of developing multiple myeloma or leukemia."[24]

Darby and Doll published a follow-up report in December 1993.[25] Doll presented the results of the study to the veterans, admitting that the incidence of leukemia was slightly higher in the test veterans but attributing this to chance and assuring them that participation in the nuclear tests had "no detectable effect on expectancy of life nor on incidence of cancer." The scientists acknowledged that the relatively high number of leukemias and multiple myelomas is "confusing" and there may well have been small hazards of both diseases associated with participation in the tests, but a connection is certainly not proven. "I do hope that participants in the tests will find the results as reassuring as we do," Doll said; "Many were worried that they had been exposed to substantial risks as a result of their participation. I hope that they will be reassured."[26]

Several veterans stomped out of the meeting. John Armstrong, secretary of the British Atomic Test Veterans Association, said, "this is the very last thing we hoped to get. I have been fighting for this for ten years, and seen hundreds of my fellow servicemen get cancer!" Ken McGinley, a founding member of the Veterans Association, expressed disbelief: NRPB is a government body and cannot be considered independent, he said, adding that the government would have a lot to lose if the study had found a link between the tests and ill health.[27]

"On account of the trouble we made," Alice concludes, "there was an inquiry, but the study was given over to Doll and Darby, and—the usual story—it turned up nothing." Not quite nothing—Tom Sorahan says—for it depends on how you interpret the data, how you read the risks. "Just so," says Alice, "these things keep turning up in situations which, according to the official story, should be safe."

Applied Epidemiology: The Ready Reckoner

Alice has been more in demand in compensation cases in the United States, where she has had considerable success. In 1980–81, she was star witness in the first case of compensation involving radiation and cancer won by a Rocky Flats worker against the U.S. Department of Energy.

Rocky Flats, located on a bleak, windswept stretch between the Rocky Mountains and the city of Denver, had a dismal safety record, though there had been, as usual, no serious studies of workers' health. It would later be ranked as the worst facility in the entire DOE complex. Leroy Krumback worked there from 1959 through 1974, when he died at age sixty-five of colon cancer. His widow Florence remembered her husband's accounts of contamination on the job, how he'd come home with his hands rubbed raw from scrubbing with Clorox, which was supposed to remove the plutonium. Her attempts to receive compensation for his death dragged on until she gave up, in 1979.

By 1980 public sentiment in the area was turning against the facility and there was pressure to shut the plant down. Alice had been invited to lecture in Denver by Dr. John Cobb, professor of Community Health at the University of Colorado medical school. Cobb was fascinated by what Alice had discovered at Hanford and invited her to talk on Mancuso/Stewart/Kneale. "I'd just given my lecture, and Cobb and I were walking back from the hospital, and we got to talking about these compensation cases, and he said, 'what a pity it is that the MSK analysis hasn't reached the stage of being able to help individuals seeking compensation'; I said, 'oh, but it has.'

"I said, 'what you've got to do is find out what dose the person received, how old he was when he got it, do this for every year he was exposed in order to obtain a cumulative dose, and calculate how long an interval elapsed between exposures and death. You modify the dose he received with these variables to come up with the *effective* dose. On the basis of this, you can estimate how much his work exposure has increased his risk.' I call our system the *Ready Reckoner* after that little printed book you got in math classes as a child which would help you do conversions—you know, translate yards into miles and pence into pounds, and old forms into new forms.[28]

"I said, 'give me this man's doses and I'll tell you whether he should get compensation.' Cobb pricked up his ears and the next day he came back with this man's records and said, what a pity it was he hadn't known about this when he'd been approached by a lawyer and asked to help with the Krumback case. And I said, 'well, leave the records with me, I'll put them through the mill and we'll see what we come up with.' So I submitted this man's case to our calculations—I did it by way of my thank you letter—and sure enough, we found that he was three times as likely to have gotten his cancer from exposure as from natural causes, which meant he was highly eligible for compensation.

"So Cobb went and called Krumback's lawyer, who was no longer in practice, and was told that too long a time had elapsed since the case had been filed. But apparently the judge had thought the case was so important he put it in a special category of *sine die,* meaning there's no end or limit to the date—it's like murder, you know, you remain at risk of being caught for being a murderer forever. The DOE had wanted to dismiss the case, but the judge said 'no, we're not going to dismiss it, we're going to put it on the shelf.' He had done this on his own authority.

"So Cobb calls in a new lawyer, a young activist, Bruce DeBoskey, who retrieves it"—and this was the celebrated Krumback case. "I was able to prove with the *Ready Reckoner* that the radiation Krumback had received was more than enough to cause his cancer."

Krumback's records showed that he had received 45.67 rems of whole body exposure over a period of fifteen years, which the DOE and the DOE contractor, Dow, claimed was safe, well within permissible levels. But Alice calculated that the actual *effective* dose was much higher because Krumback had received a substantial portion of it while over the age of forty, when his sensitivity to radiation was greater. His *effective* dose was more like 222 rems.[29]

"They put us through a terrible grilling," Alice recalls; "and, of course, they brought in the A-bomb studies. They knew this case would open the floodgate to other cases, so they fought tooth and nail. But we won."

On June 3, 1981, Colorado granted Florence Krumback $21,000. "It was a paltry sum, not enough to cover medical expenses," Alice says, "but it was a first." The decision was challenged again and again and went all the way to the Colorado Supreme Court. "It went right up through the high courts and we made legal history by winning that case. This was the first time that anyone ever won a case against the DOE for compensation in this industry.

"After that, DeBoskey starts lining up other cases and the next time I'm around—DeBoskey tried to catch me whenever I flew over Denver—he says, 'we've got four more compensation cases'; and I keep a little file in Birmingham and I say, 'this is a good one,' 'this is a bad one,' 'don't touch this one.'

"And the DOE starts sending their top lawyers from Washington, and they try every trick in the book. They of course won't fly two witnesses over from England so these clever Washington lawyers make quite sure they fire at me all the statistical questions, which George could have easily answered, but I sometimes can't. And though we win some cases, we lose others.

"There was one case when they put up a figure on the board and I recognized it, but it seemed to me all wrong and I couldn't translate it. Fortunately, it's late in the afternoon and things are coming to a halt, and I go back to the hotel room and finally realize they'd done a mirror image version of my graph and turned it upside down so it looked all wrong; though it was correctly labeled, nobody would have presented the graph this way, and that's why I was flummoxed. But with a night's rest I managed to turn it back on them. Lots of these sorts of narrow shaves, and the lawyers found I was a fairly tough customer."

("Do you enjoy those sorts of encounters," I ask. "Well, of course, to some extent," Alice replies; "If you dislike it, you don't do it. Once I'm in the battle, I'm going to try and win, aren't I? But do I like doing this? No, actually legal cases are a nuisance, a distraction, and you have to spend far too much time on them, so I've done my best to cut them down, not encourage them.")

The Krumback victory was unprecedented in the annals of radiation litigation; it was, as DeBoskey says, "the first assault on the citadel, the first case where a nuclear worker demonstrated that exposure to *permissible* levels of radiation caused cancer—the first time a court upheld that this is not acceptable, men and women are dying." It was also DeBoskey's first workers' compensation case—he was two years out of law school—and his work with Alice forged a lasting friendship (though he'd heard "reckoner" as "retina" and thought she'd been talking about the "ready retina" during much of the trial, which he hadn't questioned since he thought her method provided a new way of viewing cancer risk and compensation.)[30]

DeBoskey also found the testimony of Karl Morgan enormously moving—"the sight of these two white-haired scientists, brave veterans, pioneers in this struggle." Morgan was brought in as expert witness and testified in August, 1980 that on the basis of his twenty-eight years experience as top health officer at Oak Ridge, he would have shut down Oak Ridge if exposures had been that high.[31] "The DOE brought in its heavy hitters to refute them—Ethel Gilbert was one; Eugene Sanger, who'd been implicated in radiation experiments in the fifties and sixties, was another. I can still recall one of Alice's more memorable lines. The DOE studies were being held up as having found no cancer effect, and she retorted, 'the best way not to find something is not to look for it.' It just shut them up—that kind of common sense biting retort that had the last word."

But in terms of drama and personal interest, DeBoskey found the

Gabel case the most compelling. "Don Gabel had graduated from high school in Denver and got a job in Rocky Flats as a janitor, when one day he saw an announcement that he could get 'hot pay'—ten cents an hour more—so he took a job that involved working with radioactive materials. He was working with his head right by a pipe that had radioactive gas going through it. When he asked his boss about it, he was told, 'no, no, don't worry about your head; it's only your body you need to worry about.' " Gabel got brain cancer at age thirty and died at thirty-one.

(The pipe was a ventilation duct from the furnace where the plutonium was being fired, and Gabel had spent nearly half his time on the job with his head six inches from it.)

In an emergency hearing held shortly before he died—his speech slurred from brain surgery, his head shaved, a plate in his skull—Gabel described working conditions at Rocky Flats, the frequent contaminations, the nonchalant attitude and inadequate protection, the use of leather gloves worn through to holes for handling radioactive materials. DeBoskey recalls his efforts to get information about the pipe: "the records had disappeared. We asked for permission to go out to the plant and look at the pipe and measure radiation levels from it—but the pipe had disappeared. We arranged to have Gabel's organs tested for plutonium, but within an hour of his death, his widow got a call from someone at Rocky Flats requesting her husband's brain so that they could test it. She consented to release it; they came and got it—then the brain disappeared! The DOE told her it had been lost. We did manage to test the rest of his organs, and they were found to contain some *five thousand times* the plutonium as the average Coloradan."[32]

At this point, DeBoskey brought in Alice. "The DOE was basing its case on a study done by a DOE researcher at Los Alamos, George Voelz, who had found an excess of brain cancer at Rocky Flats, eight brain cancers where four would have been expected, and eight other nonspecified neoplasms.[33] Alice got curious about the nonspecified neoplasms, and found out that they too were brain cancers, as she'd suspected—which brought the number up to sixteen where four would have been expected." DeBoskey cross-examined Voelz: "he had to admit there'd been an error, and what with the missing brain and the fiddled study, we were able to get Mrs. Gabel compensation."

"They were grueling cases," recalls DeBoskey, "and we did not win them all. We won three, settled three, which is a form of victory, and lost four or five. The more we got into it, the more sophisticated the DOE opposition got. At a certain point, Alice didn't want to do anymore."

Meanwhile, Gabel's widow remarried and was no longer presumed to be dependent on her deceased spouse, "so she didn't get any benefits—even though she subsequently got divorced and was left sole support of three kids. She got very little money out of the case—and ditto for me.

"When we lost, I got nothing, but even when we won, there was very little money involved. Once when I grumbled about this to Alice, she snapped at me: 'if you don't like it, you don't have to take this kind of work.' And she was right. Her fee was ridiculously low; she charged me $300 for all of her work on Krumback, where someone of her skills could have been charging $300 an hour."

The lawyers and witnesses working for the DOE contractors had no worry about money, since their fees came out of the public's tax dollars. DeBoskey figured at one point that for the ten cases he'd prosecuted, the government had spent $2.3 million defending itself, while paying compensation in the range of $400,000.[34] "That showed how important these cases were to it, how worried the DOE was about precedent setting."

"There ought to be compensation for cancer," Alice maintains. "If you are working in a nuclear facility, it should be acknowledged that there is a cancer risk." She believes that the *Ready Reckoner* can reliably ascertain whether a worker has got cancer as a result of exposure on the job and, moreover, that the industry could have saved itself trouble by using it. "We said that about 5 percent of their workers would die as a result of their work—I don't, myself, think it's the appalling threat people seem to—but the industry wanted the figure to be zero. They think if they admit a cancer risk, they'll be responsible for compensating every single case of cancer their workers get, and since about 20 percent of the population gets cancer, that would mean compensating 20 percent of their workforce, which would be a huge cost. But if they adopted our system, they could sort out people who'd got cancer on the job from those who hadn't. If you know the dosage at every stage of a man's life, the date of diagnosis and death, then you can reckon with some accuracy whether he got his cancer from his work. Those who have more than doubled their risk would be entitled to full compensation; those who'd increased their risk by smaller percentages could get smaller contributions. It would remove from the industry the feeling that they were going to have to pay huge sums of money, and everyone would be satisfied that justice had been done.

"But no—out would come the DOE lawyers, on the attack. You can't get the industry to admit that there is *any* risk and certainly not as long as they have the international committees backing them."

Though there have been various efforts to get workers' compensation for nuclear workers, the industry remains adamant.[35]

U.S. Veterans: Compensation Denied

Even in the United States, where the Freedom of Information Act allows better access to information and Alice has a higher profile, her work to secure compensation for veterans has met with little success.

Alice was drawn into an important case in 1984, by Dr. Wally Cummins, an aide to an Oregon congressman. Cummins had been approached by a veteran who had been stationed at Hanford and had alerted him to cancer among troops who'd been stationed there. Cummins found Alice at nearby Portland State University, where she was spending a sabbatical, and they devised a study. There had been some twenty-three thousand troops in Hanford; they had helped build and maintain the facility and then were phased out in 1962. A comparison of these men with soldiers who had served at Fort Lewis, Washington, and had not been exposed to radiation would give the researchers a perfect case and control group, since the men had comparable health, age, and occupational characteristics. Death certificates could easily be obtained from the Department of Veterans Affairs (VA), or so it was assumed.

So enthusiastic was Alice about this study that she committed $42,000 to it, drawing on the Right Livelihood Award. Cummins kept the study going by raising grant funds of $225,000 and getting concerned scientists like Professor Rudi Nussbaum, Alice's sponsor at Portland State, to contribute their time.[36]

At first the VA showed every evidence of cooperating, but suddenly Cummins was told that the VA had no money to make the data available. The agency's cooperation was crucial, but it was refusing to release the death certificates. Even worse, $205,000 appropriated for the study by the U.S. Congress and released to the VA seemed to have disappeared. Cummins and Alice decided to carry on anyway. After November 1991, the study and efforts to secure the death certificates of the veterans were funded by Cummins, out of his own pocket. "It's nearly bankrupted me," he said.

Repeated requests to the VA that the funds be released have received no reply. "They are doing their best to kill it. We could finish it in a year and a half," said Cummins in March 1997, "with the $205,000 already appropriated. We have all the data in hand for 1952 to 1962. We are ready to go. But the VA doesn't want the study finished."

"The VA is trying to postpone the day of reckoning," says Alice, "but it will come. There are 500,000 radiation-exposed veterans. The government is afraid that if they give compensation to any one of these, it will open the floodgates."

Rocketdyne: Industry Resistance

In the fall of 1996, Alice was invited to be part of a study of a top secret DOE facility near Los Angeles, the Rocketdyne division of Rockwell International, scene of one of the worst nuclear accidents to have occurred before Three Mile Island. A local activist group had been fighting the facility for years; they succeeded in shutting it down and even getting a DOE grant of more than $1 million to do an epidemiological study of the effects of nuclear contamination on workers' health. The core of the group are women who live down the mountain from Rocketdyne in the Simi Valley, who became concerned about their own and their children's health problems. They teamed up with Physicians for Social Responsibility and Committee to Bridge the Gap, a Los Angeles group that works on environmental and nuclear issues, to form the Rocketdyne Clean-up Coalition.[37]

Alice was called in and she and George found a cancer effect at dose levels that were considered safe and found also that these extra cases were largely concentrated among workers who were over fifty years of age when exposed. ("Just as we would have predicted—though even we were surprised that we were able to get this out of so small a sample!") She went to a meeting later in 1996, "expecting to find the whole thing written up, with our input in it, to discover instead that the report had been circulated to the industry, who had proceeded to hold a conference and collected together a whole lot of scientists and abstracted from their comments what you could only describe as reasons for not going forward with the publication. It was a typical industry maneuver, trying to put off the evil day. I was the only person at this meeting willing to say, 'I can't imagine why we are listening to this silly stuff; the report is ready to go.' We had all been sworn to secrecy until December 31, but I said, 'I for one am not going to keep this secret past that date because I don't believe there should be secrecy, and I don't believe the industry should be allowed to influence the report.' I said I'd spill the beans come January 1 if they didn't agree to the report being published.

"So they promised to bring it out in January, and they finally brought it out in September. It's important that this sort of thing be public knowledge and out on the table. I'm sure I was right."

And so the story goes: Alice is called in by activists and radiation victims, her investigations turn up cancer in excess of what would be predicted by A-bomb data, the studies are handed over to official bodies for investigation, the official studies invoke the A-bomb data to discredit her findings. Government agencies stall and refuse to cooperate, journals are slow to publish papers, the industry maneuvers. Time passes. "It's a long, slow business," she says.

Part 4
A Message to the Planet

Chapter 14

Epidemiology and Alice Stewart

"I have two of the ingredients for success in radiation epidemiology—longevity and persistence. Sheer doggedness. I've hung on and here I am, still quietly going on."

"The reason people don't believe in radiation is, it's out of sight, out of mind—then, twenty or thirty years later, someone drops dead. We are dealing with something so imperceptible to the senses and with such late effects—sometimes third and fourth generation effects—that we are very far from solving the mystery."

Alice Stewart sees herself as a kind of medical detective, practicing a Sherlock Holmes type of medicine.

Epidemiology is the search for the causes of a disease, for its etiology. What makes this so difficult with cancer is that the disease may take decades to manifest, and human populations tend to be highly mobile and subject to many influences. "You can't experiment on human populations, you can't give people doses of radiation to see how they react—or you're not supposed to. You can only *observe* the health of people who have been exposed to radiation. You can't control your experiment as you can in a laboratory. You can't eliminate the 'noise.' Yet you must somehow tease out, from the tangle of human life, that a particular exposure has led to a particular cancer, years later."

Background radiation complicates the question: "you're not looking for a spot of trouble against a spotless backdrop; you're looking for a spot of trouble in a very messy situation. You can expect a fifth to a quarter of the population to die of cancer in the normal course of things—so how do you sort out the naturally occurring cancers from those brought about by influences that are over and above?

"You're looking for something very small—small and rare—and an event that may be delayed for years. This is why it's so difficult to establish proof and why it's been possible for the other side to get away with murder, so to speak."

Defense of Epidemiology

"We did it in the Oxford Survey: we picked up a very small effect, a one in two thousand chance—that a single, nonrepeated exposure to a slight dose of ionizing radiation before you are born is sufficient to increase the risk of an early cancer death. We observed something very rare, very delicate—it was a needle in a haystack. And that's frightfully important! It might not have been observed if we hadn't, almost by accident, stumbled on it.

"Nobody was prepared to believe that you could handle a disease as rare as leukemia by the survey method. There was a feeling that epidemiology could only deal with big, gross effects. The leukemia incidence was considered too small, so it hadn't been much studied by the Social Medicine people. We had no competitors for our study.

"But to do it right, you must have large numbers and a proper system for collecting data. This is why epidemiology had to wait until the nineteenth century, for the reforms of William Farr, who realized the importance of keeping records on people and initiated the system of vital statistics. Farr, who was in the General Registrar's Office, which has since become the Office of Population Censuses and Surveys, realized that it wasn't enough simply to record dates of birth and death—you must record the cause of death and systemize the language for describing it.[1]

"Vital statistics are the alphabet of epidemiology, its *sine qua non.* John Snow couldn't have tracked down the causes of the 1848 cholera epidemic if Farr hadn't collected data on cholera death rates.[2] We couldn't have done our Oxford Survey without the proper public health records. Today every advanced society has a huge collection of vital statistics that tell us about health and social conditions. We know a lot about human populations—we could know more, if we used these records imaginatively."

But epidemiology is underappreciated and underutilized, dismissed by "pure" scientists as a crude science. It is still thought to be useful only for detecting large effects—the "noise" in the system is said to interfere with picking up subtle effects. "In laboratory work you try to set up conditions for maximum control and keep the field clear of interference. You can't do this in epidemiology. You can never get the situation free of noise in the way you can when you design an experiment—you have to bring in the noise, hear it, interpret it. You must keep questions open while you fill in the picture. This is very difficult when, as a scientist, you've been taught to seek control."

Alice feels that it's important to think of epidemiology as an *observational science,* a science with its own means and methods, rather than as something deficient by the standards of laboratory science: "then you'd see what it can do. I think we've been so overtrained in laboratory sciences that we've grown dull to the requirements of observational sciences. The observational sciences have taught us a lot—how else did we learn about the tides and the shape of the moon? By observing and recording—patiently, carefully. The epidemiologist is like the astronomer who learns from keeping records of the movements of the stars or the moon, knowing perfectly well that he can't interfere with them or control them in any way. You can't control the moon and the tides or experiment with them; you can't *prove* that their movements are related, yet you know that they are."

This only *looks* simple, but it's far from simple, and it amazes Alice that "people think they can just count a few things, make a few correlations—that they can just step into epidemiology from another subject, be a physicist or something and then become an epidemiologist overnight. Epidemiology requires a great deal of skill and training."

Alice recently reviewed a book that purported to be about the science of epidemiology. "On the basis of this book, I would have to conclude that there *isn't* any science of epidemiology, there's only a series of tools which have been used to approach various problems. This is a misconception, and it's why epidemiology continues to command so little attention or respect."[3]

A related misconception concerns the question of proof: "It's often said that epidemiology cannot *prove* anything, since it lacks the control of a laboratory science: all it can do is show correlations or associations. It's this that allows the nuclear industry to claim there's no proof of risk from low-dose radiation. I say, it's a quibble. I say, if you make observations long and steadily enough, you get what *amounts* to proof, from the weight of the evidence; and actually, that is all that proof ever is. You don't have to have absolute proof at every step of the way in order to get on with the job—often things that *seem* to be true will serve your purpose just as well." The "quibble" Alice refers to is not insignificant: it has allowed industries to go on producing toxins that have yet to be "proved" dangerous.[4]

Alice points out that scientists in other disciplines know better than to insist upon "proof": "Mathematicians know *pi* can never be an exact number, but we can still use it. It's been of enormous practical value. The trick is to get the best guess of the thickness of the ice when crossing a lake. The art of the game is to get the correct judgment of the weight of the evidence, knowing that your judgment is subject to change under the pressure of new observations."

Epidemiology is not, Alice admits, an exact science, but she insists that "it has a sharper edge than it's credited with—if you remain open to what the data are telling you. You need to keep questions open, to hear what the data are saying. The epidemiologist is like a conductor—you must hear every note, you must be able to detect a false note anywhere. If you hear a false note, you don't send the violins away: you try to work with them. You must include all types of seemingly extraneous data in the collection process—it might be the key to unraveling a mystery. Handling the noise is the greatest thing in epidemiology.

"I always sensed, in setting up a survey, that it wasn't right to put yourself in a straitjacket and look for only one thing. When you set up your study, get all the information you can, even if you don't exactly know what use you're going to make of it—see what comes in, let the field carve itself out. When you go to the source to gather the information, don't just ask about one thing. Cast the widest possible net."

Excluding the Noise

One of Alice's strongest objections to the A-bomb studies is that they've excluded the noise.

It is also one of her main criticisms of the work of Sir Richard Doll. "He has been heard to say that no survey that hasn't been completed within five years is worth its salt. This has had a very dampening influence on the whole field. You can't do it that way: the time and frame of epidemiology are simply different from laboratory work—you can't tidy it up like this. It needs a long time; it's got to have untidy edges." It's Alice's opinion that Doll "went and damaged his own ankylosing spondylitis survey by not casting a wider net."

Here is her account of the ankylosing spondylitis study, which is today a classic source of data on radiation health effects.

"A radiotherapist in Holland had used radiotherapy to treat this very painful rheumatic condition of the spine and was astonished to find that no fewer than five of his patients had either died or were suffering from leukemia. He turned to England, thinking that National Health Service would be able to set up a study, and it came to the Medical Research Council. I had agreed to help with the data collection and got called in for a meeting, and there representing the MRC were Doll and Court-Brown. We were instructed to go round to every radiotherapy department in the country and inspect the records of all patients with nonmalignant conditions who had been treated with radiotherapy—we were to pull out the ones with ankylosing spondylitis."

Alice found herself in an awkward position: "There at one end of the table were Court-Brown and Doll, lecturing us on how to recognize ankylosing spondylitis and distinguish it from arthritis of the spine and other spinal conditions. Now I must tell you that as a physician I was a tremendous snob—I'd done far more medicine than Court-Brown or Doll and I knew how to identify this condition, and here were these young men reading us symptoms from a textbook."

But Alice had deeper problems with what they were doing: "Then I thought, *why?* Why were we limiting ourselves by looking only for this one condition? Here we had the opportunity to inspect all the records for nonmalignant conditions treated with radiotherapy, so why were we looking only at this one? The MRC had gone to all sorts of lengths with this survey and instead of making use of this opportunity, here were Doll and Court-Brown confining the field of inquiry, boring in on ankylosing spondylitis. So I said, 'why don't we, while we're there, look at *all* nonmalignant conditions that have ever been treated with radiotherapy? Then we'd have a much larger sample and be able to compare the diseases and make sure we weren't dealing with a complication of this disease rather than an effect of radiation.' Well, I was more or less told to shut up.

"But I have this feeling about epidemiology: don't go cutting your information down to what you *think* you need—while you're there get all the information you can. Doll and Court-Brown limited themselves unnecessarily, and, sure enough, they had to go back and do a follow-up afterwards to make sure there wasn't some special association between ankylosing spondylitis and leukemia—when they could have figured that out from the beginning.[5] If they'd looked at three or four diseases they'd have excluded that possibility with their original sample; they'd have seen that the leukemia is what was common to all the diseases and realized that what was causing it was radiation."

Another problem Alice has with their study is that it turned up a lot of ulcerative colitis—which she thinks is an effect of radiation—and dismissed it as a complication of arthritis: "you see, the radiation treatment was administered to the spine, the colon is fixed to the spine along the back, and there's a long strip of the colon that must have been exposed. If they'd looked at other diseases, this is an association they could have established. It's very important because there are a lot of digestive disorders in the A-bomb survivors that never get associated with radiation—yet we know from the survivors that radiation destroys the lining of the gut.

"What Doll did is very common in epidemiology: an observation is made in terms of one cause and one effect, and then the person just goes looking for this association, instead of opening the question out to other possibilities. But that is definitely not the way to get the most from it. You have to include the noise rather than try to shut it out in the interests of time or tidiness. You have to be a 'snapper up of unconsidered trifles,' " she concludes, tossing off this allusion to the trickster in Shakespeare's *Winter's Tale* as though it were common fare.

"Epidemiologists must remain open to as many variables as possible so as to see the big picture. Epidemiology is by definition the big picture. People in the labs are dealing with the cells and people in the hospitals are dealing with the individuals, so it's left to epidemiology to get out there and see the whole thing."

Part of the "noise" she includes is the unheard noise, the missing diseases. "It's very typical of her way of thinking that the missing diseases have a guiding role in her work," as friend and fellow scientist Klarissa Nienhuys points out. "Others might think that no such disease is a sign that everything's okay, but Alice asks, what about those who died before the A-bomb studies were started, or those who didn't come in to see the doctor, or who left their jobs before they could get the disease?"[6]

A National Treasure

Alice had a grant of £1,000 when she initiated the study that turned up the link between fetal x-rays and childhood cancer. She did not have the resources to hire staff and consultants to make a study to assess what kind of study to set up, to weigh the pros and cons of this or that approach or method. What she did have was an instinct to cast the widest possible net.

"We told our interviewers, 'When you go to the mother, ask her questions about the child's illness, let her talk, let her go right back to the time she was pregnant and tell how the child got delivered and what illnesses he or she had before falling ill. You mustn't go and just ask the question you're interested in, you must get all the information you can. Take it down, *verbatim*. Never alter the record, take it down as is, and if it turns out to be more complicated than the form you've devised, devise a more complicated form. Never cut your information off at the source.' "

This is why the Oxford Survey of Childhood Cancer became such a gold mine, a storehouse of data capable of testing dozens of hypotheses, and why it continues to be useful to this day. It is, as Bob Alvarez says, a national treasure.

"And there it was when I needed it," says Tom Sorahan, Alice's colleague, who continues to mine it and find riches. Seeking information to test Martin Gardner's controversial theory that children of Sellafield workers got leukemia from parental radiation contamination, he found that Alice had included a question about the father's job history; he was able to show that preconception exposure to radiation on the part of the father is a more likely risk factor for childhood cancer than the child's own exposure.[7] More recently, he used survey data to test associations between paternal smoking and childhood cancer. "Tom said to me one day, 'Alice, did you keep records about smoking?' And we had done so, as an afterthought. I remember we said 'while smoking is in the air, as it were, we must tuck it in,' and that's where it is on those early forms, which we'd already printed, stuck on as a label, an afterthought." Tom demonstrated that as many as one in seven childhood cancers may be caused by fathers who damage their sperm while smoking—that fathers who smoked ten to twenty cigarettes a day had about a 25 percent excess rate compared to nonsmoking fathers, and heavy smokers, defined as those who smoked more than twenty cigarettes a day, had about a 40 percent excess rate.[8]

In December 1996, these findings got the Birmingham researchers coverage on Channel 4 and BBC evening news, and in the *Guardian* and *Times*.[9]

"It is such a monumental set of data," says Tom. "Even though Alice has published well over one hundred papers from it, you can never actually come to the end of it. She collected a lot of incredibly useful data without actually knowing at the time what they might mean. She seems to have an instinct for what's going to be useful."[10]

Alice takes great pride in the Oxford Survey—in the sheer physical fact of it. It exists today in the form of twenty-three thousand white manila envelopes, each containing the records of a live and a dead child, stored in a fusty, out of the way corridor connecting two temporary buildings on the edge of the Birmingham campus. She explains that when she came to Birmingham, "I had the utmost difficulty finding a place for it. I finally noticed this corridor, which I passed by regularly because it was on my way to the cancer registry." Though to the uninitiated, these shelves of dusty old files that stretch on for a hundred yards or so may not look like a particularly inspiring monument to a life's work, Alice surveys them with pride. "When I come into this corridor and see these records and I think, 'you generated all these,' I have a secret feeling of self-congratulation."

Alice has shown that epidemiology is a much finer instrument than it's credited with being, though to do this required that she use a retrospective method that was and still is out of favor. "*Retrospective* was a dirty word. Textbooks still tend to give it a second place.[11] But in certain situations it's far and away the best, as the Oxford study shows. Where it certainly has a very important place is in establishing whether there's anything worth going after.

"Doll knew this when he began his investigations of lung cancer. He and Bradford-Hill began with a retrospective study, comparing people dying of lung cancer with those dying of other diseases, and found a connection between lung cancer and smoking. But then, instead of carrying on with the retrospective method, he switched over to the prospective method and followed smokers forward in time. He used the retrospective approach the way it's generally used, to spot some trouble, and then left it as soon as possible in favor of the preferred method.

"With our study of prenatal x-rays, we used the retrospective method and found something, then decided it was the way to carry on—so much had been revealed by it, why not continue with it? By definition the prospective study can only take one thing on board, but we found not only the association between x-rays and cancer; we collected data on the way social class, maternal age, prenatal and postnatal infections, inoculations all bear on the cancer question."

Time and the Epidemiologist

"Truth is the daughter of time," Alice is fond of saying, and it's a saying that's especially true in radiation epidemiology. Cancer takes time to manifest; radiation takes time to show its effects.

"The longer the follow-up, the better the chance of seeing a cancer effect," says Alice. "I remember Mancuso saying, 'we haven't had enough time, we haven't had enough time.' Well, we *still* haven't had time with the nuclear industry—it's still relatively young. You need studies of duration to see the cancer effect, and you need even longer studies to see intergenerational effects.

"Steve Wing's follow-up studies on Oak Ridge workers showed this. He follows his population for a longer time, and it made all the difference between what he found and what previous researchers found."

Oak Ridge workers had been studied through 1977, and researchers had turned up no excess cancer; but by following them for just seven additional years, Wing allowed cancers with longer latency periods to show up. His results suggest that earlier negative results in other DOE epidemiologic studies should be treated with caution and probably be regarded as incomplete. Since most of the workers had received doses well below what is permissible, he said this "should lead us to question" the five rem limit for workers: his findings indicate that radiation effects are about ten times greater than claimed by the Hiroshima studies.[12]

Steve Wing came upon these discoveries without any prior knowledge of Alice Stewart's work. The Oak Ridge study fell his way because he was an epidemiologist at the University of North Carolina, Chapel Hill, to which the DOE contracted part of the Oak Ridge study. His findings were not good news to the DOE. "We weren't expected to turn up anything and when we did come in with positive findings, I quickly realized they would not be welcomed by the higher-ups. I went through some interesting times getting them published. When my paper finally appeared, it was accompanied by an editorial, 'There's No Free Lunch,' which argued, if you want nuclear technology, some people have to die."[13]

Wing had heard of Alice Stewart, but only as the dreaded enemy. "Well, I hope Alice Stewart doesn't get wind of this" or "what would Alice Stewart say if she saw this data?" were the sorts of comments he heard: "I had no idea why she was viewed this way and didn't pay much attention." But after the study was published, he got a letter from Rudi Nussbaum, professor of physics at Portland State, who pointed out that if he'd been familiar with Alice's work, his findings would have come as no surprise. "Rudi had sent this letter to the journal we'd published in, hoping it would print it, which of course it did not—but they did send it to me. I went to do some reading of Alice's work, and there it all was, just as Rudi said.

"I subsequently met Alice at a conference in Europe. I was totally taken by this fierce enemy. She's so wonderfully unassuming and down to earth and goes to such lengths to make you comfortable, to behave in a way that denies the inequalities of society. She gave me an unpublished manuscript and signed it, 'To Steve, from the dreaded Dr. Stewart.' "[14]

The Age Effect

There is another way that time features centrally in Alice and George's epidemiological surveys: it was George's method for testing sensitivity over time that enabled him to find the age effect in the Hanford data. "If you understand this varying sensitivity of individuals to radiation over time, you find an age effect; if you don't, you don't," explains Alice. George and Alice realize that a dose is not a simple number but that its effect varies according to whether you're a fetus or a child, young or old, male or female, healthy or sick, well or badly nourished or weakened by previous exposures to radiation or chemicals. This is one of Alice's arguments with health physicists, whose calculations assume, for all practical purposes, that "radiation hits cardboard." Health physicists concentrate on the dose a person receives but take no account of the varying responses of varying human beings.

This is another way of excluding the noise.

"It's much easier *not* to ask these sorts of questions, to pretend that all workers are homogeneous," says Dave Richardson, who has been working with Steve Wing and has just completed his Ph.D. thesis on the age effect. "But Alice's work showed from the beginning a sensitivity to age difference. You can see this as starting with her work on the fetus: she found a huge difference in sensitivity even within the nine month gestation period, when she discovered that exposure in the first trimester was more dangerous than exposure later in the pregnancy and that sensitivity to radiation dropped off after birth. Then, when she and George looked at an adult population, the Hanford workers, they discovered that sensitivity to radiation increased with time."

Richardson explains, "this is what marks Alice and George's work out as different from anything published today—a totally different sophistication about how effects are not uniform but change through time and are modified by age. There's very little other literature that looks at these effect modifications so thoroughly. It's much easier to ignore these complexities, which is why people go on publishing as though Alice and George don't exist."[15]

Alice and George

Alice's sensitivity to the way age undermines resistance to cancer has to do with her experience as a physician. "I'm a thoroughly grounded doctor—a lot of real practice came my way during the war, a lot of varied experience, in diagnosis and treatment."

It is one of her strongest convictions that epidemiology needs to be done in teams. "It needs a doctor, someone with a feel for the data, someone who's dirtied their hands, so to speak, not just done the textbook stuff, but someone steeped in medicine, with ideas about how bodies work, about how toxins affect human tissues. And it needs someone who understands numbers. Physicians lack the special skills in mathematics and computers to make the crucial statistical inferences, and statisticians lack the understanding of illness.

"It needs two experts—separate representation at equally high levels—in medicine and statistics.

"I think George Kneale and I have worked very well together. George is a high-powered statistician who knows about computers and methods of analyzing statistics, at the forefront of techniques and so forth. He is a thoroughly grounded mathematician. And he—he's a genius. He's immensely logical, very knowledgeable, and he steers an accurate course through the data. He's imaginative, but he's also sensible, so that his imagination doesn't take him off into any wild areas. He gets his ideas so clear in his head that he can write a whole paper with all the mathematics and everything and even the sentences in the right place, top to bottom, A to Zed, all in the right order.

"I couldn't have done anything without George. I have only a rudimentary knowledge of his field. I like to think he's the power engine of our ship and that I occasionally flick the steering wheel. I may have a few ideas, but he's the one who really works them out. I give him an idea and he puts it through his tests to see if it's any good, and if it's good, he can take it all sorts of new places."

(George Kneale's work is acknowledged to be "very difficult, but outstanding, excellent," by more than one fellow scientist. Rosalie Bertell calls it "exquisite.")[16]

"But he has at his elbow someone feeding him ideas, someone with a feeling for the medical implications of the data and strong views about the medical meaning of the numbers.

"We are, I think, a rather fruitful combination."

When Bob Alvarez asked George what his job was, he said, after a long pause, "It's my job to prove that Dr. Stewart's theories are wrong. I am, in effect, trying to disprove her. Hence the strength of our long association." (George to this day refers to and addresses Alice as "Dr. Stewart.")

Second Class Citizens

Epidemiology has come a long way since Dr. John Ryle established the Social Medicine Institute at Oxford. It has seen, since the forties, a slow but steady expansion and now includes the study of diseases with social origins as well as communicable diseases. But as late as the sixties, universities were still refusing to incorporate it into their curricula. "Today epidemiology is a bit more accepted—medical students have at least some introduction to biostatistics; but it remains a second class citizen," says Alice.

The reward structure of modern medicine doesn't favor epidemiology as a research area, notes Robert Proctor in *The Cancer Wars;* the emphasis on high-tech, heroic interventions leaves it low in prestige and funding. Whereas molecular genetics and cancer therapies are lavishly supported, epidemiology and preventive medicine are starved of funding.[17]

So too has its ally, industrial medicine, remained a second class citizen. A 1983 survey of U.S. medical schools showed that only 66 percent of responding institutions taught occupational health and only 54 percent included it as a required course.[18] It has developed more in the United Kingdom than in the United States, on account of Britain's stronger trade union movement. But the closest an epidemiologist ever came to winning a Nobel Prize was Sir Richard Doll, whose name was put in for his work on lung cancer and smoking.[19]

"Epidemiology still fails to attract physicians. It attracts statisticians and sociologists, mostly biostatisticians with a crash course in medicine; only a small minority of epidemiologists is medically qualified. Physicians flee it—there's a big salary differential. You lose money by going into it. And you lose status—you play second fiddle to the statisticians, who are second class citizens themselves."

Alice herself gave up a brilliant career. "There I was, headed for a practice in Harley Street. My friends thought I was mad to chuck it all in, and I well know the material rewards I gave up—I've had two brothers and a sister who are doctors, so I know what the going rate is. But I knew I didn't want to be a Harley Street physician. You have to have dinner parties—you need a wife. Besides, I was interested in the research end." She admits that most physicians would not want to work under the conditions she has. "I've been underpaid by comparison with my skills, seriously underpaid, with no special privilege. Most physicians don't want to go down in salary, down in status and authority—why should they give up a lead part, when they could have nurses and students following them about?

"But I've seen my brothers become bored with their subject and eventually give it up—one took to farming and another to building boats—whereas I've had continued interest. I count myself very lucky."

Alice faults physicians for not seeing the potentials of epidemiology. "The medical profession is very gravely to blame for not putting more of its brains into epidemiology. I'm absolutely certain that the present system of teaching epidemiology just puts medical students off. They come to it too late. They should be introduced to the subject before they get gripped with the fascination of clinical medicine—which *should* grip them. I know the excitement someone has coming from clinical medicine. But if you could get to students before they're seized with this passion and teach them the excitement of looking at disease through the lens of the group, if you could attune them to think of the larger picture every time they looked at an individual disease, you could get them to see the fascination of epidemiology.

"It still hasn't dawned on people how much can be learned by studying groups of people—by taking a telescope to the job instead of a microscope."

Pioneer and Pariah

"One of the founders of radiation epidemiology?" Alice repeats my question. "Well, yes, you could call me that.

"I wasn't a very gifted scholar, but when I found myself as a clinical student, I seemed to have the right material. It was as though I was a craftsman and I was waiting for the day when somebody would give me some good wood and then suddenly I realized I was a carpenter. I had a gift for diagnostic problems.

"I came into epidemiology almost by accident after I'd been a practicing physician, and a very successful one, flying high. I came into the field at the right moment, while it was still young, before it existed as a discipline. It was a great moment to do this—there were so few people there before me, which gave me the freedom to do something imaginative. And then along came the war, which created all these unique opportunities and material to play with.

"If it hadn't happened this way, I doubt it would have happened. I'd have been put off by the way it is taught today. I wouldn't have taken a mathematics degree, and as for public health, I'd have had the same snobbery most physicians have. I'd have been put off by the rules and requirements—no room for that marvelous inventiveness." As epidemiology became established, it congealed into more rigid notions of how surveys ought to be done, into conventions modeled on the laboratory sciences. (It is worth noting that in other new areas of scientific inquiry, women were allowed shaping roles, as Dorothy Hodgkins had in crystallography and Ida Ban had in ophthalmology.)

Alice was present at the inception, a founding member of the Society of Social Medicine and the International Association of Epidemiology. She was editor of the *British Journal of Social Medicine.* Yet when Social Medicine came together as a group and the Royal College of Physicians for Public Health and Social Medicine was formed—out of the group she had helped found—not only was she not on the list of founding members, she was not invited to join.

"From the moment I said something bad about radiation, it was as though I'd trod on somebody's corns. It's as though the medical profession didn't want to hear about radiation, basically didn't want to know. I was never on another Medical Research Council committee. From the time I started studying the nuclear workers in America, I was completely cut off from epidemiology in Britain." She was making one of the most important epidemiological discoveries of the century, but the British medical establishment didn't acknowledge the Hanford work as epidemiology. Finally, now that radiation health research has been transferred to the U.S. Centers for Disease Control (CDC), it is no longer possible to miss the connection. "CDC stands for the control of diseases, the charting of epidemics, which is of course what epidemiology is about. And now that the nuclear workers records are where they belong, at the CDC, here we are at the forefront. Time has come round," she says.

Alice speculates about her marginalization in the United Kingdom. "It's difficult to know how these things work, what goes on behind closed doors, but it's as though there was an undercurrent that I was all the while swimming against—there were no invitations, no offers, although what we were doing was known throughout the world. There was never anybody in England to stand up for me and say, 'here's somebody who needs a grant, deserves a prize.' It was known that Doll didn't think much of us, and he has been the most powerful and influential epidemiologist in Great Britain."

(Sir Richard Doll today expresses admiration for the Oxford Survey—in fact, a recent article of his supports its findings, but he was initially critical of it. Of the Hanford studies, he said in 1995, "I think she has gone off the rails in her more recent radiation work. . . . Her methodology is not scientifically valid.")[20]

Alice was finally, in 1985, made a fellow of the Royal College of Social Medicine and Public Health. "Perhaps it was that the radiation studies were transferred to the CDC; perhaps it was that George Knox, the head of our department at Birmingham, stepped in. I don't know the inner workings of these things." And in September 1996, as she was nearing her ninetieth birthday and the Three Mile Island grant was running out, Birmingham University made her a professor.

"I got a notice from the finance people saying I'd no longer get any pay, and then the next week I got another notice saying I'd been appointed honorary professor, term of five years. They emphasized this was to be honorary and without salary, but it turned out to be rather useful because when I went to check out a book, the library said, 'oh, you've been struck off the list'; however my secretary Anne Walker went back and found out I'd been put back on the list. And then the next thing that happened was, I heard someone knocking on my door and I said, 'come in,' but they weren't knocking, they were removing the notice that said 'Dr. Alice Stewart' and putting up the sign that said 'Professor.' Apart from these things, and a few people congratulating me, there's been nothing."

Asked what the promotion to professor meant to her, she replied, "Not much—I don't identify with the title 'professor'; I've always been comfortable as Dr. Stewart. But I did appreciate that someone made an effort on my behalf."

Over the years, it's been the pleasure of the process that's kept her engaged. "I like a medical puzzle. I'm always happiest when I've got a whole lot of data I'm trying to get arranged to see if I can't see something in it. Friends often say, 'every time I come to see you, you're sitting there adding up the same row of figures.' To many people it looks very dull, but to me it's absorbing. I've had this great gift, to be never lacking interest. I didn't go into medicine to save lives—I just thought it was interesting. I was very glad it was something that would help people rather than make me a fortune, but I did it for the sheer interest."

The one thing she has regretted is the lack of students. "I don't think I missed out on the research side, I've had marvelous research opportunities. But I did miss out on making a contribution to the teaching of epidemiology. I would have liked to have had a say in how the subject is taught." She feels that not having students has deprived her of the opportunity to test out her ideas on others and to work them through in a public forum. "It would have been good to have an audience that would help me sort things out. When I taught medicine, it was the students I learned from."

Some of her friends feel that this isolation has cost her dearly. Because she's worked so much on her own, and in close cooperation with a man who is not comfortable with people, she has cut herself off from scientists who might have been supportive. A related problem is the inaccessibility of style of many of their articles. A friend comments, "George invented a new methodology with the Hanford studies, maybe superior, but not carefully explained—this is a main reason their work was discounted. Their papers are convoluted and difficult to read to the point where other specialists in the field have simply not been willing to invest the time it would take to follow them. Alice never took the time to sit down and write a clear review paper of all her work, to bother explaining why she and George followed very different methods of analysis. So people dismissed their work and their enemies had an easy victory." The same friend, upon returning from a conference full of establishment figures, including members of the NRPB, ICRP, UNSCEAR, NRC, and RERF, was struck by this when he tried to bring up her work: "The greatest crime done to Alice was in forcing her out of academia into isolation; her writing style never recovered—it's cost her the clarity that would have enabled her to get her message across. That has been a tragedy. And she has not been her own best friend."

Alice has been concerned by the lack of followers. "This does worry me—what with nobody asking us to conferences, no money to carry on what we do, no junior doctors to train, we have nobody who can carry on what we've done." But recently there have appeared on the horizon Steve Wing and Dave Richardson—"It's wonderful to have people like them carrying on. It *is* a legacy—Wing is in his forties, though he looks in his twenties, and Richardson is barely thirty."

In April 1997, she traveled to Chapel Hill, North Carolina, to consult with them. Wing had just got a government grant to carry on research on the Hanford workers—"he won it, even though he said in his bid that he'd be following the Kneale method." The grant comes from the National Institute for Occupational Safety and Health (NIOSH), one of the agencies mandated with the nuclear worker studies after they were removed from DOE auspices. "I went over to see to a smooth transfer of the records, from the system we'd been using to his.

It's very hopeful for the future, a government scientist working with the dread Dr. Stewart."

Vision and Doggedness

The citation on the Right Livelihood Award, 1986, credits Dr. Alice Stewart with *vision:* "For vision and work forming an essential contribution to making life more whole, healing our planet and uplifting humanity." And *vision* is a word that more than one person has applied to her.

Dr. Jack Geiger, coauthor of the Physicians for Social Responsibility study *Dead Reckoning,* credits her with "uncommon scientific vision": "Every so often—not very often—an epidemiologist comes along and unlocks the door to expose a major threat to health that no one has seen before. Alice Stewart has done this repeatedly. She is striking not only for the elegance of her work but for her independence and absolute tough-mindedness."[21]

Mancuso describes her as being like a good musician: "What Alice Stewart has is a little difficult to explain—but she's like a musician who has a rare gift of sound."[22] Her gift for including the noise, casting the widest possible net, keeping the questions open and letting the data carve out the questions, is a capacity like the "cultivated attentiveness" Evelyn Fox Keller describes in Barbara McClintock, the ability to "listen to the material," "let the material tell you."[23] She works more like an artist doing a painting that comes slowly into view than a laboratory scientist who insists on the step-by-step certitude of a controlled experiment. "You have to keep the big picture in your head and move from that outline inwards, to fill in the details—and then as you fill in the details, keep checking back to see if it's fitting with your original idea and keep moving back and forth and be ready to change under pressure of new observations. Keep the picture open, keep the edges unfixed." She takes whatever time is necessary to move back and forth between the part and the whole, checking the detail against the pattern until the whole shape emerges, resisting the often falsely reassuring certainty of the experimental sciences.

She is also noted for a kind of maniacal patience. According to Dr. Molly Newhouse, one of her oldest friends and a pioneer in industrial medicine, "she has this ability to see a problem and seize on it, long before other people are aware of it. Then she sticks to it long enough for the world to catch up. She's very single minded. When she's doing one thing, she's doing that. Tremendously hard working and tremendously single minded."[24]

Dr. Morris Greenberg, a senior public health official at the Health and Safety Commission in Britain, who got to know Alice through the Ramazzini colloquium, describes the special quality she and George have: "They're not heroes as one popularly imagines it: it's a quality of perseverance, it's doggedness that does it." He adds, "It hasn't been easy for Alice. She's had all manner of personal problems to deal with, and she has never got the academic acceptance that one might have expected for her. Yet she carried on, making her case in a sensible, unhysterical manner. She is a polite, gentle soul who doesn't express herself aggressively, just firmly. I respect also that she doesn't complain. Her work has definitely not been overturned—she hasn't been proved wrong. It's been quite an achievement."[25]

John Goldsmith, M.D. and M.P.H., professor of epidemiology at Ben Gurion University, one of the founders of the International Society for Environmental Epidemiology, calls Alice Stewart "the outstanding radiation epidemiologist of the century. Her instincts as to radiation effects are often non-conventional, but rarely wrong."[26] Karl Morgan states, "in time there is no question that she will be held up as one of the greatest epidemiologists of our century."[27]

Not everyone has such high praise. In the 1996 Channel 4 documentary about Alice and two other pioneer women scientists, "Our Brilliant Careers," Sir Richard Doll described her methods as "a bit slapdash. She was very enthusiastic, she got a great deal of cooperation throughout the country, but she tended to accept results at their face value without detailed checking to test their accuracy." Of the Hanford study, he said, "She used a technique which was not considered appropriate, and I think that analysis was a barmy analysis, yes, it just wasn't a scientific analysis." About low-level radiation, "I would be prepared to say that she'll never be proved to be right." Asked how he perceives her, he replied, "Well, I don't really perceive her," then muttered not quite inaudibly, "except when I have to."[28]

It later emerged that these remarks had been excerpted from an interview that was much longer and more complex, in which Sir Richard Doll said many favorable things about Alice; he felt he'd been badly misrepresented. But Alice, watching the program as it aired August 19, 1996, didn't know this. "I really—I'm quite shocked! I knew he didn't approve of me, but that he'd say this! It explains a lot, the cold shoulders, the lack of offers or invitations. I see that I haven't made it up."

A tale of professional rivalry, scientific disagreement, political disagreement, or simple dislike? Alice will never know.

But most people who know her have the highest regard for her, though they praise widely different qualities: vision, bold imaginative leaps, practicality, perseverance, patience.

Dr. John Gofman calls her the grande dame of radiation health science. "She's prevented untold numbers of premature deaths from miserable diseases like leukemia and cancer, and all history will owe her a great debt."[29]

Chapter 15

The Good Doctor

"What we've got here is a message for anyone who is interested in the cause of cancer."

The whole of this remarkable woman's work, since the Oxford Survey turned up its revolutionary discovery about fetal x-rays and childhood cancer, has been a search for the cause of cancer. "This is the mystery, and it's kept me busy my entire life. To me the whole thing has been an effort to tease the story out, to understand the etiology of childhood cancer."

From the start she felt that their discovery about fetal x-rays had put them on to something momentous. "I regarded it as very important, a fixed point to steer by, like the polar star from which the science of navigation comes. I can remember saying to myself, if a mariner had set out without knowing where to go and what to do but was fortunate enough to discover the North Star, one fixed point in the firmament, he could fill in the other geographical components—if he could only hang on. We'd found our fixed point, our polar star, the sure thing we could navigate by.

"I think that's what kept me going on a subject nobody else seemed to be interested in. I'm quite certain that's what we're doing—from this one finding we're gradually building up all sorts of things."

Early on, she and George found a strong correlation between cancer and infection sensitivity. During the time the child is incubating cancer, he or she becomes increasingly sensitive to infections. By the end of the silent phase of immune-system cancers—cancers of the reticuloendothelial system (RES)—he's more than three hundred times more infection prone than a normal child.[1] This correlation alerted Alice and George to relations between cancer and the immune system that were developed in subsequent theories about Sudden Infant Death Syndrome (SIDS), Burkitt's lymphoma, and cancer and inoculations.

Alice wishes this part of their research were better known: "I would like to be remembered as someone trying to puzzle out the causes of childhood cancer, rather than a radiation expert, which I'm not. The x-ray story has attracted so much attention that people haven't heard about the rest of it—though it's all there in the literature."

It may be there in the literature, but it is tucked away so obscurely that it is in danger of being lost.

Unmaskings

One of the ironies of Alice's story is that although she is best known for discovering the link between fetal x-rays and childhood cancer, hers is not really a radiation-focused theory.

"I can remember the moment I realized this, working on the Oxford Survey—I suddenly realized, *supposing there had been no antibiotics?* I woke up in the middle of the night with one of those brain waves my mother used to get, knowing—it came to me in a flash—*of course* this is why we're seeing leukemia, it was on account of the antibiotics! The advent of antibiotics, by reducing the chance of death by infectious diseases, had allowed us to see the true incidence of leukemia. It's as if the tide had gone out and you could now see what the bottom of the ocean looked like. The true state of affairs was revealed for the first time.

"When you are developing a cancer of the immune system, you lose immunological competence: you have no resistance, you react badly to minor infections, minor infections become major infections. This means that before antibiotics and other drugs were introduced, even a common cold might be sufficient to remove a leukemia death from the official record. In the old days, the leukemia was there all right, it just didn't get a chance to show: infections, measles, mumps, pneumonia, whooping cough, other respiratory diseases, would kill the child first. The odds against surviving long enough to see the disease were so great that childhood leukemia was a rarity.

"Then along come antibiotics and up goes the leukemia figure two and a half times. They kept the child alive long enough for the leukemia to show.[2]

"I can remember my father saying, you rarely see a case of leukemia in a child, and when you do, you always notice that it takes the child who's never been ill. And this was a curious thing we found in the early days of the Oxford Survey—so many of the mothers claimed that the child who died had seemed to be the healthiest of their children. You can see why: if he'd ever met with a serious infection in the days before antibiotics, he'd have died."

When she and her co-workers began looking at the rise of leukemia, David Hewitt's review of official statistics turned up an increase in leukemia mortality that had kept pace with a drop in general mortality.[3] A later study by George Kneale confirmed that during the years when pneumonia deaths of children are up, there is a drop in leukemia, though leukemia incidence rises again in the following nonepidemic years.[4] In the year 1918, the year the Spanish influenza swept the world, there was a massive disappearance of leukemia across the world: "never was the leukemia death rate so low as it was in 1918. Why? because no one incubating leukemia would have survived the influenza."

The most important factor in the rise of childhood leukemia, Alice insists—more important than the use of medical x-rays—is antibiotics. "The effects of man-made radiation on increased leukemia rates cannot be known until the story is seen in this context, of the control of infectious diseases. Thanks mainly to above-ground weapons testing, we've doubled background radiation worldwide; at the same time, childhood cancer has more than doubled. But how these two phenomena are related is difficult to determine. Radiation has its effect, of course, but how much the rise of leukemia is caused by the new sources of radiation and how much by the unmasking effect, is hard to say. It's impossible to measure precisely, since there's been this other enormous change, the eliminating of major competing causes of death."

Cancer, Infection, the Immune System

Alice believes that most childhood cancers originate from *in utero* exposure to background radiation or medical x-rays because hardly anything else gets through the placental barrier. "The uterus is the best protected organ in the human body. And it's got to be well protected to allow nine months for getting everything to do with normal development not only right, but in the right sequence. Fetal development requires tremendous protection against outside interference, and you can be pretty certain that in normal circumstances, nothing harmful gets through the barrier except the one thing that human beings can't protect against—and that is radiation. Chemical carcinogens play a part, of course, but as promoters—secondary, not primary, causes of cancer."

Alice sees *in utero* mutations as the cause not only of childhood cancers, but also as unrecognized causes of miscarriages, stillbirths, and deaths in the early years. "You notice that there is a high death rate in the early stages of life—these are the children with congenital weaknesses or defects. Somehow nature says, you won't be strong enough to survive, so you're bumped off early. After that you're presumably programmed for a certain life span, a span which is partly determined by the strength of the material. Eventually the material wears out, like metal fatigue in an airplane—we die of heart failure or the failure of some other crucial organ. Natural selection is interested only in life to the point of reproduction. We're not meant to live forever and it would be a disaster to do so—though of course human beings may have other ideas.

"But what can kill you off in the interval between the congenital weaknesses and the final wearing out—what determines whether you survive—is whether you can recognize self from non-self in a situation where every moment, with every breath and every bite you take, you're taking in foreign matter and using it to replace your own tissues. So here you are, with this job of taking in non-self which has got to become a part of self. You've got this surveillance mechanism that's got to recognize self from non-self, and it must on no account destroy normal cells. And you can see that there are two great enemies—infections and cancer. Vulnerability to both increases with age: it's as though in the process of the wearing out, you become less able to muster the forces to defend against foreign invasions.

"It's the immune system that performs this function of recognizing self from non-self at the cellular level. It's a surveillance system geared to recognize foreign proteins—recognize and destroy them. Much is known about the way it works: cells chase around gobbling up foreign proteins, other cells come along and absorb what the first phalanx misses, and so on. One theory of cancer is that in order for a mutation to survive to become a cancer, to form a clone of cancer cells, there must be something wrong with the surveillance system.

"In infections and cancer you've got two very similar processes, though they seem to be so different. An infection is caused by a living organism with protein foreign from you, getting inside your body by some trick and multiplying. Now these organisms may get inside you but not multiply. We probably—many of us—have within us tubercular bacillus; perhaps we've had a slight attack of influenza when the tubercular bacillus was struggling to take over; perhaps we even have a scar on our lungs or little lesions. Disease happens when the foreign organism gets the upper hand and succeeds in multiplying.

"How does the immune system work to repel foreign invaders?

"We know that though you're born with a certain natural immunological competence, you've got to acquire a lot more during your lifetime, and you do this by constant contacts with infections that you learn to overcome. The system has to learn to detect and destroy foreign proteins, so it must be a very delicate mechanism: it's got to be able to recognize the difference between you and foreign molecular elements that are not part of your cells and stay away from the ones that are you. You have to be very careful not to kill yourself.

"If your resistance is working, your body finds ways of localizing infection—you may get a boil or a carbuncle instead of a generalized disease, and that shows that your body has managed to contain it, to bring it down to size. Localized forms of infection are less dangerous than diffuse forms, just as localized forms of cancer, or tumors, are less dangerous than leukemia or lymphomas. Even if infection really takes over and becomes acute and you have high fever and infection everywhere, the body usually finds a way of settling the problem relatively quickly.

"But cancer presents different problems. Cancers are like infections in that something foreign is trying to get the upper hand, but in the early days the cancer cell may be only very marginally foreign. Something which is half you is different from a foreign organism. It's developed an apparatus it shouldn't have, to circumvent the body's defense mechanisms.

"I always say it's rather like having a delinquent child in the house—you can't treat him the way you'd treat a burglar. It's partly yours, a part of you—partly self, partly non-self—and very destructive: it can do a great deal more damage. The process is slow and subtle.

"Now what is cancer? Cancer starts with a mutation and develops when the mutant cells gain the upper hand. When radiation passes through matter it alters the molecules through which it passes, knocking away their electrons and creating chemically unstable particles. Radiation is a mutagen, that is, it creates a mutation. Not every mutation develops into a cancer—the body has extraordinary resistance. Ninety-nine times out of a hundred or 999 times out of a thousand, nothing happens. You can have lots of mutant cells without being harmed. Perhaps you've even been hit by radiation and had a mutation while we've been sitting here talking. (You realize, we only seem to be sitting quietly here, we're actually whirling through space and being bombarded with background radiation.) But most organs contain more than enough cells to spare and maintain normal function—you can afford to lose lots of cells and nothing happens. Obviously, a hit is more likely to cause damage when you're the size of a tadpole, consisting of a few clusters of cells, each of which is going to develop into a part of your body. If you should happen to be hit by radiation when you're this size, there will almost certainly be a trail of trouble.

"There are three ways radiation can affect cells. One possibility is that cells may be killed outright, end of story—unless many cells are killed or a crucial cell in a developing embryo is affected, in which case you die. The second possibility is that cells may be damaged, but in a way that is harmless: a few malfunctioning cells will not affect an organ where most cells are behaving normally. But in the third case, cells may have their DNA damaged and yet survive and go on to spawn defective daughter cells.

"It looks as though the mutagen has to strike the cell at the moment when it's dividing for this kind of damage to occur. It's at the moment of cell division that the risk of DNA damage is greatest. When the cell divides, the nucleus unfolds to form a spindle with chromosomes on either side, each one of which contains duplicate sets of controlling genes in a particular sequence which it is important not to alter. The genes at this moment are not locked back in the nucleus where they're usually tucked away; they're out there and vulnerable. If a hit occurs while this delicate process is occurring, it's going to be difficult to get each part of the intricate system to reassemble properly.

"Provided that the DNA damage is limited to one of the helix crossbars, there will usually be full repair; if there is destruction of both members of a pair there will be attempted repair, but probably at the cost of permanent gene damage. You have to have a double hit that doesn't repair and doesn't kill the cell for a mutation to occur.

"The slightly damaged cell may succeed in reproducing itself, in setting up a little clone. The mutation gets by the body's surveillance mechanism by being sufficiently like the normal cells to pass without bringing the immune system down on it. The cell loses control of its own growth processes and stops obeying the instructions of the body as a whole; it becomes liberated from the usual checks on replication and proliferation and becomes a cancer."

Alice's main quarrel with the nuclear establishment is this: "they say that below a certain dose—which they call a threshold—you get chromosomal repair or radiation damage. They say repair reduces risk. I say it might *increase* risk. I use the example of a cracked plate: you can repair the plate, but you'll still have traces of damage, and all you need for cancer is a small trace. Every time it is stressed, it becomes more liable to break. It will not react to disease or physical injury as well as an undamaged cell, and when it reproduces, it will pass on its defect."

Another analogy Alice finds useful is that of the aftermath of a train wreck: "you may reassemble the parts and get the train back on the track and get it to its destination all right, but some of the passengers happen to be dead. Think of the carriage as the chromosomes containing the genes. Because the chromosomes manage to reassemble themselves, this doesn't mean you haven't done some damage to the DNA within."

Alice's guess is that even after the damaged cells manage to clone themselves, a lot of these mutations get to this stage and no further: in fact, all the forces of the body are working to contain them. She sees evidence for this in the fact that precancerous conditions of the cervix often reverse themselves even when women fail to come in for follow-up, though in theory they should develop into cancer; but she also points out that such reversals occur only in women under fifty, and the younger the woman, the better her chances. "Our Hanford studies show that the odds against the mutation developing into a cancer are very high when you're about twenty, but that they decrease with age and the weakening of the immune system—the age effect again."

Mutations may produce tumors, on the one hand, or birth defects, on the other hand. If it's a germ cell that's been damaged—a cell in the ovaries or testes—and that germ cell later forms a child, the child will suffer from an inherited abnormality. "Or the damage may skip a generation before you see anything—and not only one, but two or three generations. If the defective gene is recessive, it has to wait to meet up with the same damage in another parent before it manifests the damage;[5] it may take generations to manifest—but it will be there. Once you've fed defective seeds into the gene pool, they lurk as potential trouble for the future."

Theory of Childhood Cancer

Alice sees childhood cancers as *delayed congenital defects,* or late effects of *in utero* mutations. She believes that the *in utero* initiation of these cancers accounts for the tissues they affect and the forms they take, producing developmental effects different from those leading to adult cancers.

Early in development, an embryo consists of a mass of primitive, undifferentiated cells called *blast* cells, each of which is a stem cell undergoing rapid division. Because its cells are rapidly dividing, embryonic tissue is vulnerable to mutagens. "When you're the shape and size of a tadpole, you're going to be especially vulnerable to mutagenic hits. If a hit occurs in a tissue that's going to form an essential component of the body such as the lungs or intestine, you won't survive. This is why children do not get cancers in these organs—these children don't get born. But an embryo has tissues that are not needed later, that are necessary to its development but have no future; some of these remain even as new systems develop. It can survive a mutation in a vestigial tissue."

Childhood cancers develop from three groupings of cell systems in the embryo: the autonomic nervous system, which is involuntary and will soon become subservient to the central nervous system; the reticulo-endothelial system (RES), the blood-forming and bone marrow tissues, which are the basis of the immune system; and the Wolffian ridge, a large, semi-circular structure attached to the back. The Wolffian ridge shrinks down to become the kidney, and from it come the genito-urinary system, lungs, intestines, and limbs. Cancers of the Wolffian ridge become Wilms' tumors; cancers of the autonomic nervous system become neuroblastomas and cerebellar tumors; and cancers of the RES are leukemia and lymphoma.

Virtually all children's cancers develop from these types of tissue: vestigial, the immune system, or the brain. About 50 percent of children's cancers are either leukemias or lymphomas—RES neoplasms—whereas only 5 percent of adult cancers are. Half of the remainder are brain or neural tumors (neuroblastomas); half the remainder of these are Wilms' tumors; and the rest, about 10 percent, are other types of cancers.

If the Wolffian ridge undergoes a mutation, the mutation will, as the embryo develops, take the form of a tumor surrounding the kidney, Wilms' tumor. The tumor develops on the outer rim of the kidney and sits on top of it, leaving it still able to function. If it's the brain tissue that's affected, the mutation may become a brain tumor. It is significant that 99 percent of the brain tumors of children are in the hindbrain, the cerebellum, which develops earlier than the forebrain and is larger in the embryo and therefore more likely to receive radiation hits. (The brain tumors of adults are mostly in the forebrain, the cerebrum.)

"You can survive the early effects of brain tumors—in fact, hardly any of the nervous system matters much until you are born—so the cancer can go unnoticed. And you can survive the early effects of leukemia because the immune system is not needed until a month after birth—your mother's is doing the work for you until then, longer if you're being breast-fed, in which case you receive immunity from your mother for six to nine months. You don't need your immune system until you go off hers, but then you need it very quickly."

Which tissue is affected determines the age at which the cancer appears. Since the cancer partakes of the characteristics of the tissue and reflects its growth rate, each kind of cancer has a time when it typically manifests. Leukemia has its peak instance before the age of five, but slow-growing tissue makes for slow-growing cancers: bone cancer has a peak in young adult life. Hodgkin's lymphoma peaks at around fourteen to twenty years of age—this is the slowest growing of the RES neoplasms.

"The cancers that peak latest, I'm going to guess, are testicular and early breast tumors. Nobody has ever looked at them this way, but I think that early breast and testicular cancer may be delayed congenital effects, manifesting later because of the late unfolding of the sex organs."

Sudden Infant Death Syndrome: An Untested Theory

When, in the early days of the Oxford Survey, Alice began looking into childhood leukemia, David Hewitt's study turned up this odd and unusual peak age of two to four years. Alice had long been wondering, why these peak ages? And why did children get only lymphatic leukemia and rarely myeloid leukemia?

She then noticed another curious thing—that twice as many children who died of leukemia under six months were born in the first half of the calendar year. There seemed to be no discernible pattern in children who died when they were older than six months, but Oxford Survey data revealed that children who died from leukemia under six months of age were *twice* as likely to be born between January and June than in the second half of the year, between July and December.

"Why? There is no seasonal difference between the first and second half of the year—they're both half winter and half summer. In Britain, the population regularly records more deaths between October and March than between April and September. But the number of deaths between January and June is not different from the number between July and December. So how to account for this difference?"

She soon worked out that the difference was whether a child who had survived to one month of age was moving into warmer or colder months. "You can see that all one-month-old infants who are born in the first half of the year will experience more warm weather before six months of age than during the next six months, whereas all one-month-old infants who are born in the second half are moving into the cold. The children born in the first half of the year, January to June, are moving into the warmest months, whereas those born from July onwards are moving into winter, where they have more exposure to danger from infection. Since those babies moving into summer are less likely to meet with an infection at the most crucial times, they have a better chance of surviving six months—to die of leukemia. So it looks like children born in the first half of the year get more leukemia.

"And exactly the opposite is happening on the other side of the calendar. If you're born in the second half of the year in Britain, you have three times as much exposure to the colder months by the time you're six months of age than if you are born in the first half of the year. So just before you're coming up to die of leukemia, you get nipped in the bud with an infection. It looks like it's good to be born in the second half of the year because you're less likely to die of leukemia, but it isn't really: it just means you've died of something else before leukemia has a chance to show."

But what could that something else be? "So I said, let me see if I can't find a cause of childhood death that is commoner in the winter than in the summer (and so affects babies born in the second half of the year), something which has not been affected by the advent of antibiotics and which is concentrated in the first six months of age. In other words, I was looking for the obverse or mirror image of the leukemia deaths." Alice is not called Alice for nothing, as she's fond of saying: "I see everything through the looking glass."

And there it was, a cause of death that hadn't decreased with antibiotics: in fact, it had become more prominent—Sudden Infant Death Syndrome (SIDS), known as "crib death" in the United States, "cot death" in England. It kills between 6,000 and 7,000 babies a year in the United States alone, between 2 and 3 cases per 1,000 live births, or one in every 350 babies.[6] Nobody knows what SIDS is, nor why it's on the rise. What is known is that it occurs more often in winter than summer and that it occurs mainly between four and six months of age.

Alice believes that these sudden infant deaths are masking myeloid leukemia. "My theory is this: the reason we aren't finding myeloid leukemia in children is that the child with myeloid leukemia is dying of a sudden unexplained death, if he hasn't already died of anoxia during the second stage of labor. Most SIDS deaths occur within one and six months of age, which is just when the child is losing its mother's immunity and achieving its own. While the normal child is gradually acquiring his own immunity, the child with leukemia is gradually losing immune competence. Since you get from your mother defenses against infection in the form of passive immunity for one month or more, the weakness in the system doesn't get put to the test until you go off your mother's immune system."

Myeloid leukemia is more acute than lymphatic leukemia. It has a shorter latency, manifesting between one and three years of age rather than two to four years, and it involves the red blood cells as well as the white. Children who are incubating myeloid leukemia are—like all pre-leukemics—more infection sensitive than normal children. But they are also born with a defect in their hemoglobin: they have something wrong with their red cells as well as their white cells.

"While in the womb," Alice explains, "the fetus produces fetal hemoglobin, which is geared to receiving oxygen through the placenta; but soon after birth this is replaced by adult hemoglobin, geared to receiving oxygen through breathing. At birth you have both kinds of hemoglobin present, enabling you to breathe both through the placenta and through the new apparatus of the lungs; then you gradually get rid of the fetal hemoglobin. But children who are incubating this kind of leukemia don't make the changeover from fetal to adult hemoglobin and are left with too much fetal hemoglobin. This hemoglobin fails to take up oxygen from the lungs, so that when they go into a deep sleep, or have the first effects of a respiratory infection, the oxygen level falls to a fatal level and they're liable to go into anoxia—shortage of oxygen.

"There have been studies showing that children who die of SIDS have an exceptionally high ratio of fetal to adult hemoglobin, something like sixty-five to fifteen—though this is difficult to measure after death, and it's not something all hematologists accept."[7]

Alice's theory is that SIDS children have difficulty replacing passive immunity with active, and fetal hemoglobin with adult, and the two effects combined might be sufficient to cause a sudden death. SIDS children die when they're sleeping, and the mechanism of death seems to be respiratory obstruction—purple bruises are sometimes present, tiny bleeding points called "petechiae," perhaps resulting from the infant's attempts to take deep breaths against some obstruction in the airway.[8] Since the child is so sensitive to low air pressure situations, it may take only the shallow breathing of deep sleep or the slight respiratory blockage of a cold to cause death.

"SIDS deaths are more common in winter than summer, which is when the immune-compromised child is more likely to succumb to infection. They often occur in a family situation where an older child brings an infection home, or where everyone in the family has a cold and the child goes to bed with sniffles and doesn't wake up. You have no defense of your own, so you meet with an infection and go out like a light. This also fits our findings on the Survey that the firstborn get more leukemia—because the firstborn doesn't meet with infections brought home by an older sibling and so has a greater likelihood of living long enough to manifest leukemia.

"It is also known that SIDS children have an easy delivery with a short second stage of labor. The second stage of labor is when the baby becomes dependent on its own hemoglobin for breathing and when any defect in its system could be fatal. This fits my theory—these babies would *have* to have got into the world fairly easily because if they'd had a difficult labor, they'd have died. It also fits with something else found by the Oxford Survey: that leukemia is uncommon in the second twin. Second twins have difficult second stages of labor—they have the longest time before they breathe—and a defect in the system would kill them."

Alice believes that these missing myeloid leukemias may account also for many miscarriages and stillbirths. She notes that 5 percent of stillbirths are totally unexplained and occur—during the second stage of labor—to apparently healthy babies.

Alice's theory of SIDS has been there in the literature since 1975, but nobody's picked it up.[9] This is the more remarkable, since it could so easily be tested. "There's a blood test done on all children shortly after birth—the same test should be used to look at the proportion of fetal to adult hemoglobin. Then when the mother gets a follow-up exam at four weeks, do a second test for proportion of fetal to adult hemoglobin—then monitor the population for all causes of death in the next eleven months.

"According to me, you'd expect children who died of SIDS to have shown a high proportion of fetal hemoglobin at one month of age. You can't test for this after death, since the blood count can only be diagnosed by flowing blood, but you could monitor children while alive—and you could easily establish whether SIDS children have a disproportionate amount of fetal hemoglobin.

"I tried to launch a study of SIDS in America through the Childhood Cancer Research Institute, but there wasn't enough funding, and nobody in England has shown the slightest interest. I simply can't understand why. No one knows anything about this mysterious syndrome—they're stuck—so why not test my theory. As long as SIDS remains a mystery, my theory is as good as any other."

She adds, "I think it would be a great comfort to the mothers of these children—the children often come out in bruises, so they look battered. It would be a comfort to these mothers to know they didn't murder their children—they were victims of an inevitably fatal disease."

The sort of monitoring Alice recommends might also provide a way of detecting leukemia at one month, and "you might then invent something to stop it, something to boost the immune system—inoculations, for example." Alice and George found evidence in Oxford Survey data that inoculations act as cancer inhibitors. "In the Oxford Survey, you'll remember, we pitted a cancer case against a live control, and we discovered quite an extraordinary thing—that our controls had been inoculated more often than our cases. Perhaps they were alive *because* they'd been inoculated? George put the data through his tests and we found that 20 percent of children incubating cancer had been lucky enough to have inoculations at a moment when they seemed to arrest the cancer. The inoculation had to occur at just the right moment, and the critical moment seems to be late in latency—as though the child, about to develop leukemia, happens to get an inoculation and it nips the cancer in the bud.[10]

"Inoculations don't prevent you from getting cancer, but they may be a way of stopping it, by giving a boost to the immune system. They don't affect the initiation of the disease, but they may interrupt an otherwise ongoing process. This is consistent with the rest of what we know: if leukemia and other cancers weaken resistance to infection, you can imagine that by boosting the body's immune system, you might strengthen resistance to cancer and reverse the process. There is undoubtedly a tremendous amount to be gained just by boosting the immune system in every possible way you can think of. But you have to hit it at exactly the right time.

"We applied for funding to some drug companies to test this, but again—nothing. It looks as though this is another Stewart theory that's going to have to go untested. I seem to be constantly bumbling into situations where nobody thinks the way I do."

Burkitt's Lymphoma

Another of Alice's theories concerns Burkitt's lymphoma as a contained form of leukemia. Burkitt's lymphoma is the most common kind of children's cancer in parts of the world where there's widespread exposure to malaria. Alice suggests that it may be a form of lymphatic leukemia that gets contained by a highly developed immune competence.

"In tropical Africa and other places where there's lots of malaria, Burkitt's lymphoma accounts for about 90 percent of childhood cancers. It's a rather mild form of cancer, easy to treat, and the children don't suffer a lot. In England and America, Burkitt's lymphoma is very rare—it appears in areas of the world where children have lots of attacks of malaria. Everyone thinks it's a special viral infection, but I think it's the children showing exceptionally strong resistance to the mutations that cause leukemia because they've developed such high resistance to malaria.

"These are children who have not only had repeated attacks of malaria, but who are born of parents who've had repeated attacks, so that maternal levels of immunological competence are exceptionally high. The progress of the mutation, instead of running rapidly to a diffuse disease as it does in our part of the world, where it becomes either lymphatic or myeloid leukemia, turns to a localized form because abnormally high levels of immunological competence hold it in check. The child's resistance seems to allow for a containment of the cancer process: the children get enormous tumors, and they occur in places where you'd expect to see infections.[11]

"A further reason for seeing Burkitt's as a localized lymphatic leukemia is the appearance, in these parts of the world, of a localized form of myeloid leukemia, virtually unheard of anywhere else—chloromatous myeloid leukemia. Normally in children, the myeloid cells are so vigorous that they become diffuse throughout the body, but here they collect into localized tumors called chloroma, tumors on the scalp, greenish in color, consisting of myeloid cells." Alice recalls that in the first few years of data collection for the Oxford Survey, there were two cases of chloroma: "they were probably the last cases in England, left over from those days when children got multiple infections; they soon disappeared when the infection rates fell off."

According to Alice's theory, if malaria were to disappear from an area, the ordinary form of leukemia would start appearing instead of the localized form. And sure enough, a study in the Suez area found that when the malaria rate came down, so did the incidence of Burkitt's lymphoma, and the usual form of leukemia appeared instead.[12]

"It's one of my two outrageous and outstanding theories, the other being SIDS. There's a brief account of both in 'Childhood Cancers and Competing Causes of Death,' a paper I wrote in a temper when I heard Doll was dismissing the whole of the Oxford Survey as having done nothing but turn up a doubtful x-ray effect."[13]

Free of Charge

As Klarissa Nienhuys suggests, "Alice and George were finding out things about the immune system that were quite ahead of their time, that may in the long run be more important than the radiation discoveries."

Alice's theories of childhood cancer, of SIDS as an effect of leukemia, of Burkitt's lymphoma as a contained leukemia, and of inoculations as cancer inhibitors, all came out of Oxford Survey data. "It shows how important it is just to go on monitoring the situation, to continue gathering data. The Oxford Survey should have continued. It wasn't costing much, it could have ticked quietly away. If you are interested in diagnostic problems, which I obviously am, then you want to get to the root of the thing—you don't want to leave it halfway. I feel we were within a few years of answering questions about SIDS and inoculations that might have solved the mystery of cancer.

"Ever since I woke up in the middle of the night realizing how important antibiotics were in the rise of leukemia, everything we've discovered has pointed to the centrality of the immune system to the prevention of cancer."

Other sorts of correlations were turning up. "Early in the Survey, we found an excess of Down's syndrome among the leukemic children—thirty times higher than normal—and this fits with the rest. Down's syndrome prevents the normal development of the blood-forming tissues; these children have deficient immune systems that make them extremely susceptible to infections—and to leukemia. Before antibiotics they rarely survived infancy. Other congenital diseases (ataxia-telangiectasis and Bloom's syndrome) have the same genetic defect that prevents normal development of the immune system, and these also have strong associations with infections and leukemia—though only after the advent of antibiotics did these children live long enough for these associations to be recognized."[14]

Another project Alice hoped to launch, which could have been tested had the Oxford Survey been allowed to continue, was on the effects of ultrasound. "Ultrasound came in 1975 and our data stopped after 1980, so it's necessarily left unsettled. Once the use of ultrasound becomes universal, it will be impossible to study, because once it's uniform, there will be no control group—as is the case with background radiation. We have data on some of these transitional years, it would have been nice to have gotten more—it would have been important to look closely at what happened as ultrasound came on the scene. I have no hunch about whether there's a cancer effect. I wouldn't be all that surprised to find one, though I wouldn't be surprised not to. Non-ionizing radiation doesn't shatter the cell the way ionizing radiation does."

Even with the Oxford Survey's termination in the early 1980s, it continues to offer riches. There is Tom Sorahan's discovery about paternal smoking and childhood cancer, which Alice calls "of key research interest and very good news for the Oxford Survey. I've always known there were lots of uses to be made of the data. The records are there for new theories to be placed on them. I've always known that our discovery about fetal x-rays was like the mariner's North Star, the fixed point in the firmament."

It is poignant, hearing Alice toss out her cutting-edge theories to audiences of activists and students who are unlikely to do much with them. As she was expounding her theory of SIDS to an audience at Portland State University, she was asked, "Why aren't you doing a study of SIDS now?" "I'm too old," she replied; "I couldn't get a grant in my own name. Even if I were younger, it's not as if I had a big department and an honorary position anywhere and could gather a team of people around me. But you see, I've alerted your interest. I think there's a prize there for somebody."

Someone from the audience commented, "we get these ideas free of charge." Alice responded, "we give all our ideas free of charge. We put out these ideas on the table—for others to develop. The most I can do is throw the pebble into the pool and create circles and leave it to somebody else to get on with the work."

Chapter 16

Pioneer and Pariah

"I've had a marvelous time, I've had a lot of fun, I've never been short of interest—though we did get kicked around pretty badly."

"It is very difficult for the Alice Stewarts of the world to survive," says Sheldon Samuels, health consultant to the AFL-CIO, one of the speakers who presented Alice Stewart with the Ramazzini Prize. "All people are creatures of their institutions, especially scientists, who must generate resources for computers, labs, staff, equipment, publication, in order to function. Any scientist who doesn't follow the mainstream has difficulty. She has a struggle not only to be independent but to find the support—social, intellectual, and material—that a scientist needs."[1]

"People don't realize how dependent scientists are; they have no idea," says Rosalie Bertell. "Who pays the scientist? Government, industry, the university—and they don't fund controversial projects. Scientists need grants, equipment, resources. People have no idea what it's like when you have no money, no staff, no support. Activists come to you asking you to write a report for them, read a government document, pull a case together, go to court with them. Most of the time you're living on the edge, with barely enough to keep yourself going."[2]

"Most of us work in research institutes, universities, and government," says Steve Wing; "we have to, on a daily basis, accept many assumptions in order to survive and get on with our work. The study of radiation and health has been very strongly marked by its attachment to the development of nuclear weapons and by the fact that for fifty years, almost all research into radiation and health was funded by the governments and the industries who had an interest in the development of nuclear power. Science is presented to us as though it objectively looks at things, looks at what's really there, but I think that's a myth: scientists, like other people, see what they look for and see what the methods they bring to material allow them to see."

This is why the push for change in science, as Wing and others suggest, usually comes from outside. "It's people from outside the scientific mainstream who are in a position to challenge the assumptions we've made ourselves comfortable with and to lead us new places."[3] But the innovators are rarely honored in their times—in fact, they're more likely to be reviled as quacks and charlatans.

"Yet we managed to carry on," says Alice. "We met constantly with opposition; we kept the thing going on a repertory basis rather than a full-scale production, but we managed always to stay in business."

Marginalization

That baptism by fire her first day at Cambridge in 1925, when she walked into the lecture hall to the sound of two hundred or so male students stomping in time with her steps, was good preparation for what was to come. Forced down to the front row to take a seat with three other women and a Nigerian, she recalls that "from that moment on, I had no illusions."

Though Alice was never again physically "stomped" by her male colleagues, this was not the last of the insults she was to endure. Marginalized by mainstream British medical research circles, starved of funding, blackballed, blacklisted, cold-shouldered, she has borne with more than her share of slights, snubs, and rebuffs.

The Medical Research Council (MRC) funded a portion of her childhood cancer study for Scotland, but it never again gave her a substantial grant. Funding from other sources in England was nonexistent; it was support from the United States that enabled her to carry on, though just barely. She was never asked to be on a MRC committee or invited to take part in any of the reviews or commentaries that shape medical research.

"After Hanford, there was a general tendency to push us out of the story, drop our names from references. Everyone knows who the experts are: Doll, Darby, and Gilbert; everyone knows that they agree with each other and everyone agrees with them, and there's a general feeling that the subject has been done and there's only one dissident voice, and that's us, so there's no need to take us seriously. We never got invited to official meetings; we never got asked to give our point of view. We had no funding at all for some time, so we couldn't afford to take ourselves to meetings, let alone hold meetings of our own, so we never had a forum for our views."

Then, come retirement age, it was made clear that she was no longer welcome at Oxford and it became impossible for her to continue the Oxford Survey. After she left, she was never invited back to Oxford to speak, though by then she had an international reputation and was inundated with invitations from around the world. "For a long time after I went to Birmingham we were marginalized, literally—put in a Portakabin [trailer] not belonging to anybody. It was a nice quiet place to work but not exactly in the corridors of power. There was no way of meeting people; we might just as well not have been there."

("Follow the sign that points to the laundry and then dive through the hole in the hedge," she'd direct visitors.)

She has found out about conferences on her subject being held at Oxford and Birmingham, on her front doorstep, as it were, to which she was not only not asked to speak, but not even invited. Sue Rabbit Roff, author of *Hotspots: The Legacy of Hiroshima and Nagasaki,* describes attending the 1995 Roentgen Centenary Conference of Radiologists and Radiographers, held at Birmingham: "there were more than three thousand present from all around the world—Alice had not been informed that conference was on."[4]

She has encountered rudeness that takes the breath away. She was once present at a conference where the audience had been told not to read her research—"I was put on a list of proscribed people, banned!" When she requested access to the Hiroshima data to test out her theories, she was turned away by the keepers—literally—though she knew other researchers were allowed access. She has not been invited to be part of any of the investigations of cancer clusters in Britain nor asked to testify at compensation cases in Britain.

She has been called barmy, senile, gone round the bend, off the rails.

"But I just keep out of earshot, I don't listen—it doesn't concern me." Alice feels it is useful, this skill learned as the middle child of eight. Besides, this was only part of the story, for "suddenly I woke up one day and realized that people all over the world, people everywhere, knew what we'd been doing." And so she found herself in this odd position of being nonexistent in official circles and in demand throughout the world.

Nonexistent on the One Side, Honored Speaker on the Other

That situation was epitomized as she was awarded the Right Livelihood Award in 1986, an award established in the late seventies, when Jakob von Uexkull, a young Swedish baron, approached the Nobel family with the idea of expanding the Nobel Prize to honor those who have made contributions to the betterment of society. The Nobel family was intrigued by von Uexkull's suggestion but legally unable to act upon it under the terms of the will setting up the prize endowment. So the baron raised his own money and endowed his own prize. Today the "Alternative Nobel," as the European media call it, is endowed for more than $2 million. The awards, first instituted in 1980, are given in the Swedish Parliament in Stockholm every year on the day before the Nobel Prize presentations.

On December 8, 1986—the year of the Chernobyl accident—Alice received an award of $25,000 (£16,000) for "work on practical and exemplary solutions to the most urgent problems of today." Rosalie Bertell was also presented an award.

Alice was delighted with the money and the recognition. "It gave me a platform and a little money to go on with work that was threatened." But even as she was being honored, she became aware that she was being snubbed—by the British government.

"In the course of the few days in Stockholm, I became aware that Rosalie Bertell was being treated as a VIP—the Canadian embassy had put a car at her disposal, there was a dinner held for her, while there was nothing of this sort being done for me. I happened to have lunch with a woman I'd been to school with at St. Andrews, who was on the staff of the British embassy and told me the Right Livelihood people had written the British ambassador about the prize and hadn't even received a reply—not even a reply. It was very rude, really. It's quite a nice prize, and it's not every day that someone gets this sort of prize. Our embassy is supposed to be out there making friends so that we can hold our own in the world, what with this shrinking empire of ours, and not even a reply.

"There was only one mention of the Right Livelihood Award in the English press, in the *Yorkshire News*, and that's because the Right Livelihood has its headquarters in Bradford. But I don't think it was mentioned anywhere else in England. There was not a breath of official recognition from England."

Alice has had more recognition from the United States, partly because of its greater openness and interest in anti-nuclear issues and partly because—she suspects—there is no powerful, patronizing figure in the United States who shapes official responses toward her. Yet even so, she has had only one invitation to spend a sabbatical at an American university—and that was from Professor Rudi Nussbaum of Portland State University.

Nussbaum, who taught physics for thirty years and published widely in nuclear physics, invited Alice to Portland in 1984. During this time, she was wined, dined, interviewed, and honored in a way she had rarely been. As she was introduced as the annual convocation speaker, about to address an audience of faculty and students, Alice heard the first public acknowledgment ever that she'd made a significant critique of the A-bomb data. "This was the first time I'd heard myself referred to publicly as someone who's got something to say on the survivor studies."

In 1981 Alice was invited to Hiroshima for the Peace Memorial Conference—not by the Radiation Effects Research Foundation (RERF), but by the peace groups. "On Hiroshima Day, August 6, they have a big open air gathering in the Peace Park. A river runs through this park—the dying had thrown themselves into the river, it was full of bodies—and for this ceremony, by the side of the river and on the bridges, thousands gather after dark. It's summer and there are vendors selling floating candles, little boats made to carry candles, of very fine folded paper, and you write on the paper a message to a dead person and then float it off on the river, like a slow farewell. This custom grew up gradually—at first, any observance was forbidden; people weren't even allowed to talk about the bombing—and as you float the candles down the river, it's quite uncanny, the effect. I can't tell you how moving it is—I cry just recalling it."

While in Hiroshima, she decided to visit the RERF, the custodian of the A-bomb data and imprimatur of international safety standards. "They knew I was coming but pretended there had been some confusion between the Japanese and English. After I'd spoken with them, I was taken down the hall and left to wait for a cab alone. Now this didn't strike me as odd because it's natural in Britain, but then somebody came along, asked who I was and what was I doing sitting there, and expressed astonishment at my being left alone because they never do that to a visitor in Japan, just leave you without an escort. And he was very upset that a visitor like me had been left alone that way—it was as though they'd hurled an insult at me. He stayed with me because he thought it was impolite to leave me alone, and he escorted me to a cab.

"When I got back to England I decided to put in writing the request that I'd made while I was there, for some data. I never got a reply.

"It's very odd because whenever I talk to a RERF researcher individually, they'll agree with me and say 'yes, that makes great sense.' Then they say 'but'—and of course, what's important comes after the 'but'—'but it's an untestable hypothesis.' Individually, they seem to listen to me, but as a group, it's like a stone wall. I must seem to them like a nasty buzzing hornet. They can't stop the noise but they don't think the noise is of any interest; they'd like the noise to go away, but I buzz on."

The snubs have been real, though they seem to be more keenly registered by Alice's friends and colleagues than by Alice. It's from her friends that one learns what Alice has stayed out of earshot of. It was Rudi Nussbaum who suggested, "get Alice to tell you about the RERF incident"; it was her longtime friend Faith Nichols who urged me to inquire about the Right Livelihood Award. Alice will tell you about these things if asked, but she does not go around advertising them.

It was one such insult that caught Klarissa Nienhuys's attention. She had organized a seminar at Groningen University and invited a speaker on low-level radiation.

> He started out with a general introduction, talking in a rather scientific way, a bit dull, but that suddenly changed when he came to talk about Stewart and Hanford. What he said and how he said it, the words he used, were very personal, with very little content as to what the research was about. He said that when Dr. Stewart had found these results on prenatal radiation and childhood cancer in the fifties, no one had believed her, though finally in the seventies it was acknowledged that she'd been right; but by then she'd become so frustrated that she'd gone to work with Mancuso and had produced ridiculous results about workers at Hanford. The implication was that this scientist had more or less flipped out and now these two crackpots, Stewart and Mancuso, had found each other.

Nienhuys had met Alice the year before, at the Gorleben hearings, March 1979, and "the idea that she was nuts was completely inconsistent with my memory and experience of how she behaved, how she talked about the subject, and how she talked about other scientists in the field. I was shocked to hear this man make such out-of-line statements, apparently based on second- or thirdhand gossip." The attack piqued her interest: "what could be in her work, in her being, that a person like this felt it necessary to make such statements? It could only mean something very antifemale or that the subject was symbolic in some way." Intrigued, she made efforts to get to know Alice and found her "friendly, warm, and human," someone "stimulating and nourishing me as an intellectual as hardly anyone ever had before. I felt delighted, and that is always the case when I meet her."

A few years later, at a March 1985 meeting at Rotterdam University, "Nuclear Energy: Prestige or Necessity?" Nienhuys was witness to another insult. Alice gave a paper, and the chair of the meeting

> behaved in such a rude way that several persons in the audience approached her after the meeting to apologize for his behavior and next day the organizers had several calls complaining. Alice had made a point based on Mancuso/Stewart/Kneale findings, and this man interrupted her with, 'Well, *Mrs.* Stewart, of course you cannot refer to the Hanford study because the Hanford study is the failure of bad epidemiology and we all know that.'[5]

But it's from Alice's friends and supporters that one hears these things. From Alice, one hears, "I've had quite a fight to do things but never enough to overcome such strengths as I have."

Sex and the Scientist

Alice Stewart came on the scientific scene in the 1950s, a decade that was a nadir in the history of women in the sciences.[6] She was pushed to the margins, left with minimal space and funding to carry on.[7] "My predecessor had a professorship, my successor had a professorship, and I had a readership. The world is designed by men, for men, for their convenience—it's not something you can fight."[8]

But she also had enormous advantages. Both parents were physicians who were able to give her practical as well as moral support. Her father was Cambridge educated and well connected professionally, and he had strong ideas about giving his daughters opportunities equal to his sons. He sent Alice to a secondary school where she made friendships that became important to her at Cambridge and throughout life. Her Cambridge education, her Oxford position, gave her an entre into the elite circles of medical research in Britain.

As it happens, she probably wouldn't have gotten very far without such advantages. Women in science needed them, as Hilary Rose suggests. Rose finds highly privileged class origins, as well as the support and encouragement of a scientist father or husband, in the backgrounds of nearly all the women in her study. These were practically a sine qua non for a woman in science in the early twentieth century.[9]

Alice had the backing not only of a scientist father but of a mother who had become a physician and managed to stay professionally involved despite the demands of a large family. She also had the benefit of both parents' good advice—to stay on the path that had been cleared by other women. "I became a Consultant at the Royal Free, the first teaching hospital to admit women; I managed to get a position at the Elizabeth Garrett Anderson, again in a context of women helping women, again, thanks to my parents' advice. They were right—if I'd moved off that path, I'd have had my head chopped off. I was benefiting from the efforts of women who came before me."

Nevertheless, marriage imposed handicaps that would have defeated a woman of lesser gifts: and not only did Alice marry, she had children—and raised them on her own. "I was well aware that the only successful women scientists were either single or had no children. There was tremendous prejudice against married women. Most women who went into science dropped out when they married or had young children, and found themselves unable to get back into a career later. You weren't allowed even to hold a house job [residency] if you were married; you were cut off from experience, right from the beginning."

What gave her the break she needed was the war, which dissolved the prejudice against women in medicine: "suddenly the whole of society was on our side, wanting women to be doctors." Then even the disadvantage of having children turned to her advantage because it meant she couldn't be called up for service and had a crack at positions left empty by the men. This got her to Oxford, and once there, she set about doing work that earned her distinction. But her role as a caretaker did not end when her children were grown, for she took on responsibilities with her grandchildren that continued to cut her off professionally: conferences were luxuries she couldn't afford.

Hilary Rose suggests that women in science often respond to the prejudices they encounter either by withdrawing or by identifying with the power structures and becoming "honorary men."[10] Alice has little use for women who make themselves honorary males.[11] Her sense of her mother's difficulties, of her own struggles to balance motherhood with profession, make her identify strongly with women. When she came to direct projects of her own, she made special efforts to hire women, to give them breaks; and her friendships with women have lasted her life.

Moreover, she refused to attempt to ingratiate herself with the men who might have helped her. Something—the bad luck of Ryle's death, her own outspokenness, her refusal to play the game?—left her without male mentoring in midcareer and badly undefended. Sue Rabbit Roff speculates, "I think it would be very interesting to look into the question of male mentors/patrons—why did Alice lose hers in midcareer? Was there no one to protect her in the old boys' network after Ryle died? Would it be worth looking at the careers of other women such as Sarah Darby, Ethel Gilbert, Valerie Beral who have been working in the same field to see to what extent they have been 'sponsored' by senior male colleagues?"[12]

Male mentoring would have made Alice's life easier, and she might have found it, had she tried. More than one friend comments on her refusal to try. Nienhuys comments, "I know what she says about committees and all, but as the director of an institute she *should* have been sitting on committees—at least on those committees that allocate research. Maybe she wasn't invited, or maybe she wouldn't have been let in if she'd tried, but the Alice Mary Stewart that I know didn't bother to try."[13] As such, she absented herself—or was absented?—from the decision-making processes that are so crucial to the shaping of scientific knowledge.

"She was not prepared to go to the right places and be nice to the right people," says an old friend. "She could always think of something more interesting to do than to go to a vice-chancellor's garden party where you had to dress up and be nice to people. Also, undoubtedly, she was forthright in her opinions and that annoyed people. She felt that people should automatically welcome her ideas—she didn't prepare the ground enough, didn't lobby enough. It's quite incredible that she's not in *Who's Who,* since she's one of the great figures of epidemiology. It was partly the gender issue and partly a rather prickly personality."

"I didn't care about being pretty," said Alice when I commented on her looks in the old photos. "I didn't want to learn how to be nice to men, how to find a husband. I only thought: I want to be independent, I want to earn my own living."

"In a way, self-sufficiency might be Alice's strength and weakness as a woman making her way in a man's world," speculates Roff. Perhaps, as more than one friend has suggested, Alice was independent to a fault. Or it may be that independence is by definition a fault in a female, in a world where a husband can say to his wife as he leaves her for another woman, "she needs me more than you do."

A prickly personality? or a refusal to play the male power game within the only rules it provided—to act as deferential female or assume the role of honorary male? Was her withdrawal motivated by pride or by a sensible decision that there could be no winning within these terms, that it would be better to make a separate space and terms of her own?

Gender—or Genius?

What Alice did was withdraw from the corridors of power and find ways of devoting herself to the questions that interested her. She found—or created—work that drew on her special skills. Her skills are by no means those of a recluse, however; they are, on the contrary, highly social—whether she was mobilizing student volunteers for the TNT survey and keeping them amused while they lived and worked at the factory or enlisting the assistance of every health officer in England in the administration of the Oxford Survey. "The way she could motivate so many people to help her carry out the work, going to all the medical officers of health in the country and getting them to agree to help and give her staff, was extraordinary," says her colleague Tom Sorahan, who has worked closely with her in recent years and can testify that this skill has lasted through her life. Sir Richard Doll refers to her unusual capacity to rally the support of others; I have seen her orchestrate—gracefully, graciously—large social gatherings in her nineties.

By the time she designed a survey to investigate the rise of leukemia in children, Alice had given birth to two children and had a strong sense that she ought to "ask the mums." Then, what guided her through the complexities of the American nuclear industry was a no-nonsense way of seeing what was before her—what are these YYYYs and NNNNs? she asked; it was this question that led to an understanding of the healthy worker effect. Her revolutionary critiques of the A-bomb data are similarly grounded in common sense and an understanding of the body: the survivor population *couldn't* have returned to normal after five years; radiation *couldn't* have affected young people more than old. Again and again she points out flaws in the way the Hiroshima data are interpreted, basing her objections in "what any practicing physician would know."

Part of her genius as an epidemiologist derives from her ability to keep questions open and tolerate uncertainty, a capacity that well suits her to epidemiology, which, as an observational science, is less interventionist than the laboratory sciences.[14] She has a sensitivity to epidemiology as an approach with its own special qualities, a feeling for her subject. It is tempting, Keller suggests, to see such empathetic attentiveness as characteristic of a "feminist" science,[15] and Alice does acknowledge that her way of going about problem solving may be related to her gender.

Steve Wing describes her genius for culling from different areas: "she takes information from different areas of science and brings it together in a way that's not linear . . . it's a synthetic mode of thought, an artistic mode."[16] More than one fellow scientist credits her with lateral thinking,[17] and she herself says, "I'm not a thinker who sticks to one line—I make sudden side leaps and bring in connections." But it's not just areas of scientific inquiry that she'll bring together: she's as likely to reach for Shakespeare to make a point. Such leaps and lateral movements have been described in recent creativity literature as the synthesizing capacity of the creative imagination. Further aspects described by this literature as characteristic of genius include the ability to tolerate contradiction, to risk following intuitions and go after questions not defined within conventional paradigms, and a high level of self-confidence that allows risk-taking and the toleration of social opprobrium.[18]

Then, too, there is the unusual trajectory of Alice's career, her apparently serendipitous move from subject to subject—something else that's often seen in the careers of women, though a bit more dramatically in Alice's. She keeps "bumping into controversies," as she says. "First was the discovery about fetal x-rays and childhood cancers. Then came the fuss over nuclear worker risk at Hanford—I had no idea what I was getting into—and that embroiled me in a third major controversy, on the A-bomb data. Each time, it came as a real surprise, almost as a shock. Almost by accident I fell into the middle of the most important development of the twentieth century, radiation. I didn't go out looking for it. I was just an epidemiologist prepared to take up a problem and go after it."

She feels that the controversy has been good for her. "Perhaps the best thing that happened to me was that nobody believed me. I think, actually, this is almost the best thing that can happen to a researcher, for there to be a very severe and persistent criticism of what you're doing—it means that someone is taking an interest in you. Each time, this pushed us to go on and on, to get more data and prove our positions. And because we stuck with it, the results we've gotten have been rich. It's important to have opposition—it's like playing with a better tennis partner. It gets in a sense more exciting, the more opposition there is—if you can hang on. It's great fun. I think I even thrive on it."

Insofar as gender worked against her, it was like that "little misfortune" she has always believed you should hope for. She has more than once seen bad luck turn to opportunity: the bad luck of not getting MRC funding allowed her to devise a survey of her own; being squeezed out of Oxford left her free to respond to the most interesting challenge of her life, the call from Mancuso; the bad luck of obscurity left her free to go her own ways. "I'm a great believer that life can be too kind to you. Other people see me as maligned and so forth, but I don't hear the insults. For every one who belittles me there's someone who comes along and says, 'you're marvelous, come be on television.' I think there must be some strong ego at the center of this, some inner conceit. Perhaps it's just the sheer self-confidence of the Englishman, lucky to be born an Englishman. If I had been born with less self-confidence, it would have been a lot harder."

On the whole, Alice feels that she's been helped more than hindered by being a woman. "If I'd been a man, I'd never have stood it—the pay was too low, the prospects too bad, I'd have had my eye on the prize. As a woman I got to slip through the cracks, and here I am, like a drop of water. Since nobody took me seriously, I was left to go my own ways."

Double Meanings

Alice's story does not yield a simple meaning, like "woman triumphs over impossible odds" or "single-minded pursuit of a goal leads to success"—because the success is by no means assured, the recognition is nowhere near what it should be, the resistance to her findings is still fierce—and the story is not over.

Obscurity has suited her temperamentally and spared her the busywork of "sitting on important committees and so forth. I may work in a hut, but I seem to get around as much as the people who are working in palaces and I think I go about as a much freer agent. The people on committees, they get the first sight of everybody's publication, they know exactly what's going on, but they've quite forgotten how to take up a pencil and paper and do the job themselves."

On the other hand, it has also left her without power to command respect and attention. Occasionally she wonders if she shouldn't have stood up for herself and Social Medicine after Ryle died. "If I'd looked to my own interest instead of lying low, if I'd said, no, you mustn't do this to the department, you must give it a professorship, you must give me a proper salary—perhaps people would have had to attend to us. As is, they were able to discount us."

Opposition has pushed her to build support for her theories, to refine and develop them. But it has also brought the Oxford Survey to a halt and is impeding her research on the nuclear workers. Because there's a tendency to ignore her in radiation literature, she's found herself with few followers—though she is very pleased that young researchers like Steve Wing are finding their way to her.

Nevertheless, she considers herself "extraordinarily lucky": "There's this sort of popular idea that I've suffered greatly as a result of my beliefs, worked under appalling conditions and so forth—it is a total myth. I've had a very comfortable scientific life. I've had excellent colleagues and a fascinating subject. I've had a steady salary, a readership from Oxford, a pension. I've never had support commensurate with my training or talents, it's true, but I make more than a secretary. It's a reasonable income—not by American standards, not even by most men's standards, but enough to keep me independent of the world and to have most of the things I want. I have simple taste, and once you get obsessed by this work, you haven't time to spend money. Fortunately, I have grandchildren, so I get rid of the money all right.

"And fortunately I've been able to keep some connection with a university to maintain some respectability. And there's always been, just when I thought I was coming to the end of it, something new to go on to—just as I'm about to retire, someone puts at my feet decades worth of data to play with! I've been lucky to have the right job at the right time. I know few people who are at my level of official recognition—that is, invisible—who are also known all round the world.

"If somebody stops me tomorrow, I shall say I have had a very good run for my money."

I would agree that Alice has been lucky—lucky to have the parents she had, to have the confidence and connections that come with privilege; lucky to be born with the Naish family energy and the gene for good cheer (not everyone in her family had it); lucky to combine brilliance with good sense. I think that her grounding in a woman's experience gave her an approach to scientific questions that enabled her to go about problem solving in unconventional ways. Her understanding of the body, derived from experience as a physician and a mother, her capacity to persevere in a line of inquiry that was actively discouraged, her intuitive leaps, her gift for lateral thinking, her formidable diagnostic skills combine uniquely in a vision that has taken her through the looking glass, a vision that promises to transform the paradigm of knowledge in her field.[19]

Then, too, she has had the gift of interest. Alice's persistence, which is legendary among her friends, her insistence on going back over and over the material, comes not from a drudge-like compulsiveness, but from a pleasure in the process. She is a scientist with a highly developed sense of social responsibility, but she's also a medical detective with a keen scent on the trail, a pleasure in the pursuit. Whereas she's seen two physician brothers lose interest in their subject, she has had this gift of being continually engaged. It was said of Honor Fell, a pioneer in cell biology at Cambridge, that she was married to her work; the same could be said of Alice, who describes research, the romance of discovery, as the great love of her life. But she's also had flesh and blood marriage and romance, to say nothing of strong ties to children and grandchildren and to an extended familial and social network. She has dared more and risked more by way of human commitments than do most scientists, female or male.

Through it all, she maintains faith that "the truth will come out sooner or later. Not that we'll be proved to be absolutely right—I wouldn't think that for a moment. But that the records will be there, it will be written down on paper so anybody can get it. Someday, someone will work the story out. You have no idea how comforting that is." She has confidence that her work has made a difference. "I'm absolutely certain on this point—and it's historically rather interesting if I'm right—that none of these studies of low-level radiation would have happened if it hadn't been for the Oxford Survey. Everything would have been lovely in the garden and there would have been no need to look at anything, and the whole thing would have slipped quietly by."

She is pleased by the part she's played in this controversy: "It's been a good thing there's been this public debate. I'm impressed with how the public has educated itself, particularly the American public. There's been a growing awareness that radiation is more dangerous than advertised and quite a lively and effective citizens' movement. People have become highly informed, which gives the lie to the idea of the public as incapable of understanding anything. They've recruited educated people into the discussion, which has improved the balance. Now the expertise isn't all on the one side."

Alice is known far and wide as a killer Scrabble player and is especially famous for one maneuver: "People who play Scrabble with me say, she's so maddening because it looks like you're winning, and then she turns and wins on the last move. The great thing in Scrabble is to go out on a triple. You put down all your letters in a position that triples the score, and if it's the last move, you get fifty more points besides. So you can be right at the bottom, in a hopeless position, and win on the final twist. And I'm rather hoping that the last move in my game will be a triple, and the other players will have to pay all their points to me."[20]

For herself, she is "quite content with the idea that what we are doing will be better understood by the next generation than by the present. I think I'll prove right on that score. I'm conceited enough to think I will." But she is also aware that time is running out for her personally and she'll need others to carry on the work. She is worried about George Kneale and indignant on his behalf: "It's absurd—he's a leading figure in this field, and he ought to be holding a good position, not facing the prospect of unemployment."

There is also a dry, ironic side to her (as she read over a version of this chapter, she commented, "I get rather tired of hearing what a cheerful chap I am"). She is capable of being bleak, even outraged, though less on her own behalf than on behalf of humanity. Alice did not start out with anti-nuclear opinions, but she has come to fear that "an almost irretrievable mistake has been made. I'm afraid we've gone down a wrong path, and the whole of society is going to pay. I suspect that in twenty years the world will realize this, only they'll be twenty years further on in the mess, which they could be stopping today if they would only do it. There will be a totally irrevocable buildup of a pool of recessive genes into the population. It won't be noticed until it becomes common; then we'll never root it out, we'll never get rid of it."

The Radicalizing of Alice

Alice Stewart did not start out with a mission, but in the course of her work she has developed one. She has been radicalized by "the strength of [her] data" and by the behavior of the nuclear industry, which has been—as she says—"very stupid."

She feels at home in anti-nuclear circles, where "there are more women than men. It was bad enough in British medical circles, but the nuclear industry is worse—it's totally male, except for a few honorary males. I can remember meetings where I'd make myself speak up and say something quite small, and I can remember being quite frightened. But women have been much the keenest ones to be activists about radiation." Alice feels a responsibility as a scientist and physician to support them: "If the women who aren't qualified are showing that it's important, then the women who *are* qualified ought to help them."

But she is sometimes made uneasy by this alliance. She insists that she is not herself an activist—she is first and foremost a scientist. She bases her positions on her own research; she is not a publicist of other people's opinions—she feels it is important to make this distinction. She insists on keeping her politics apart from her science and holds to the belief that science can be objective, neutral, and value free. Though she has a first-hand knowledge of the misuses and politicization of radiation science, she maintains what seems to some of her friends a touching and somewhat old-fashioned faith in the possibility of pure science.

Nevertheless, she has been driven to take strong stands. "I didn't use to feel I had a message to the world. But I do now. I think it's important people should understand they've been misled. I think it's very important.

"The nuclear industry has a very special responsibility because radiation is the most powerful mutagen known to man, a poison that no one can feel, smell, see, or detect until people drop dead. Yet the industry is putting up background levels of radiation all the time—it's lunatic! And people have no choice but to live in this setting. You can choose whether or not to have an x-ray (you should of course be careful to have as few as possible),[21] but you cannot choose whether to have background radiation.

"We have, in fact, doubled the background radiation rate. The figure stands at least double of what it would be if there were no extra, man-made radiation from weapons testing, accidents, power stations, waste. Now if you've doubled a zero risk, it doesn't matter. But if you've doubled *any* risk—and our studies show that there is a risk—that's something very different.

"Even more than the cancer threat is the genetic damage—that's what you ought to be really afraid of, the possibility of sowing bad seeds into the human gene pool. You'll need three generations before you really see trouble, possibly four. It won't show up till then because it's initially more likely to produce recessive gene damage than dominant gene damage. But the terrible thing about recessive genes is that once they're into the gene pool, you can't get them out.

"If only we'd been intelligent enough to put our resources into discovering some other form of energy. Instead, we've pushed ahead on this side of the science to the point that it looks as though we can only get our energy from the fission mechanism." Alice maintains that "if we had a Manhattan Project, a cooperative effort, a powerful country to say this is so important we're going to get all our experts and give them the maximum amount of opportunity for realizing their full potential, we could have viable solar power in less than twenty years. If I were God tomorrow, I should take every job away from the nuclear physicists. I'd take all the money that's been going into this corner of knowledge and put it into getting energy direct from the sun. That's what I'd do."[22]

Taking the Game

Alice Stewart may find herself in good company, with other scientific innovators who met with opposition in their day but whose findings were validated by later generations. "The strength of the opposition often appears to be proportional to the freshness of their ideas," writes Ralph Moss, former cancer researcher and author of *The Cancer Industry.* "It sometimes takes many years for the establishment to acknowledge that a pioneer was right and it was wrong. Usually all the contestants in the battle have passed away before that happens."[23]

Rudi Nussbaum says, "Those people, the very few, who dared to question the current canon—whether it was Copernicus, or Galileo, or Einstein—had a great price to pay for asking these questions, but they felt it was their duty as scientists. Alice's work questions some basic assumptions. Her critique is fundamental, it isn't about a detail—and when someone questions basic tenets, that's a heretical attitude, and heretics get burned at the stake."[24]

Steve Wing says, "She has been independent enough to question what are for many unquestionable beliefs. She has been willing to disagree with a scientific establishment that has great power, prestige, control of resources, and she's been willing to risk her whole career, her life's work, to disagree, to carry on, in the face of those powers which control the universities, the journals, the medical schools, the international radiation protection agencies."[25]

Fortunately for Alice, it hasn't been the search for glory or recognition that's kept her going, nor even the mission: it's been the pleasure of the puzzle, the unraveling of the mystery. And engaged it does keep her: "People used to say, what do you do with your spare time?—and the joke answer was, 'go to the lavatory.' " She never takes a vacation, but then she never needs to: "My work gets me to interesting places and people. I never feel the need to go for a holiday because there's so much variety in my life."

Not that she minds recognition or shuns it when it comes her way. "Well, of course rewards are always very nice, I don't reject them when they come, but I'm not the slightest bit upset that there haven't been too many." She's delighted when she finds out that people on the far side of the globe know of her work and is pleased at the recent interest from mainstream media: there have been, in the past few years, several write-ups in high-profile places—the *Financial Times* (1 January 1995), the *Times Higher Education Supplement* (28 July 1995), *Ms.* (August 1996), and the Channel 4 documentary (August 1996). She wonders where these stories have come from. "It reminds me of the day I suddenly woke up and realized that all these activists seemed to know about my work. How did they find out? I certainly didn't tell them."

But she feels about glory the way she feels about money—it's nice to have but not worth working for. "Anybody who thinks they're going to be happy getting a great reward for what they've done should remember the famous phrase of John Webster, which goes as follows: 'Glories like glow-worms, far off shine bright, but look to near, have neither heat nor light.' It's a nice quotation, it's from *The Duchess of Malfi*. I take great pleasure in it. Glories are, from my point of view, nice additions to the work you're doing, but you must never lean on them—they're just nice little passing bright lights. And I do think that people who get satisfaction out of their life—it doesn't matter whether it's from work or family—it comes from the inside."

Chapter 17

Endings

"I've never been able to write about it, or even review it. I feel I'm in the middle of a story and there's no point at which I can step back from it because no sooner do I step back than the scene changes."

Alice Stewart's story is that of a woman scientist whose unpopular positions cost her standing, salary, access to funding. It's also the story of a woman who's had "a marvelous time," enjoyed a full and interesting personal and professional life, and had the satisfaction of seeing her work shape a critical scientific debate. It's the story of a physician and researcher who has put her expertise to the service of the global community, a story that holds out hope for those who are striving for a more inclusive and holistic approach to the health of individuals and society and who seek to focus attention on the cause and prevention of disease.

Alice has lived long enough to see radiation science move in her direction, to see researchers confirm that the biological effects of radiation are more dangerous than once believed,[1] to watch regulatory bodies respond to these findings. She has lived to see the A-bomb studies reevaluated in her terms. In August 1981, a major survey of the bombs' effects compiled by a team of Japanese scientists and social workers found "irreversible injury" to human cells, tissues, and organs that continued to plague victims after many years, causing more leukemia deaths, blood diseases, nervous system disorders, and loss of immunological competence than previously acknowledged.[2] Even some RERF researchers are coming round to her view: in 1995, I. Shigematsu et al. discovered a number of non-malignant effects.[3] Sir Richard Doll acknowledged, "it looks like she'll be proved right on this issue—the A-bomb studies are turning up effects other than cancer."[4]

New research is coming out of Harwell, the heart of radiation research in Britain, demonstrating at the molecular level that radiation alters chromosomes in ways that are more complex and far-ranging than once believed. A 1997 study by Eric Wright, head of experimental hematology at the Medical Research Council's Radiation and Genome Stability Unit, Harwell, reports that radiation damage to a cell may show up only after the cell has divided several times, producing a genomic instability that might result in broken or misshapen chromosomes and mutated genes and early cell death. His conclusions confirm what Alice has been saying all along, that the injured cell may cause trouble down the line; and his findings—like her own—make the current regulatory system seem inadequate. Whatever the A-bomb studies tell us, Wright asserts, the extrapolation from a group exposed to a large, acute dose of radiation to a group receiving small, chronic doses may not be adequate to understanding this kind of damage.[5] A related discovery has been made by a team at the Paterson Institute for Cancer Research, Manchester, indicating that sperm cells exposed to radiation produce inherited damage in offspring that leaves them more vulnerable to further carcinogenic insult—a finding that shows a potential mechanism by which paternal irradiation may increase susceptibility to cancer in the next generation and lends plausibility to Martin Gardner's theory.[6]

In May 1994, Alice Stewart had an unprecedented meeting with the head of the Department of Energy. Six months earlier, Hazel O'Leary, President Clinton's Secretary of Energy, had startled the world by blowing the whistle on her own department, disclosing that the United States had set off 204 unannounced nuclear explosions and conducted radiation experiments on hundreds of human beings without their consent.[7] Her revelations lead to hearings with radiation victims throughout the United States in 1994 and 1995. To one of these hearings, at the Press Club in Washington, DC, Alice was invited by Bob Alvarez, who'd been appointed by O'Leary as Deputy Assistant Secretary for National Security and Environmental Policy. "Alvarez dragged me over to meet her. She seemed to know who I was and what I was doing and why were being introduced, and cameras started clicking. People knew I was a friend of the activists and this was a photo opportunity—here I was shaking hands with the head of the DOE. We stayed and chatted quite a bit.

"Then I got a message saying that Dr. Tara O'Toole, head of the medical section of the DOE, would like to meet with me. Could I come advise them on the standards for compensating radiation victims—I who had barely set foot inside the building. I thought this would be an excellent opportunity to publicize my method of measuring radiation risk, so there I was explaining the *Ready Reckoner* to a group of DOE lawyers—probably those same lawyers who'd been taking shots at me in Denver—and they were astonished. I can't think why, since it's all there in our published papers. They were fascinated; we had difficulty breaking off.

"In return O'Toole asked me, could they do anything for me, did I want to be paid a fee? I said that wasn't necessary, but I could do with their help in getting some of the A-bomb data. And I was amazed, the day after I arrived back in England, to find a letter of invitation from the head of the Radiation Effects Research Foundation. The DOE must have put pressure on people in Hiroshima."

This was how Alice and George got access to the multiple injury data, which they used to establish that the survivor population is a selected, not a representative population, selected in favor of resistance to radiation.

Alice turned ninety on October 4, 1996. She celebrated the event on the lawn at Fawler, among a gathering of 135 or so friends and family.

In July of that year, she presented the multiple injury findings at the International Workshop on Radiation Exposures by Nuclear Facilities at Portsmouth, where Steve Wing and David Richardson presented their corroboration of the age effect in the Oak Ridge population.[8] In October, she was brought by Physicians for Social Responsibility, Committee to Bridge the Gap, and Simi Valley activists to consult on the study of Rocketdyne, near Los Angeles.

In November, she gave a paper at an international conference at Ben Gurion University, Beer Sheva, in Israel. Dr. John Goldsmith, one of the conference coordinators, had heard her at Portsmouth and was so impressed that he invited her to be keynote speaker. She gave a version of the multiple injury findings. "My paper met with friendly reception. A member of the audience from the DOE said he was fascinated by what I'd said and asked for a copy of the paper. So too did an American who is forming the next BEIR committee, BEIR VII, ask me for a copy and asked who I could recommend from England to be on the new committee."

In March 1997, Alice was featured at a conference held by the Rocky Flats Health Advisory Panel of the Colorado Department of Public Health study of radiation effects on the population near Rocky Flats, funded by the DOE. In July, she was awarded an honorary doctorate from Bristol University. This is the first honorary doctorate she has received in England, and it pleased her that it came from the British College of Physicians who, decades before, had made her their youngest woman member. Later in July, Alice returned to Bristol to address the Conference on Low Level Radiation, where she made contact with another establishment figure, Professor Bridges, who chairs the Committee on Medical Aspects of Radiation in the Environment (COMARE), a committee that has investigated cancer clusters around British nuclear installations and found "no evidence" of increase. Bridges, a Medical Research Council researcher, listened to her views with great interest.

In the first week of September 1997, Alice gave a seminar at the University of Dundee, Scotland. In the middle of the month, she returned to Los Angeles to consult about the Rocketdyne workers. At the end of the month, she testified in Reading in a case concerning Greenham Common women.

When I asked her in October 1998 what had happened in the year since I'd seen her, she said, "not much," then thought for a moment, "except"—in March, a lecture at City University of New York; in April, a meeting at the Westfalishe Wilhelms-Universitat in Munster; in July, a lecture for COLA at Greenwich University. In September 1998 she was honored, along with Karl Morgan, at a lively and well-attended conference in New York, organized by Dr. Helen Caldicott and the STAR Foundation (Standing for Truth About Radiation) at the New York Academy of Medicine.[9] Her presentation of the multiple injury findings met with a standing ovation.

At ninety-two, life is full. She continues to spend weekends at Fawler and midweek at Birmingham, where she and George have retained their library privileges and part-time secretarial assistance. She doesn't like it that she and George have been squeezed into a very small office and that they still have to scrounge for funding. But she is pleased by the interest of individuals working within government agencies like the DOE and COMARE, agencies that have long ignored her: she feels it's evidence of the recognition she hoped would eventually come, the triumph of truth in the course of time.

She has developed a new line, writing book reviews for the *Times Higher Education Supplement.* She tells me she'd like to start a new project but is not sure she should—"my projects tend to take thirty years."

Notes

Chapter 1

1. Keith Schneider, "Scientist Who Managed to 'Shock the World' on Atomic Workers' Health," *New York Times* 3 May 1990: A-20.

2. Renate Barber's "That's Alice" was enormously helpful in getting me started on this project.

3. Richard Doll and Richard Peto, "The Causes of Cancer: Quantitative Estimates of Avoidable Risks of Cancer in the United States Today," *Journal of the National Cancer Institute* 66 (1981): 1191–1308. Robert N. Proctor describes the influence of Doll's work: "It was endorsed by the Office of Technology Assessment, the NCI, and countless government and academic treatises," by popular magazines and the *New York Times; Cancer Wars: How Politics Shapes What We Know and Don't Know about Cancer* (New York: Basic Books, 1995), 68–69.

4. James Younger, "Nuclear Plant Cleared in Leukemia Cluster," *New York Times* 8 March 1994.

5. Evelyn Fox Keller, *A Feeling for the Organism* (San Francisco: W. H. Freeman, 1983).

6. Alice Stewart et al., "Preliminary Communication: Malignant Disease in Childhood and Diagnostic Irradiation *in utero*," *Lancet* 2 (1956): 447; "A Survey of Childhood Malignancies," *British Medical Journal* 1 (28 June 1958) 1495–1508.

7. A 1953 AEC pamphlet claimed that low-level exposure "can be continued indefinitely without any detectable bodily change." Catherine Caufield, *Multiple Exposures: Chronicles of the Radiation Age* (Chicago: University of Chicago Press, 1989), 120.

8. Cf. a 1981 statement made by State Department attorney William H. Taft: the "mistaken impression" that low-level radiation is hazardous has the "potential to be seriously damaging to every aspect of the Department of Defense's nuclear weapons and nuclear propulsion programs. . . . it could impact the civilian nuclear industry . . . and it could raise questions regarding the use of radioactive substances in medical diagnosis and treatment." Brian W. Jacobs, "The Politics of Radiation: When Public Health and the Nuclear Industry Collide," *Greenpeace* (July–August 1986): 6–9. The quotation is on p. 7.

9. Richard Rhodes, *The Making of the Atomic Bomb* (New York: Simon and Schuster, 1986), 443ff; Stephen Hilgartner et al., *Nukespeak: The Selling of Nuclear Technology in America* (San Francisco: Sierra Club Books, 1982), 22; Mark Hertsgaard, *Nuclear Inc.: The Men and Money behind Nuclear Energy* (New York: Pantheon, 1983), 11–13.

10. Nuclear physicists agreed not to submit any papers to scientific journals; editors were asked to watch for manuscripts that might touch on sensitive areas and send them for review by a committee that would decide whether they should be published or not. Violations were reported to the Office of Censorship; Hilgartner 25–30; Monica Braw, *The Atomic Bomb Suppressed: American Censorship in Occupied Japan* (Armonk, NY: M. E. Sharpe, 1991), 107.

11. Hertsgaard 14.

12. Hertsgaard 12–13; Anna Gyorgy and friends, *No Nukes: Everyone's Guide to Nuclear Power* (Boston: South End Press, 1979), 99.

13. Karl Grossman, *Cover-up: What You* Are Not *Supposed to Know about Nuclear Power* (Sagaponack, NY: Permanent Press, 1980), 17, 152–55.

14. Caufield 148–50; Hilgartner 41, 75; John Stauber and Sheldon Rampton, *Toxic Sludge Is Good for You: Lies, Damn Lies and the Public Relations Industry* (Monroe, ME: Common Courage Press, 1995), 34–35.

15. Amory B. Lovins, "The Origins of the Nuclear Power Fiasco," *The Politics of Energy Research and Development; Energy Policy Studies,* ed. John Byrne and Daniel Rich (New Brunswick, NJ: Transaction Books, 1986), 3: 7–34; and W. Greene, "Cold War Costs," *Boston Globe* 6 May 1992.

16. Peter Bunyard, *Nuclear Britain* (London: New England Library, 1981), 28–30; Hertsgaard 11; Hilgartner 53.

17. Hertsgaard 103–5.

18. There hasn't been a new nuclear power plant ordered in the United States since 1978; there hasn't been a low-level waste dump sited since 1984. Wolfgang Rudig, *Anti-Nuclear Movements: A World Survey of Opposition to Nuclear Energy* (Harlow, UK: Longman Group), 1990.

19. In 1992, the land mass of Department of Energy facilities comprised more than the states of Delaware and Rhode Island combined; the total of nuclear plant workers ever employed exceeded 600,000. H. Jack Geiger et al., eds. *Dead Reckoning: A Critical Review of the Department of Energy's Epidemiologic Research,* Physicians for Social Responsibility, Physicians Task Force on the Health Risks of Nuclear Weapons Production, 1992, 17–18.

20. Bernardino Ramazzini, an eighteenth-century Italian physician, is known as the father of occupational medicine. His 1713 treatise, *Diseases of Workers,* is the first comprehensive documentation of occupational disease.

21. Keith Schneider, "U.S. Hints It May Yield Data on Weapon Workers' Health," *New York Times* 28 October 1988.

22. Rudig 53.

23. Tom Athanasiou, *Divided Planet: The Ecology of Rich and Poor* (New York: Little Brown, 1996), describes the Council for Energy Awareness's campaign for nuclear as "clean, safe, power" (248).

24. Marvin Resnikoff et al., *Deadly Defense: Military Radioactive Landfills* (New York: Radioactive Waste Campaign, 1989); Center for Defense Information, *Vital Signs* (New York: Norton, 1995), United Nations, Freedom House.

25. "For several years, DOE has had a budget for waste management and environmental restoration of $5 billion or more, and overall expenditures at this level or higher are expected for about 30 years"; Arjun Makhijani, Howard Hu,

and Katherine Yih, eds., *Nuclear Wastelands: A Global Guide to Nuclear Weapons Production and Its Health and Environmental Effects,* Special Commission of International Physicians for the Prevention of Nuclear War and the Institute for Energy and Environmental Research (Cambridge, MA: MIT Press, 1995), 255. The estimated cleanup cost of weapons production is, according to a recent study, between $216 billion and $410 billion; this must be added to the $5.5 trillion spent by the United States on nuclear weapons between 1940 and 1996. *Atomic Audit: The Cost and Consequences of U.S. Nuclear Weapons Since 1950,* ed. Stephen I. Schwartz (Washington, DC: Brookings Institute, 1998).

26. Helen Caldicott, "The Weapons, the Power, the Waste," *Nation* 29 April 1996: 11–16; Harvey Wasserman, "In the Dead Zone: Aftermath of the Apocalypse," *Nation* 29 April 1996: 16–20.

27. Bernard Lown and Eugene I. Chazov, "Physician Responsibility in the Nuclear Age," *Journal of the American Medical Association* 274, no. 5 (2 August 1995): 416–19; M. J. Elders, "News Report," *Lancet* 1 (1994): 41; Victor Sidel, "Farewell to Arms, the Impact of the Arms Race on the Human Condition," *Physicians for Social Responsibility Quarterly* 3 (1993): 18–27. The legacy is only slowly being understood. In July and August 1997, U.S. media had stories about a National Cancer Institute study ordered by Congress in 1982 that estimates that the populations of the western states were exposed to doses of radioactive iodine at least ten times larger than those caused by the explosion at Chernobyl and one hundred times greater than were being estimated by the government in 1959. The NCI estimates that fallout has caused ten thousand to seventy-five thousand thyroid cancers, 70 percent of which have not yet been diagnosed. Matthew Wald, "U.S. Atomic Tests in 50's Exposed Millions to Risk," *New York Times* 29 July 1997: 8; "Thousands Have Thyroid Cancer from Atomic Tests," *New York Times* 2 August 1997.

28. I am deeply indebted to those who've shared this material with me. See acknowledgments.

29. "The linear dose-effect model is believed to be conservative and quite suitable for purposes of radiation protection"; Jacob Shapiro, *Radiation Protection: A Guide for Scientists and Physicians* (Cambridge, MA: Harvard University Press, 1990), 358–59. These are the premises of mainsteam radiation studies, such as E. Russell Ritenour, "Overview of the Hazards of Low-Level Exposure to Radiation," *Health Effects of Low-Level Radiation,* ed. William R. Hendee (Norwalk, CT: Appleton-Century Crofts, 1984), 13–18.

30. Schneider, "Scientist Who Managed to 'Shock the World.' " Clive Cookson, "Unsung Heroine," *London Times* 19 Jan. 1995; Gail Vines, "A Nuclear Reactionary," *Times Higher Education Supplement* 28 July 1995; Amy Raphael, "Alice Stewart, Rebel Scientist," *Ms.* July–August 1996; "Sex and the Scientist: Our Brilliant Careers," Channel 4 television, 19 August 1996. Other informative articles include Len Ackland, "Radiation: How Safe Is Safe?" *New Scientist* no. 1873 (May 1993): 34–37; Christine Cassell, "Profiles in Responsibility: Alice Stewart," *Physicians for Social Responsibility Quarterly* 1, no. 1 (March 1991): 53–57; Judith Johnsrud, "On the Trail of Childhood Cancers," *East/West* (November 1987): 43–47; Virginia Myers, "Nuclear Enemy No. 1," *Oldie* (October 1992): 22–23.

Chapter 2

1. Thomas Neville Bonner, *To the Ends of the Earth: Women's Search for Education in Medicine* (Cambridge, MA: Harvard University Press, 1992), 51–56, 161–65.

2. "As Lucy lay dying, confined to her bed and nearly blind, I asked her questions about her life," Nora Naish told me, explaining how she came to write this biography. "She was more than glad to talk, and I was more than glad to listen." Nora, the ex-wife of Alice's younger brother, John, is a retired physician and the author of several novels, including *The Butterfly Box, The Magistrate's Tale,* and *A Time to Learn,* the fictionalized version of "Dr. Lucy."

3. Arthur F. Abt, M.D., *Abt-Garrison History of Pediatrics* (Philadelphia: W. B. Saunders, 1965), 103.

4. W. F. Bynum, *Science and the Practice of Medicine in the Nineteenth Century* (Cambridge: Cambridge University Press, 1994), 79.

5. According to Bynum, the number of deaths from typhoid fell to 100 per 1,000,000 in 1900, as opposed to 350 per 1,000,000 in 1870. Overall death rates in England and Wales fell from 20.8 per 1,000 in 1850 to 18.2 in 1900 (223).

6. Abt 151; B. Seebohm Rowntree, *Poverty: A Study of Town Life* (London and New York: MacMillan, 1901), 110, 122; George Newman, *Infant Mortality* (1906). Cited in "Dr. Lucy."

7. "Dr. Lucy."

8. The infant death rate of England and Wales in 1891–1900 was 153, in 1901–10 it was 128, in 1911–15, 110, and in 1919, 91 (Abt 166).

9. The Board of Guardians was the governing board of the Poor Law institutions, public institutions (hospitals, asylums, workhouses) that constituted a kind of rudimentary welfare system that had been in place since the sixteenth century; Bynum 61, 185–86.

10. Talk given by Professor John Emery to the Aesculpian Society of Sheffield, 1993.

11. Alice's brother, John Naish, remembers her as a "maternal force close in primeval energy to Brecht's *Mother Courage,*" though he adds that "despite her dominance in all things domestic there was an unspoken assumption that in all grave matters of finance and education, the last word was her husband's." Ernest was, "despite his gentle voice and impeccable manners, forever remote and somewhat terrifying" (*A Physician's Eye,* Mellen Lives, vol. 6 [Lewiston: Edwin Mellen Press, 1997], 14–15, 18).

12. Lucy Naish, "Breast Feeding: Its Management and Mismanagement," *Lancet* (14 June 1913): 1657–78.

13. In England and Wales, sixty-two thousand people died of the influenza, more than in the great cholera epidemic of 1848–49 ("Dr. Lucy").

Chapter 3

1. Humphrey Milford, *St. Leonards School, 1887–1927* (London: Oxford University Press), 1927.

2. Rita McWilliams-Tullberg, *Women at Cambridge: A Men's University—Though of a Mixed Type* (London: Gollancz, 1975), 207–8. As recently as 1974, only 7 percent of all University appointments at Cambridge were held by women. Less than 3 percent of professors were women, less than 7 percent of Readers (216).

3. Michael Holroyd, *Lytton Strachey: A Critical Biography,* vol. 2, *The Years of Achievement, 1910–1932* (New York: Holt, Rinehart, and Winston, 1968), 387.

4. Virginia Woolf, *A Room of One's Own* (New York: Harcourt Brace, 1957).

5. Kathleen Raine, *The Lion's Mouth* (1977; New York: George Braziller, 1978), 18, 25.

6. Rhodes calls Rutherford the Newton of atomic physics (230). See also Caufield 23–24; Shapiro 5.

7. C. P. Snow, *The Search* (New York: Charles Scribner's Sons, 1934), 87–89.

8. Rhodes 27–28.

9. Jessie Stewart studied with Jane Harrison, the well-known Cambridge scholar of Greek religion, and became her biographer and the author of *Jane Ellen Harrison: A Portrait from Letters* (London: Merlin Press, 1959). Edward Shils and Carmen Blacker, eds., *Cambridge Women: Twelve Portraits* (Cambridge: Cambridge University Press, 1996), 63.

10. Julian Trevelyan, *Indigo Days* (London: MacGibbon & Kee, 1957), 16, quoted in Philip Gardner and Averil Gardner, *The God Approached: A Commentary on the Poems of William Empson* (London: Chatto and Windus, 1978), 20.

11. Raine 44–45.

12. M. C. Bradbrook, "The Ambiguity of William Empson," *William Empson: The Man and His Work,* ed. Roma Gill (London: Routledge and Kegan Paul, 1974), 2–12. The quotation appears on p. 3.

13. Alan Rusbridger, "French Letters of a Cambridge Don," *Guardian* 13 November 1991.

Chapter 4

1. F. H. Hinsley and Alan Stripp, eds., *Code Breakers: The Inside Story of Bletchley Park* (London: Oxford University Press, 1994).

2. Alice Stewart et al., "Some Early Effects of Exposure to Trinitrotoluene," *Half a Century of Social Medicine: An Annotated Bibliography of the Work of Alice M. Stewart,* ed. C. Renate Barber (West Sussex, UK: Piers Press, 1987). First published in *British Journal of Industrial Medicine* 2 (1945): 76–82.

3. Alice Stewart et al., "Chronic Carbon Tetrachloride Intoxication," *Half a Century of Social Medicine,* Barber. First published in *British Journal of Industrial Medicine* 1 (1944): 11–19.

4. Alice Stewart, "Pneumoconiosis in Coal Miners," *British Journal of Industrial Medicine* 5 (1948): 120.

Chapter 5

1. F. A. W. Crew, *Measurements of the Public Health: Essays on Social Medicine* (Edinburgh: Oliver and Boyd, 1948), xi. Social Medicine had a longer history in Europe, going back to the second decade of the century in Germany and the Netherlands (thanks to Klarrisa Nienhuys for this observation and for help with much of the background for this chapter).

2. Leslie Witts, J. A. Ryle's obituary, *British Medical Journal* (11 March 1950): 612–13.

3. John Pemberton, "Possible Developments in Social Medicine," *British Medical Journal* (11 December 1943): 754–55. The data mentioned in text appear on p. 754.

4. S. L. Leff, *Social Medicine* (London: Routledge and Kegan Paul, 1953), 9.

5. John A. Ryle, "Social Pathology," *Social Medicine: Its Derivations and Objectives,* ed. Iago Galdston (New York: The Commonwealth Fund, 1949), 55–75. The quotation is on pp. 71–72. First published by the New York Academy of Medicine, Institute on Social Medicine in 1947.

6. John A. Ryle, *Changing Disciplines: Lectures on the History, Method and Motives of Social Pathology* (London: Oxford University Press, 1948), 112.

7. John A. Ryle, "Social Medicine: Its Meaning and Its Scope," *British Medical Journal* (20 November 1943): 633–66. The quotation appears on p. 634.

8. Ryle, *Changing Disciplines* vi, 114–15.

9. Ryle, "Social Medicine" 633.

10. Ryle, "Social Pathology" 71–72.

11. Ibid. 72.

12. Ibid. 68–69; Ryle, *Changing Disciplines* 15–18, 115.

13. Alice Stewart, "Problems of Tuberculosis in Industry: A Study of the Shoemaking Trade in Northamptonshire," *Half a Century of Social Medicine,* Barber. First published in *British Journal of Tuberculosis* 47 (1952): 122.

14. J. W. Webb, D. Hewitt, and Alice Stewart, "The Scope and Limitations of the Civilian Medical Board Records: A Report to the Medical Research Council [1951–53]," *Half a Century of Social Medicine,* Barber; also Stewart, "Learning from Ryle," *American Journal of Public Health* 85, no. 10 (October 1995): 9–10.

15. Alice Stewart, "A Report on the Work of the Institute of Social Medicine, Oxford, 1943–50," *Half a Century of Social Medicine,* Barber.

16. The system of "visible tape" is described in Alice Stewart, "Visible Tape: A New Method of Assembling Medical Records for Computer Analysis," *Proceedings IEA Meeting August 1968,* 469.

Chapter 6

1. In 1930, a white newborn male born in the United States had a 180 chance in 100,000 of dying of leukemia; by 1955, his chances had risen to 600 in 100,000. Shapiro 350.

2. David Hewitt, "Some Features of Leukemia Mortality," *British Journal of Preventative Social Medicine* 9 (1955): 81.

Experience with Atomic Radiation (New York: Dell, 1982), 130, quoting Karl Morgan, "Citizen's Hearings for Radiation Victims," Washington, DC, 11 April 1980, 88.

18. E. G. Knox, Alice Stewart, George Kneale, and E. P. Gilman, "Pre-natal Irradiation and Childhood Cancer," *Journal of Radiation Protection* 7 (1987): 4. This paper took ten years to publish.

19. G. W. Kneale, "Excess Sensitivity of Pre-leukemics to Pneumonia: A Model Situation for Studying the Interaction of an Infectious Disease with Cancer," *British Journal of Preventative Social Medicine* 25 (1971): 152–59.

20. S. Jablon and H. Kato, "Childhood Cancer in Relation to Pre-Natal Exposure to Atomic Bomb Radiation," *Lancet* 2 (1970): 1000–1003.

Chapter 7

1. Molly Newhouse went on to teach occupational medicine at London University and London School of Hygiene and Tropical Medicine and became a worldwide authority in risk from asbestos exposure.

2. Thomas McKewon is author of *The Role of Medicine: Dream, Mirage, or Nemesis* (Oxford: Blackwell, 1979), an important study that points out that the betterment of social and environmental circumstances—that is, prevention—has played a more crucial role in the improvement of public health than have clinical interventions.

3. "Alice had been frozen out when I arrived at Oxford, she'd been treated very badly, she'd had her support cut down to a minimum; when I came here I did my damndest as Regius professor to try to improve the situation. . . . But I knew from the attitude of people who had been here when I came that there was just no chance of getting her a professorship." Interview with Sir Richard Doll, 9 October 1998.

4. Georgina Ferry, "No Smoke without Fire," *Oxford Today* 5, no. 3 (1993): 11–12.

5. Medical Research Council, Conference on Cancer of the Lung, 6 February 1947, report presented to MRC meeting 21 March 1947; A. Bradford Hill, Memorandum, "Proposed Statistical Investigation of Cancer of the Lung," MRC, 10 November 1947, presented to MRC meeting 21 November 1947, bears out Alice's recollections. The landmark study produced by Doll and Bradford Hill showed that people who smoked twenty-five or more cigarettes a day were about fifty times more likely to die of lung cancer than nonsmokers. "A Study of the Aetiology of Carcinoma of the Lung," *British Medical Journal* 2 (1952): 12-71-86. Doll does not, to my knowledge, go around claiming the discovery as his own; he acknowledges that Bradford Hill invited him to be part of the study and says in the 1993 interview that his original hunch was that the problem was auto exhaust fumes (Ferry 12).

6. On the Channel 4 *Power List* program on 31 October 1998, Sir Richard Doll was listed as number 122 among the most powerful (or influential) people in Britain.

7. *New Scientist* 78 (20 April 1978): 132.

3. MRC records show no grants awarded to Alice for this proposal and n grants denied, so the rejection she recalls would have been for an informa application.

4. Alice Stewart, J. W. Webb, D. Giles, and David Hewitt, "Preliminar Communication: Malignant Disease in Childhood and Diagnostic Irradiation *i utero,*" *Lancet* (1956): 447. For the stir it created, see Alwyn Smith, *The Scienc of Social Medicine* (London: Staples, 1968), 126.

5. Alice Stewart, J. W. Webb, and David Hewitt, "A Survey of Childhoo Malignancies," *British Medical Journal* 1 (28 June 1958): 1495–1508. See als "Low Dose Radiation Cancers in Man," *Advances in Cancer Research,* G. Klei and S. Weinhouse, eds. (New York: Academic Press, 1971), Chap. 8.

6. For the anti-nuclear movement, see Rudig; Peter Stoler, *Decline and Fail: The Ailing Nuclear Power Industry* (New York: Dodd, Mead, 1985); Ruth Brandon, *The Burning Question: The Anti-Nuclear Movement since 1945* (London: Heinemann, 1987).

7. Anthony Serafini, *Linus Pauling: A Man and His Science* (New York: Paragon House, 1989), 186–207.

8. W. M. Court-Brown and R. Doll, "The Hazards to Man of Nuclear and Allied Radiations," Presented by the Lord President of the Council to Parliament by Command of Her Majesty, Medical Research Council, June 1956; "Leukemia and Aplastic Anaemia in Patients Irradiated for Ankylosing Spondylitis," Medical Research Council Special Report Series, no. 295, 1957.

9. Interview with Sir Richard Doll, 9 October 1998.

10. W. M. Court-Brown and R. Doll, "Incidence of Leukemia after Exposure to Diagnostic Radiation *in utero,*" *British Medical Journal* 2 (1960): 1539–45.

11. Interview with Sir Richard Doll, 9 October 1998; and R. Doll and R. Wakeford, "Risk of Childhood Cancer from Fetal Irradiation," *British Journal of Radiology* 70 (1997): 130–39. The quotation appears on p. 137.

12. "The negative findings of Court-Brown and his colleagues gained a remarkably wide acceptance," summarizes Smith 131. Doll's study continues to be invoked to refute Alice's. Steven R. Wilkins, "Truths and Fallacies Concerning Radiation and Its Effects," *Health Effects of Low-Level Radiation,* Hendee, 23–32.

13. Brian MacMahon, "Prenatal X-Ray Exposure and Childhood Cancer," *Journal of the National Cancer Institute* 28 (1962): 1172.

14. I. D. Bross and N. Natarajan, "Leukemia from Low Level Radiation: Identification of Susceptible Children," *New England Journal of Medicine* 287 (1972): 107–10. This study, like Court-Brown and Doll's, looked for only one type of cancer, which was a mistake, according to Alice, because it perpetuated the false impression that leukemia was the only childhood cancer caused by radiation.

15. Karl Morgan, "Radiation and Health," in *Energy in Central and Eastern Europe; Nuclear Power and Energy Efficiency: Two Options* (Budapest: Panos Institute, Friends of the Earth European Coordination, Greenway, 1992), 83–104. The quotation appears on p. 99. Also see Robert del Tredici, *At Work in the Fields of the Bomb* (London: Harrap, 1987), 132–33.

16. Caufield 233.

17. Harvey Wasserman et al., *Killing Our Own: The Disaster of America's*

8. Doll and Peto 1191–1308. Jeremy Laurance, "Clean House Could Cause Cancer," *London Times* 13 March 1992; Younger.

9. Raine 125.

10. Empson wrote of these years in the "South-Western Combined Universities," "A Chinese University," *Life and Letters* 25, no. 34 (June 1940): 239–45.

11. Professor Wang Zuo-liang, memoir, quoted in *William Empson, The Royal Beasts and Other Works,* ed. and introduction, John Haffenden (London: Chatto and Windus, 1986), 27.

12. Ibid. 39. With thanks to Haffenden for sharing with me parts of his forthcoming biography.

13. Quoted on blurb of the 1968 U.S. edition of *Some Versions of Pastoral.* New York: New Directions, 1968.

14. John Haffenden, introduction, *Argufying: Essays on Literature and Culture,* by William Empson, ed. John Haffenden (London: Chatto and Windus, 1987), 3.

15. Terry Eagleton, *Literary Theory: An Introduction* (Minneapolis: University of Minnesota Press, 1983), 31, 45–52.

16. Gardner and Gardner 30, quoting A. Alvarez.

17. Gardner and Gardner 27.

18. Raine 125–26.

19. Bradbrook 2–12.

20. Not included in the Empson file is an item in *Lancet* by Alice Stewart, "Gene-Selection Theory of Cancer Causation," (2 May 1970): 923–24, that begins with a quote from Empson's *Collected Poems:* "How small a chink lets in how dire a foe."

21. Rusbridger.

22. Christopher Ricks, *William Empson, 1906–1984, Proceedings of the British Academy* 71 (1985): 538–54.

Chapter 8

1. Thomas Mancuso, "Study of the Lifetime Health and Mortality Experience of Employees of ERDA Contractors," Final Report, September 1977, 13, 30. Department of Industrial Environmental Health Sciences, Graduate School of Public Health, University of Pittsburgh, prepared for ERDA, 1.

2. Michael D'Antonio, *Atomic Harvest: Hanford and the Lethal Toll of America's Nuclear Arsenal* (New York: Crown, 1993); and Rhodes 497–500; see chapter 1 of the present work, notes 9 and 10.

3. Alice Stewart, "The Pitfalls of Extrapolation," *New Scientist* (July 1969): 181.

4. Howard Kohn, "The Government's Quiet War on Scientists Who Know Too Much," *Rolling Stone* 23 March 1978: 42–44; idem, *Who Killed Karen Silkwood?* (New York: Summit, 1981), 290.

5. For Hueper, see Proctor, 36–48. For NCI reluctance to fund research that might offend the chemical industries, see Samuel S. Epstein, *The Politics of Cancer* (San Francisco: Sierra Club Books, 1978), 318–42.

6. A 1967 memorandum to John Totter, director of the AEC Division of Biology and Medicine, from Leonard Sagan, AEC Medical Research Branch, 20 November 1967, 3, stated the "unanimous opinion" of a group that had just reviewed the Mancuso study, that "aside from a certain *'political' usefulness,* it was very unlikely that new information on radiation effects will accrue from this study." Hilgartner 104, 249, n. 18.

7. Kohn, "Quiet War" 43; Kohn, *Silkwood* 291.

8. del Tredici 140.

9. *Effect of Radiation on Human Health: Health Effects of Ionizing Radiation,* Hearings before the Subcommittee on Health and the Environment, Committee on Interstate and Foreign Commerce, House of Representatives, 95th Congress, 2d sess., 24–26, Serial No. 95–179, 24, 25, 26 January and 8, 9, 14, 28 February 1978, 525, 633.

10. The reorganization did little to eliminate conflict of interest, as the Mancuso affair would show. For conflict of interest in the regulatory agencies, see Ralph Nader, *The Menace of Atomic Energy* (1977; New York: Norton, 1979), 277–79.

11. *Effect of Radiation* 525, 543–44, 547.

12. S. Milham Jr., "Increased Cancer Mortality among Male Employees of the Atomic Energy Commission, Hanford, Washington Facility," the Milham Report, Washington State Department of Social and Health Services, May 1974.

13. *Effect of Radiation* 495–96, 530, 626; and prepared attachment, "AEC Press Release read over phone to Thomas Mancuso" 559.

14. Robert Gillette, "Radiation Study Sparks Controversy," *Los Angeles Times* 1 December 1977.

15. Kohn, "Quiet War" 43.

16. The term *healthy worker effect* was coined by Anthony MacMichael in a response to John Goldsmith's "What Do We Expect from an Occupational Cohort?" *Journal of Occupational Medicine* 17 (1975): 127.

17. Gyorgy 92. For the Mancuso affair, see David Burnham, "Study of Atom Workers' Deaths Raises Questions about Radiation," *New York Times* 25 October 1976; Jean L. Marx, "Low-Level Radiation: Just How Bad Is It?" *Science* 204, no. 13 (April 1979): 160–64; Peter Bunyard, "Radiation Risks—How Low Can One Get?" *New Ecologist* 5 (September–October 1979): 161–65; Rosalie Bertell, *No Immediate Danger: Prognosis for a Radioactive Earth* (Summertown, TN: Book Publishing Company, 1985), 88–95.

18. Kohn, "Quiet War" 44.

19. del Tredici 140.

20. Thomas Mancuso, Alice Stewart, and George Kneale, "Hanford I: Radiation Exposures of Hanford Workers Dying from Cancer and Other Causes," *Health Physics* 33 (1977): 369–84.

21. Interview with Karl Morgan 16 December 1995.

22. Cf. Rachel Carson: a single larger dose "may kill the cells outright, whereas the small doses allow some to survive, though in a damaged condition. These survivors may then develop into cancer cells. This is why there is no safe dose of a carcinogen." *Silent Spring* (Boston: Houghton Mifflin, 1962), 232.

23. Thomas Mancuso, Alice Stewart, and George Kneale, "Delayed Effects of

Small Doses of Radiation Delivered at Slow Dose Rates," *Quantification of Occupational Cancer,* ed. R. Peto and M. Schneiderman, Banbury Report no. 9, Cold Spring Harbor, NY, 1981.

24. del Tredici 132–33.

25. Karl Morgan, "Cancer and Low Level Ionizing Radiation," *Bulletin* (September 1978): 30; and idem, "How Dangerous Is Low-Level Radiation?" *New Scientist* 82, no. 5 (April 1979): 18.

26. Morgan, "Radiation and Health" 101.

27. J. A. Reissland, "An Assessment of the Mancuso Study," NRPB-R79 (London: HMSO, 1978). ERDA's criticisms were that "the data has not been handled in the usual method. . . . The statistical techniques were adequate although not the usual ones which we used for such studies. . . . The study does not employ the standard epidemiological methods . . . this report has very little scientific value." Stewart and Kneale's review of criticisms obtained through FIOA, *Effect of Radiation* 575. The most quoted criticism of MSK was S. Marks, E. S. Gilbert, and B. D. Breitenstein, "Cancer Mortality in Hanford Workers," *Late Biological Effects of Ionizing Radiation,* IAEA, Vienna 1 (1978): 369–86. For Bertell's critique of them, *No Immediate Danger* 92–6; and idem, "Comments and Response to 'The Nuclear Workers and Ionizing Radiation,' " *American Industrial Hygiene Journal* 40 (October 1979): 916–22. For Alice's critique, see Alice Stewart and George Kneale, "An Overview of the Hanford Controversy," *Occupational Medicine* 6, no. 4 (October–December 1991): 641–643.

28. *Effect of Radiation* 557, 624–25.

29. Thomas Mancuso, Alice Stewart, and George Kneale, "Reanalysis of Data Relating to the Hanford Study of Cancer Risks of Radiation Workers," Hanford II a, *Late Biological Effects of Ionizing Radiation,* AEA, Vienna 1 (1978): 387–410; idem, "Hanford II b: The Hanford Data, A Reply to Recent Criticisms," *Ambio* 9 (1980): 66–73; idem, "Hanford Radiation Study III: A Cohort Study of Cancer Risks from Radiation to Workers at Hanford (1944–77 Deaths) by Method of Regression Models in Life-Tables," *British Journal of Industrial Medicine* 38 (1981): 56–66; idem, "Identification of Occupational Mortality Risks for Hanford Workers," *British Journal of Occupational Medicine* 41 (1984): 6–8; idem, "Job-Related Mortality Risks of Hanford Workers and Their Relation to Cancer Effects of Measured Doses of External Radiation," *British Journal of Occupational Medicine* 41 (1984): 9–14.

30. *Effect of Radiation* 620–21.

31. Ibid. 629.

32. For PSR's criticism of the SMR, see Geiger et al., *Dead Reckoning* 45–46.

33. Instead of comparing workers to the general population, MSK compared workers within the Hanford complex, looking at those who were exposed to radiation in relation to those not exposed. "We divided the workers into nine different groups according to the type and nature of job and radiation exposure," Alice explains. "The professional and technical staff exhibited a 'healthy worker effect,' except for those exposed at the highest level, where the effect all but vanished and increased exposure to radiation undermined the good health of the specially selected members of this high ranking class. Clerical staff and manual workers not exposed to radiation exhibited higher mortality rates."

Chapter 9

1. "*Hiroshima Genbaku Sensaishi*, Record of the Hiroshima A-bomb War Disaster; memoirs of the Marine Transport relief team," *The Impact of the A-bomb, Hiroshima and Nagasaki, 1945 to 1985*, the Committee for the Compilation of Materials on Damage Caused by the Atomic Bombs in Hiroshima and Nagasaki (Tokyo: Iwanami Shoten Publishers, 1985), 28–29, 31–32.

2. Naomi Shohno, *The Legacy of Hiroshima: Its Past, Our Future* (Tokyo: Kosei Publishers, 1986), 15, 65; in Joseph Gerson, *With Hiroshima Eyes: Atomic War, Nuclear Extortion and Moral Imagination* (Philadelphia: New Society Publishers, 1995), 32.

3. Yoshitreu Kosaki et al., *A-bomb: A City Tells Its Story* (Hiroshima: Hiroshima Peace Culture Center, 1972), 3, 4, 18; in Hilgartner 216–17.

4. Bertell, *No Immediate Danger* 138.

5. Caufield 62–63.

6. Wilfred Burchett, *Shadows of Hiroshima* (London: Verso, 1983); Sue Rabbit Roff, *Hotspots: The Legacy of Hiroshima and Nagasaki* (London: Cassell, 1995), 271; Robert Jay Lifton and Greg Mitchell, *Hiroshima in America: A Half Century of Denial* (New York: Avon, 1995), 46–49.

7. Lifton 45–46; Monica Braw, *The Atomic Bomb Suppressed: American Censorship in Occupied Japan* (Armonk, N.Y.: M.E. Sharpe, 1991), 119, 121, 123.

8. M. Susan Lindee, *Suffering Made Real: American Science and the Survivors of Hiroshima* (Chicago: University of Chicago Press, 1994), 46, 49–50.

9. Bertell, *No Immediate Danger* 143–44. Eventually, the effects of fallout had to be acknowledged. In 1977, the report of a team of scientists including the distinguished British atomic dissident Joseph Rotblat and the Nobel Prize winning biologist George Wald drew particular attention to the residual radiation suffered by the 37,000 people who had entered the cities after the explosions and who had been subjected to low doses of radiation but suffered "enhanced incidence of leukemia" nevertheless (Roff 227).

10. Bertell, *No Immediate Danger* 142–45, 250; Roff 59.

11. Lindee 28; Charles E. Land, "Studies of Cancer and Radiation Dose Among Atomic Bomb Survivors," *Journal of American Medical Association* 2 August 1995), 274, 5, 402–7; Itsuzo Shigematsu and Mortimer Mendelsohn, "The Radiation Effects Research Foundation of Hiroshima and Nagasaki," *Journal of American Medical Association* (2 August 1995): 274, 5, 42–43.

12. Daniel Land, "A Reporter at Large," *New Yorker* 8 June 1946, quotes Tzsuzuki 62–72; in Lifton 53.

13. G. W. Beebe, H. Kato, and C. E. Land, "Mortality Experiences of Atomic Bomb Survivors, 1950–74," Life Span Study Report 8, RERF TR 1–77, 1977.

14. "The use of A-bomb data for risk assessment is generally predicated on the assumption that the survivors, apart from their radiation dose, are representative human beings." Minutes of meeting of RERF Scientific Council, 17–20 March 1987, RERF Survival Report, 1 April 1986–31 March 1987.

15. Roff 149.

16. Lindee 150, 164–65.

17. Roff 60, 114.
18. Lindee 7, 85–92; Roff 60, 101, 114, 149.
19. Lindee 6–7. Roff's detailed descriptions of the working of the studies give a sense of the many difficulties researchers faced with the data collection and analysis.
20. Lindee 104; Roff 67, 150, 187.
21. Lindee 35, 99, 104, 107.
22. Ibid. 99, 145.
23. Though there is an important difference between the healthy survivor effect and the healthy worker effect, as Alice points out: "the selection of the survivors is accidental, made by the radiation exposure, whereas with the workers, it's deliberate, made before the exposure takes place; but it means that neither is a representative population."
24. Alice Stewart and George Kneale, "Pre-cancers and liability to other diseases," *British Journal of Cancer* (March 1978): 37, 448–57; and Kneale, "Excess Sensitivity of Pre-Leukemics."
25. Alice was concerned that the A-bomb studies assumed that the "late effects of atomic radiation are always the result of mutations and never the result of damage to bone marrow and other blood-forming tissues" and that they were putting too much emphasis on leukemia; Alice Stewart and George Kneale, "Mortality Experiences of A-Bomb Survivors," *Bulletin of the Atomic Scientists* 40, no. 5 (1984), 40, 5, 61–62; Alice Stewart, "Radiation and Marrow Damage," *British Journal of Medicine* (April 1982): 284, 1192.
26. Alice Stewart and George Kneale, correspondence, "Late Effects of A-Bomb Radiation: Risk Problems Unrelated to the New Dosimetry," *Health Physics* (May 1988): 54, 5, 567–69.
27. "Delayed Effects of A-Bomb Radiation: A Review of Recent Mortality Rates and Risk Estimates for Five-Year Survivors," *Journal of Epidemiology and Community Health* 36, no. 2 (June 1982): 80–86. Other key papers are Alice Stewart and George Kneale, "Non-Cancer Effects of Exposure to A-Bomb Radiation," *Journal of Epidemiology and Community Health* 38 (1984): 108–12; Alice Stewart, "Detection of Late Effects of Ionizing Radiation: Why Deaths of A-Bomb Survivors Are So Misleading," *International Journal of Epidemiology* 14, no. 1 (1985): 52–56; "A-Bomb Radiation and Evidence of Late Effects Other Than Cancer," *Health Physics* 58, no. 6 (1990): 729–35; Alice Stewart and George Kneale, "A-Bomb Survivors: Further Evidence of Late Effects of Early Deaths," *Health Physics* 64 (1993): 467–72.
28. George Kneale and Alice Stewart, "Reanalysis of Hanford Data, 1944–1986 Deaths," *American Journal of Industrial Medicine* (23 March 1993): 371–89.
29. Joseph Rotblat has similarly questioned the use of the survivors as a representative population and suggested that the nuclear workers would be a more appropriate study population; "The Puzzle of Absent Effects," *New Scientist* (25 August 1977); idem, "A Tale of Two Cities," *New Scientist* (7 January 1988): 106–10.
30. Roff 227. Roff describes the use of negative findings as a means of reassurance (178, 194).

31. Lindee, 147–48, points out that the U.S. media was an enthusiastic ally. In 1955 *U.S. News and World Report* produced a series of articles on the "healthy, happy" babies of Hiroshima, headlined "Thousands of Babies, No A-Bomb Effects" and including a photo caption, "No upsurge of cancer rates."

32. 1981 recalculations by scientists at Lawrence Livermore National Laboratory in California and Oak Ridge, Tennessee, indicate that the amount of radiation delivered by the bombs had been grossly overestimated, which suggests that radiation is far more dangerous than formerly believed and "some of the most important data on the effects of radiation on humans may be wrong." "New A-bomb Studies Alter Radiation Estimates," *Science* 212 (May 1981); Clyde Haberman, "40 Years after A-Bombs, Medical Burden is Unclear," *New York Times* 4 October 1985. Then in 1990, the National Academy of Sciences produced a report based on improved calculation of the radiation released by the bombs suggesting that the low-dose radiation may be even more dangerous. "More cancers are appearing than we predicted" in BEIR III, reported Jacob I. Fabrikant of the University of California at Berkeley. Eliot Marshall, "Academy Panel Raises Radiation Risk Estimate," *Science* 247 (January 1990): 22–23. Also "Hiroshima Study Shows Higher Risks of Low-Level Radiation," *New Scientist* 6 (January 1990).

33. Peter Bunyard, "Radiation Risks—How Low Can One Get?" *New Ecologist* 5 (September/October 1978): 161–63; Bertell, *No Immediate Danger* 173–74.

34. Morris Greenberg, "The Evolution of Attitudes to the Human Hazards of Ionizing Radiation and to its Investigators," *American Journal of Industrial Medicine* 20 (1991): 717–21, 720. Greenberg, as advisor to the nuclear industry on ionizing radiation, is close to the heart of the establishment; he nevertheless finds Alice's criticisms convincing.

35. Caufield 175–76. Patrick A. Green gives a detailed account of the constituency of the ICRP and the affiliations of its members with organizations associated with the commercial exploitation of nuclear energy and government regulatory agencies. "The Controversy over Low-Dose Exposure to Ionizing Radiations," Master's thesis, University of Aston in Birmingham, November 1984.

36. Bertell, *No Immediate Danger* 173–74.

37. L. E. Sever, Gilbert E. S. et al., "A Case-Control Study of Congenital Malformations and Occupational Exposure to Low-Level Ionizing Radiation," *American Journal of Epidemiology* 127, no. 2 (1988): 226–42; E. S. Gilbert et al., "Updated Analyses of Combined Mortality Data for Workers at Hanford, Oak Ridge, and Rocky Flats," *Radiation Research* 136 (1993): 408–21; E. Cardis, et al., "Effects of Low Doses and Low Dose Rates of External Ionizing Radiation," *Radiation Research* 142, no. 2 (1995): 117–37.

38. Geiger et al. 31.

39. 1995 NRPB report, *Guardian,* 20 December 1995. For the change between 1980 and 1990, from BEIR III to BEIR IV, to the no threshold hypothesis, see Roff 7.

40. MSK have shown that "the likely shape of the dose/response curve—that is, the relationship between a given dose of radiation and the number of cancers

which will result . . . is not the shape of the curve selected as 'conservative' by such prestigious bodies as the ICRP or Britain's NRPB;" their curve shows, rather, a non-linearity of dose response. "Whereas the linear model—a straight line relationship between dose and cancers—used by ICRP suggests that the cancer rate will double every 30 rads, Stewart's curve gives the overall doubling dose for low dose radiation as 15 rads." Peter Bunyard, "Radiation and Health," *Green Britain Industrial Wasteland* (Cambridge: Polity Press, 1986).

41. Patrick Green, interview by Virginia Myers, 28 August 1996.

42. Cf. E. Russell Ritenour, "History of Radiation Protection Agencies and Standards," Hendee, *Health Effects of Low-Level Radiation* 153–60: "Current radiation protection philosophy is based on the assumption that there is no completely 'safe' amount of radiation. In practical terms, however, there is certainly a level below which the measurement of biological effects *becomes meaningless*" (157–58, my emphasis).

43. Interview with John Gofman (with Dr. Vicki Ratner), 14 April 1994.

44. Alice was sent a diskette version of Hiroshima data that included records of acute injuries under four headings: flash burns, ulceration of the mouth or pharynx, spontaneous hemorrhage or purpura, and hair loss. Testing differences between 63,042 survivors who denied any involvement with any one of these injuries and 2,602 who claimed two or more injuries, she and George found significant differences: among survivors with multiple injuries, leukemia deaths were exceptionally common; the different cancer experiences of the two groups were largely the result of exposures early or late in life; cancer was not the only late effect of the radiation.

45. As of July 1999, it still has not found a publisher.

Chapter 10

1. *Effect of Radiation* 719–21, 725.

2. Ibid. 782.

3. Cf. Shirley Fry, one of Oak Ridge's leading researchers, on an Oak Ridge study going back to 1979: "We found no increase in any type of cancer linked to radiation, such as cancer of the lung or cancer of the thyroid. We found a very healthy population." Keith Schneider, "U.S. Hints It May Yield Data on Weapon Workers' Health," *New York Times* 28 October 1988.

4. Joseph Rotblat, review of *Review of the Department of Energy's Controversial Termination of a Research Contract,* Report to the Congress by the U.S. Comptroller General and *Radiation Standards and Public Safety,* Proceedings of a Second Congressional Seminar on Low Level Ionizing Radiation, sponsored by the Congressional Environmental Study Conference, the Environmental Policy Institute and the Atomic Industrial Forum, *Bulletin of Atomic Scientists* (October 1979): 41. Rotblat cited the International Atomic Energy Agency, which is mandated both with "worldwide promotion of nuclear energy, and preventing its misuse," as another instance of blatantly incompatible roles.

5. Kohn, "Unquiet War" 42–44.

6. Robert Gillette, "Radiation Study Sparks Controversy," *Los Angeles Times* 1 December 1977: 26.

7. Interview with Robert Alvarez, 19 December 1995.

8. Kohn, *Silkwood;* Jim Garrison, "The Mysterious Case of Karen Silkwood," *Ecologist* 9, no. 8, (November–December 1979): 291–95; Bunyard, *Nuclear Britain* 203–6.

9. Interview with Robert Alvarez, 19 December 1995.

10. *Effect of Radiation* 561, 622.

11. Ibid. 536–37.

12. Ibid. 544, 633. Thomas Mancuso, "Study of the Life, Time, Health and Mortality Experience of Employees of AEC Contractors," Health Physics Society Annual Meeting, III, November 1971; in *Effect of Radiation,* prepared statement, 527–62.

13. Bob Alvarez, *Atomic Bulletin of Scientists* (February 1980): 61–62. S. E. English, AEC's assistant manager for research, wrote that "the study should . . . permit a statement to the effect that a careful study of workers in the industry disclosed no harmful effects from radiation. . . . A corollary statement could presumably be made about other exposed populations." Letter to John Totter, director of AEC's Division of Biology and Medicine, 13 November 1967, *Red Alert: The Worldwide Dangers of Nuclear Power,* in Judith Cook (Sevenoaks, Kent: New English Library, 1986), 175.

14. *Effects of Radiation,* Alice Stewart's testimony, 618–33.

15. Ibid. 677, 745, 767.

16. Walter Pincus, "House Panel Told That Exposure Limit for Radiation Is Ten Times Too High," *Washington Post* 9 February 1978: 2; "GAO to look at Mancuso Investigation," *Science News* 12 August 1978; Robert Alvarez, "Mancuso Affair," commentary, *Bulletin of Atomic Scientists* (February 1980): 61–63.

17. Karl Grossman, *Power Crazy* (New York: Grove Press, 1986), 92–101.

18. For the anti-nuclear movement, see Brandon and Rudig.

19. Peter Taylor, "The Struggle against Nuclear Power in Central Europe," *Ecologist* 7, no. 6 (July 1977): 217–22; and Bunyard, *Nuclear Britain* 184–85.

20. Rudig 32.

21. Brandon 6, 34.

22. Lifton and Mitchell 67; Brandon 16–18; Braw 138; Hertsgaard 16.

23. The International Physicians for the Prevention of Nuclear War came into existence in 1981 and won the Nobel Peace Prize in 1985; Lifton 261–62; Bernard Lown and Eugene I. Chazov, "Physician Responsibility in the Nuclear Age," *Journal of American Medical Association* 274, no. 5 (2 August 1995): 416–19.

24. Leslie J. Freeman, *Nuclear Witnesses* (New York: Norton, 1983), 60.

25. Ernest Sternglass, "The Death of All Children," *Esquire* 72 (September 1969): 3; idem, *Secret Fallout: Low Level Radiation from Hiroshima to Three Mile Island* (New York: McGraw-Hill, 1981). Also, interview with Sternglass (with Dr. Vicki Ratner), 30 April 1994.

26. Interview with John Gofman (with Dr. Vicki Ratner), 14 April 1994. John W. Gofman and Arthur R. Tamplin, *Poisoned Power: The Case against Nuclear Power Plants before and after Three Mile Island* (Emmaus, PA: Rodale Press, 1971).

27. Freeman 112.
28. Caufield 174–75; Kohn, "Quiet War" 42–44.
29. Rosalie Bertell, *Ms.* September–October 1991, 27ff.
30. Irwin Bross and N. Natarajan, "Leukemia from Low-Level Radiation," *New England Journal of Medicine* 287 (1972): 107–10.
31. Irwin Bross and N. Natarajan, "Genetic Damage from Diagnostic Radiation," *Journal of the American Medical Association* 237 (30 May 1977): 2399; Wasserman 134.
32. Rosalie Bertell, "Breast Cancer and Mammography," *Mothering* (summer 1992): 29, 51.
33. Bertell, *No Immediate Danger* vii, 161.
34. Interview with Rosalie Bertell (with Dr. Vicki Ratner), 30 April 1994.
35. Sam Totten and Martha Wescoat Totten, *Facing the Danger: Interviews with Twenty Anti-Nuclear Activists* (Trumansburg, NY: Crossing Press, 1984), 21–22; and Freeman 22–49.
36. Bertell, *No Immediate Danger* 243; Wasserman 174–75.
37. Interview with Bruce DeBoskey, 18 January 1997.
38. Geiger et al., 55–60, cite Mancuso, Johnson, Dr. Greg Wilkinson, and Dr. Stephen Gough, among others.
39. *Effect of Radiation,* Bross's testimony, 906–11.
40. Marian Christy, "The Unsinkable Dr. Stewart," *Boston Globe* 8 June 1988; Schneider, "Scientist Who Managed to 'Shock the World.' "

Chapter 11

1. For the DOE's orchestrated critiques of MSK, see chapter 8 of this book, nn. 27 and 28.
2. Contamination of the Cumbrian beaches, *New Scientist* 22 and 29 December 1983; Peter Bunyard, "The Sellafield Discharges," *Ecologist* 16, nos. 4–5 (1986); Jean McSorley, *Living in the Shadow: The Story of the People of Sellafield* (London: Pan Books, 1990); Anna Mayo, "High-Level Omissions," *Lies of Our Times* (July–August 1994): 13–17.
3. Peter Pringle and James Spigelman, *Nuclear Barons* (London, Sphere, 1983); in Cook 125.
4. *The Management of Radioactive Wastes* (IAEA, Vienna, 1981), quoted in Andrew Blowers, David Lowry, and Barry D. Solomon, *The International Politics of Nuclear Waste* (New York: St. Martin's Press, 1991), 43.
5. *Daily Mirror* 21 October 1975; in Rudig 76, 94.
6. *Planning and Plutonium,* Evidence of the Town and Country Association to the public inquiry into an oxide reprocessing plant at Windscale, the Town and Country Planning Association, 1978, 68–81.
7. *The Windscale Inquiry,* Report by the Hon. Mr. Justice Parker, Presented to the Secretary of State for the Environment on 26 January 1978, vol. 1, Report and Annexes 3–5, 48–49; Martin Scott and Peter Taylor, *The Nuclear Industry: A Guide to the Issues of the Windscale Inquiry,* the Town and Country Planning Association in Association with the Political Ecology Research Group, 105–9.

8. Nicholas Hildyard, "The Windscale Inquiry," editorial, *New Ecologist* 2 (March–April 1978): 38.

9. Rudig 76, 94.

10. Wolfgang Rudig argues that the government's "low-profile" strategy managed to successfully defuse anti-nuclear protest in Britain. "Maintaining a Low Profile: The Anti-Nuclear Movement and the British State," *States and Anti-Nuclear Movements,* ed. Helena Flam (Edinburgh: Edinburgh University Press, 1994), 88–94, 70–100.

11. Douglas Black, *Investigation of the Possible Increased Incidence of Cancer in West Cumbria* (London: HMSO, 1984). For criticisms of the Black Report, see Cook 186–90.

12. R. W. Apple, "Scenic Suffolk, Land of Saxons, Sizes up the Atom," *New York Times* 1 February 1984.

13. Pearce Wright, "CEGB Attacks Credibility of Anti-Sizewell Scientists," *London Times* 28 February 1985.

14. Guy Rais, "Marathon Sizewell Inquiry Ends," *Daily Telegraph* 8 March 1985.

15. Alice Cook and Gwin Kirk, *Greenham Women Everywhere: Dreams, Ideas and Actions from the Women's Peace Movement* (London: Pluto Press, Boston: South End Press, 1984), 108–24; Brandon 75; Paula Allen, "An Activist Love Story," *The Feminist Memoir Project,* ed. Rachel Blau DuPlessis and Ann Snitow (New York: Three Rivers Press, 1998), 416–18.

16. Blowers 84, 93, 151–54, 186; Berkhout 3, 172; "The Billingham Experience," extracted from *House of Commons Environment Committee Report,* 1986.

17. This exchange is reported by Brian Jacobs, "It Could Be Worse . . . and Probably Is," *In These Times* 14–20 January 1987.

18. David Fairhall, "Pressure for a Quick Result at Thurso," *Guardian* 7 April 1986.

19. Peter Roche, "Low and Intermediate Level Nuclear Waste Transport: U.K. and U.S. Experience," SCRAM, 1990.

20. Robin McKie and Geoffrey Lean, "Island Fights Mini-Sellafield," *Observer* 11 June 1986; Peter Bunyard, "Gearing up to the Plutonium Economy," *Ecologist* 6, no. 10 (December 1976): 348–55.

21. McKie and Lean.

22. M. A. Heasman et al., "Childhood Leukemia in Northern Scotland," *Lancet* 1 (1986): 266; Rob Edwards, "Doubts Dog Fast Reactor's Future," *Guardian* 26 October 1989.

23. Roger Milne, "Leukemia Fears Raised at Dounreay Inquiry," *New Scientist* 9 October 1986.

24. Ibid.

25. *Transcript of Proceedings, the Town and Country Planning (Scotland) Act, 1972,* Outline Planning Application for a European Demonstration Fast Reactor Fuel Reprocessing Plant at Dounreay, Caithness, before A. G. Bell, Esq., the Reporter, and Prof. K. C. Calman, Assessor, at the Town Hall, Thurso, Caithness, on Tuesday, 14 October 1986, 29–97.

26. Erlend Clouson, "Nuclear Plant Seeking New Life after Death," *Guardian* 19 March 1994.

27. Pearce Wright, "Demonstrators Gather as a Two-Week Adjournment Is Refused," *Times* 5 October 1988.

28. P. D. Ewings and C. Bowie, "Leukemia Incidence in Somerset with Particular Reference to Hinkley Point," Somerset Health Authority, December 1988; Roger Milne, "Missing the Hinkley Point," *New Scientist* 30 September 1989.

29. Transcript of Hinkley Point C Inquiry, before Mr. Michael Barnes et al., at Somerset College of Agriculture and Horticulture, Cannington, Somerset, 3 March 1989, 32–97.

30. Danielle Grunberg, Talk given to the East-West Consultation, "Citizens Involvement and Public Participation," Riga, Latvia, 26 October 1990, was especially helpful in describing the issues of the Hinkley inquiry.

31. Alasdair Murray, "British Energy Scraps 4.9 Billion Nuclear Plan," *Times* 12 December 1995.

32. Brandon 118.

33. Interview with Janine Allis-Smith, by Virginia Myers, 28 August 1996; Sorley 145–47.

34. Interview with Barbara French, by Virginia Myers, 28 August 1996.

35. Interview with Jill Sutcliffe, by Virginia Myers, 28 August 1996.

Chapter 12

1. Wasserman 237–63. Peter Stoler, *Decline and Fail: The Ailing Nuclear Power Industry* (New York: Dodd, Mead 1985), 131–32. Official reassurances turned out to be premature; Steve Wing et al., "A Reevaluation of Cancer Incidence Near the Three Mile Island Nuclear Plant," *Environmental Health Perspectives* 105 (1997): 52–57, have demonstrated a cancer effect.

2. Sternglass, *Secret Fallout* 241–75.

3. J. G. Kemeny et al., "The President's Commission on the Accident at Three Mile Island," The Need to Change: The Legacy of Three Mile Island," Washington, DC, 1979; Stoler 120–24.

4. Stoler 125–26; Bertell, *No Immediate Danger* 327.

5. d'Antonio 233–63, for mid-1980s revelations about DOE; I. Amato, "Dangerous Dirt: An Eye on DOE," *Science News* 4 October 1986: 221.

6. Matthew Wald, "Nuclear Arms Plants: A Bill Long Overdue," *New York Times* 30 September 1988; Keith Schneider, "Candor on Nuclear Peril," *New York Times* 19 October 1988; idem, "U.S. Studies Health Problems Near Weapons Plant," *New York Times* 17 October 1988.

7. Joint Hearing of the Senate Governmental Affairs Committee and the House Government Operations Subcommittee on Environment, Energy and Natural Resources, 1 October 1988.

8. In an October 1988 *Times* cover story, "They Lied to Us," Ed Magnuson indicted the weapons industry and government response toward these revela-

tions: 31 October 1988: 61–65; also, Dan Charles, "Will These Lands Ne'er Be Clean?" *New Scientist* 24 June 1989: 288.

9. See acknowledgments.

10. PSR Press Conference, Denver, Hyatt Regency Hotel, 22 March 1989.

11. Alice Stewart and George Kneale, "Mortality of Hanford Workers," *Health Physics* 57 (1989): 839–44; Alice Stewart, "Healthy Worker and Healthy Survivor Effects in Relation to the Cancer Risks of Radiation Workers," *American Journal of Industrial Medicine* 17 (1990): 151–54; idem, "Late Effects of A-Bomb Radiation: Risk Problems Unrelated to the New Dosimetry," *Health Physics* 54, no. 5 (1988):567–69; "A-Bomb Survivors as a Source of Cancer Risk Estimates: Confirmation of Suspected Bias," *Low Dose Radiation. Biological Bases of Risk Assessment,* ed. K. Baverstock and J. Stather (London and New York: Taylor and Frances, 1989).

12. Chapter 15, pp. 240–44.

13. "Johnston May Back Move to Broaden Public Access to DOE Health Data," *Inside Energy* 20 November 1982.

14. M. A. J. McKenna and Elizabeth Neus, "Worse than Three Mile Island," *Cincinnati Enquirer* 28 April 1989.

15. Clifford T. Honicker, "America's Radiation Victims: The Hidden Files," *New York Times Magazine* 19 November 1989.

16. Interview with Dianne Quigley, 9 November 1996. When Alice was making these appearances, CCRI got her featured on *McNeil-Lehrer, 60 Minutes, ABC News,* and CNN (all in 1991) and in numerous local newspapers around DOE sites. Special thanks to Dianne Quigley and Jean Kasperson for supplying information for this chapter and sending tapes of many of these appearances.

17. Brian Jacobs, review of *No Immediate Danger,* by Rosalie Bertell, *MIT Alumni Association Technology Review* 90 (April 1987): 75.

18. UPI, Richland, Washington, *Regional News* 24 September 1986.

19. Interview with Larry Shook, March 1997.

20. Thomas Graf, *Denver Post* 2 May 1989.

21. Tim Flach, "Savannah River Site Health Risks Called Inevitable," *Greenville News* 10 November 1989.

22. Stuart Englert, "Expert: Low Level Radiation Increases Cancer Risk," *Idaho State Journal* 31 October 1989.

23. Statement before the Senate Committee on Energy and Natural Resources, Washington, DC, 5 October 1989; Geiger et al. 25

24. *Clean-up at Federal Facilities,* Hearing before Subcommittee on Transportation and Hazardous Materials, Committee on Energy and Commerce, House of Representatives, 101st Congress, 1st sess., on H.R. 765, a bill to establish a commission to make recommendations to assure a comprehensive national approach to environmental cleanup of contamination of hazardous substances at Department of Energy Facilities, 23 February 1989, 12.

25. Ibid. 54, 278–79; see also *Department of Energy's Radiation Health Effects Research Program and Working Conditions at DOE Sites,* Hearing before the Committee on Governmental Affairs, U.S. Senate, 101st Congress, 1st sess., 2 August 1989, 44–46, 275–79.

26. Ibid. 6–7.

27. Ibid. 198–203.
28. PSR Press Conference 1989.
29. See chapter 10, n. 38.
30. For the transfer of health research, see Makhijani, Hu, and Yih 261, 282; Schneider, "U.S. Hints It May Yield Data"; idem, "U.S. Panel Will Urge Release of Secret Records," *New York Times* 22 November 1989; idem, "U.S. Will Disclose the Health Files of Atomic Workers," *New York Times* 18 May 1990; Charles Marwick, "Low-Dose Radiation: Latest Data Renew Questions of 'Safe' Level," *Journal of the American Medical Association* 264, no. 5 (1 August 1990): 553–57.
31. Geiger et al. 60. For the Memorandum of Understanding, see "CCRI Update on Efforts for Public Participation in Radiation Health Risks," *CCRI Newsletter* December–January 1994: 1–4; "DOE Cedes Worker Studies to HHS Department," *Nuclear News* February 1991; "Watkins Agrees to Transfer Long-Term Worker Health Studies to HHS," *Inside Energy* 2 April 1990.
32. Helen Gavaghan, "Three Mile Island Group Wins Battle for Radiation Data," *New Scientist* (14 April 1990): 21; Jim Detjen, "U.S. to Release Data on Radiation Exposure," *Philadelphia Inquirer* 21 June 1990.
33. Mike Townsley, "British Scientist Wins Her Fight for US Nuclear Data," *Observer* 17 June 1990.
34. "Nuclear Safety: DOE to Release Health Records of Workers to Independent Researchers for Radiation Study," *Occupational Safety and Health Reporter* 23 May 1990.
35. Schneider, "Scientist Who Managed to 'Shock the World' "; idem, "U.S. Releases Radiation Records of Forty-Four Thousand Nuclear Arms Workers," *New York Times* 18 July 1990.
36. Dr. Jill Sutcliffe suggested that I "should not allow Gilbert's quote to go unchallenged. Certainly, I have been present when Alice has denied that she is" (personal communication, 6 November 1997).
37. Jim Carrier, "Danger All around, but Accidents Far from Exotic," *Denver Post* 12 June 1995.
38. Matthew Wald, "Pioneer in Radiation Sees Risk Even in Small Doses," *New York Times* 8 December 1992.
39. Kneale and Stewart, "Reanalysis of Hanford Data," cited chap. 9, n. 28.
40. George Kneale and Alice Stewart, "Factors Affecting Recognition of Cancer Risks of Nuclear Workers." *Occupational and Environmental Medicine* 52 (1995): 515–23.

Chapter 13

1. *Third National Cancer Survey: Incidence Data,* NCI Monograph 44, ed. Sidney J. Cutler and John L. Young Jr., Biometry Branch, NCI. "It's estimated that background radiation in Denver is roughly twice the normal level, yet Denver shows a relatively low cancer incidence," comments Alice. "But for one thing, you've got a very migratory population in Colorado, a constant movement of people in and out, so you have no idea where a given person came from, how long

they've been there. For another thing, all the mountain states have a relatively low cancer incidence—but this probably has to do with reporting, which tends not to be as thorough in rural as in urban communities. There are so many other variables that need to be looked at, the NCI study is so shot with weaknesses that you'd hardly expect it to have found a cancer effect, but it's often quoted as gospel."

2. George Kneale and Alice Stewart, "Childhood Cancers in the U.K. and Their Relation to Background Radiation," *Radiation and Health* 16 (1987): 203–20. Sylvia Collier, "Cancer Deaths Linked to Background Radiation," *Health and Safety at Work* October 1986: 46–48; Roger Milne, "Background Radiation Blamed for Child Cancers," *New Scientist* (23 October 1986): 15.

3. Alice Stewart, "Childhood Cancers and Competing Causes of Death," *Leukemia Research* 19, no. 2 (1995): 103–11.

4. Another study has turned up a cancer effect in rural areas: Paula Cook-Mozaffari et al. have found that in several regions earmarked as suitable sites for nuclear power stations mortality rates for juvenile RES neoplasms were well above average. "Geographical Variation in Mortality from Leukemia and Other Cancers in England and Wales in Relation to Proximity to Nuclear Installations 1969–78," *British Journal of Cancer* 59 (1989): 476–85.

5. The Freedom of Information Act was passed in the United States in 1966, allowing citizens, journalists, and investigators to bring government information into the open. For the United Kingdom, see David Sumner et al., "The United Kingdom," Makhijani, Hu, and Yih, *Nuclear Wastelands,* 398; for the Official Secrets Act at Sellafield, McSorley 78.

6. See the sources listed in note 2 of this chapter; and E. G. Knox, Alice Stewart, George Kneale, and E. A. Gilman, "Background Radiation and Childhood Cancer," *Journal of the Society of Radiological Protection* 8 (1988): 9–18.

7. E. A. Gilman, George Kneale, E. G. Knox, and Alice Stewart, "Pregnancy X-Rays and Childhood Cancers: Effects of Exposure Age and Radiation Dose," *Journal of Radiological Protection* 8, no. 1 (1988): 3–9.

8. A 1994 overview of the current debate on cancer clusters, Valerie Beral et al., eds., *Childhood Cancer and Nuclear Installations* (London: BMJ, 1993) omits mention of the Oxford Survey of Background Radiation. In a review of this book for *British Medical Journal* 309 (9 July 1994), Alice wrote, "Beral, Roman, and Bobrow have done a public service by providing easy access to more than 50 papers, abstracts, letters, editorials, and reports published since 1984," but "conspicuous by its absence from the published papers and the commentary is any mention of the Oxford Survey. . . . No reason is given for not mentioning this earlier, nation-wide survey in an otherwise comprehensive account of an important debate" (378).

9. Personal communications, 5–20 February 1996.

10. Other researchers report clusters around these installations. E. Roman et al. report an increased incidence of childhood leukemia in the vicinity of the plants at Aldermaston and Burghfield. "Childhood Leukemia in the West Berkshire and Basingstoke and North Hampshire District Health Authorities in Relation to Nuclear Establishments in the Vicinity," *British Medical Journal* 294 (1987): 597–602; E. Roman et al., "Case-Control Study of Leukemia and non-Hodgkin's Lymphoma among Children aged Zero–Four Years Living in West

Berkshire and North Hampshire Health Districts," *British Medical Journal* 306 (1993): 615–21.

11. Sara Downs, "Hiroshima's Shadow over Sellafield," *New Scientist* (November 1993): 25–29.

12. Martin Gardner et al., "Results of Case-Control Study of Leukemia and Lymphoma among Young People near Sellafield Nuclear Plant in West Cumbria," "Methods and Basic Data of Case-Control Study of Leukemia and Lymphoma among Young People near Sellafield Nuclear Plant in West Cumbria," *British Medical Journal* 300 (17 February 1990): 423–28, and 429–34; Leslie Roberts, "British Radiation Study Throws Experts into Tizzy," *Science* 248 (4 April 1990): 25.

13. R. Doll, H. J. Evans, and Sarah Darby of the Imperial Cancer Research Fund at Oxford published "Paternal Exposure Not to Blame," concluding that "the association between parental irradiation and leukemia is largely or wholly a chance finding." "Commentary," *Nature* 367 (24 February 1994): 678–80.

14. Younger. Doll has recently reiterated this view; Steve Conner, "Virus Blamed for Sellafield Cancer," *London Times* 20 July 1997.

15. Also cited was the 1992 study by Doll and colleagues arguing that "current predicted risks of childhood leukemia after exposure to radiation are not greatly underestimated for low dose rate exposures"; J. H. Olsen et al. "Trends in Childhood Leukemia in the Nordic Countries in Relation to Fallout from Atmospheric Nuclear Weapons Testing," *British Medical Journal* 304, no. 6822 (18 April 1992): 1005–9.

16. Chris Busby, *Wings of Death: Nuclear Pollution and Human Health* (Aberystwyth: Green Audit Books, 1995), 124; Dick Raylor and David Wilkie, "Drawing the Line with Leukemia," *New Scientist* 21 (July 1988): 53–56. The quotation is on p. 56.

17. Rosie Waterhouse, "Families Lose Fight to Sue BNFL," *Independent* 10 May 1994, *Hotspots,* Roff, 290.

18. *New Scientist* (7 January 1988): 33; Brian Cathcart, Reuters North European Service, 28 January 1983; Caroline Dewhurst, "The Macabre Secret of Christmas Island," *Birmingham Sun* 18 May 1983; Barry Wigmore and Alan Rimmer, "Justice at Last?" *People* 10 December 1990.

19. Interview with Dr. Tom Sorahan, August 1996.

20. E. G. Knox, Tom Sorahan, and Alice Stewart, "Cancer following Nuclear Weapons Tests," letter, *Lancet* 2 (9 April 1983): 815; Paul Lashmar, "*Lancet* Letter Backs A-Test Cancer Theory," *Observer* 10 April 1983; "Doctor Tells of A-Bomb Cancer Link," *Observer* 13 June 1983.

21. E. G. Knox, Tom Sorahan, and Alice Stewart, "Cancer following Nuclear Weapons Tests," letter, *Lancet* 2 (8 October 1983): 856.

22. Steve Conner, "Risk and the Radioactive Service," *New Scientist* (4 February 1988): 30–31.

23. Paul Lashmar and Robin McKie, "A-Bomb Survey Row," *Observer* 2 October 1983.

24. S. C. Darby et al., "A Summary of Mortality and Incidence of Cancer in Men from the U.K. Who Participated in the U.K.'s Atmospheric Nuclear Weapons Test and Experimental Programmes," *British Medical Journal* 30, no. 296 (January 1988): 332–38.

25. S. C. Darby, G. M. Kendall, T. P. Fell, Richard R. Doll, A. A. Goodill, A. J. Conquest, D. A. Jackson, and R. G. T. Haylock, "Further Follow-Up of Mortality and Incidence of Cancer in Men from the U.K. Who Participated in the U.K.'s Atmospheric Nuclear Weapons Test and Experimental Programmes," *British Medical Journal* 307 (11 December 1993): 1530–35.

26. Tom Wilkie, "Bomb Tests 'Gave No Extra Cancer Risk,' " *Independent* 10 December 1993.

27. Conner 31; and "Nuclear Test Report Angers Ex-Soldiers," *Herald,* Glasgow, 10 December 1993.

28. Alice Stewart, "MSK Model for Cancer Induction Effects of Occupational Exposures to Ionizing Radiation: Method of Risk Estimation for Badge Monitored Workers," Department of Public Health and Epidemiology, University of Birmingham.

29. Alice Stewart, Testimony before the Colorado Workers Compensation Department Hearings, *Krumback v. Dow Chemical,* February 1980, quoted in Wasserman et al., *Killing Our Own,* 162; Pamela Avery, "Rocky Flats Cancer Death Blamed on Radiation," *Rocky Mountain News* 4 June 1981: 4.

30. Interview with Bruce DeBoskey, 19 January 1997.

31. Testimony before the Colorado Workers Compensation Department Hearings, *Krumback v. Dow Chemical,* August 1980, quoted in Wasserman et al., *Killing Our Own,* 162.

32. Tad Bartimus and Scott McCartney, *Trinity's Children: Living along America's Nuclear Highway* (New York: Harcourt, Brace Jovanovich, 1991), 195–204. Gabel is interviewed in *Dark Circle,* the powerful documentary film on Rocky Flats produced by Chris Beaver and Judy Irving, 1983.

33. G. S. Voelz et al., "An Update of Epidemiologic Studies of Plutonium Workers," *Health Physics* suppl. 1 (1983): 493–503.

34. Bartimus and McCartney 205.

35. In 1990, President Bush set up a $100 million fund to compensate victims of the nuclear weapons program, a measure thought to be "less expensive than paying injury and liability claims"; Keith Schneider, "U.S. Fund Set Up to Pay Civilians Injured by Atomic Arms Program," *New York Times* 16 October 1990. In 1992, Rep. David Skaggs introduced a "Defense Nuclear Workers Bill of Rights Act" that would cover health reinsurance and radiation exposure compensation. President Clinton's Secretary of Labor Robert Reich investigated the possibility of compensation for nuclear workers, but this came to nothing.

36. Interview with Dr. Wally Cummins, 16 March 1997; "The Hanford Veterans Cancer Mortality Study: A Brief Prepared for Senator Kerrey"; Janet Goetze, "Scientist to Study Camp Hanford Health Data," *Oregonian* 26 May 1998.

37. Jon Wiener, "Citizens Arrest D.O.E. and Rockwell," *Nation* 18 September 1995: 274–80.

Chapter 14

1. Major Greenwood, *Some British Pioneers of Social Medicine* (London: Oxford University Press, 1948), 61–82.

2. Abraham M. Lilienfeld and David E. Lilienfeld, *Foundations of Epidemiology* (New York: Oxford, 1980), 35–37; Thomas C. Timmreck, *An Introduction to Epidemiology* (Boston: Jones and Bartlett, 1994), 70–72, 417–43.

3. Alice Stewart, "The Grim Reaper's Trusted Scythe," review of *Plagues: Their Origin, History, and Future,* by Christopher Wills, and *Investigating Disease Patterns: The Science of Epidemiology,* by Paul D. Stolley and Tamar Lasky, *Times Higher* 19 July 1996.

4. Alice adds, "Statistical significance is what you are aiming for, a 95 percent chance of certainty—but absence of statistical significance is not necessarily refutation."

5. W. Court-Brown and R. Doll, "Leukemia and Aplastic Anemia in Patients Irradiated for Ankylosing Spondylitis," MRC Special Report No. 295, HMSO, London, 1957; P. G. Smith, R. Doll, and E. P. Radford "Cancer Mortality among Patients with Ankylosing Spondylitis Not Given X-Ray Therapy," *British Journal of Radiology* 50 (1977): 728–34; P. G. Smith and R. Doll, "Age and Time Dependent Changes in the Rates of Radiation Induced Cancers in Patients with Ankylosing Spondylitis following a Single Course of X-Ray Treatment." *Late Biological Effects of Ionizing Radiation* 1 (1978): 205–18, IAEA-SM-334/711, International Atomic Energy Agency, Vienna; P. G. Smith and R. Doll, "Mortality among Patients with Ankylosing Spondylitis after a Single Treatment Course with X-Rays," *British Medical Journal* 284 (1982): 449–60.

6. Personal communication, 3 February 1996.

7. *CCRI Newsletter* December–January 1994: 12–13.

8. T. M. Sorahan, R. J. Lancashire, Peck I. Hulten, and Alice Stewart, "Childhood Cancer and Parental Use of Tobacco: Deaths from 1953 to 1955," *British Journal of Cancer* 75 (1997): 134–38.

9. Nigel Hawkes, "Smoking Fathers May Sow Seeds of Childhood Cancer," *Times* 17 December 1996.

10. Interview with Tom Sorahan, August 1996.

11. See, for example, Shapiro 351; David A. Savitz, "Basic Concepts of Epidemiology," *Health Effects of Low-Level Radiation,* Hendee, 47–56, 54.

12. Steve Wing et al., "Mortality among Workers at Oak Ridge National Laboratory," *Journal of the American Medical Association* 265, no. 11 (20 March 1991): 1397–1402; Keith Schneider, "Study Links Cancer Deaths and Low Levels of Radiation," *New York Times* 20 March 1991.

13. William R. Hendee, "There's No Free Lunch: The Benefits and Risks of Technologies," *Journal of the American Medical Association* 265 no. 11 (20 March 1991): 1437–38.

14. Interview with Steve Wing, 17 December 1995.

15. Interview with David Richardson, 15 September 1996.

16. Interview with Rosalie Bertell, by Virginia Myers, February 1995.

17. Proctor 265–70.

18. U.S. Congress, Committee on Government Operations, *Occupational Illness and Injuries,* Washington, DC, 1989, 24–25; Proctor 327, n. 76.

19. Proctor 267.

20. Doll initially had reservations about a "recall bias" in the Oxford Survey,

but see R. Doll and R. Wakeford chapter 6, n. 11. Clive Cookson, "Unsung Heroine," *Financial Times* 19 January 1995.

21. Quoted in Ann Hornaday, "Straight Talk about Chernobyl," *Ms.* August 1986.

22. del Tredeci 140.

23. Keller 197–207; and "A World of Difference," *Reflections on Gender and Science* (1985; New Haven: Yale University Press, 1995), 158–76.

24. Interview with Jill Boniske and Susan Robinson, *A Second Opinion: Alice Stewart on Low-Level Radiation,* documentary forthcoming in 1999.

25. Interview with Morris Greenberg, by Virginia Myers, 30 July 1997. See also Morris Greenberg, "The Evolution of Attitudes to the Human Hazards of Ionizing Radiation and to Its Investigators," *American Journal of Industrial Medicine* 20 (1991): 717–21.

26. Personal communication, April 1997.

27. Interview with Karl Morgan, 16 December 1995.

28. "Sex and the Scientist: Our Brilliant Careers," Channel 4 television interview, aired 19 August 1996.

29. Interview with John Gofman (with Dr. Vicki Ratner), 14 April 1994.

Chapter 15

1. Alice Stewart and George Kneale, "Pre-cancers and Liability to Other Diseases."

2. The first paper to suggest that the increase in leukemia was a direct consequence of the falling death rate from pneumonia and other infections was Alice Stewart, "Aetiology of Childhood Malignancies, Congenitally Determined Leukemias," *British Medical Journal* 1 (1961): 452–60. In 1969, Alice argued that "the most likely explanation of the rise in leukemia was that a reversion towards the normality of death rate was occurring from sulphonamides and from the widespread use of antibiotics"; "The Pitfalls of Extrapolation," *New Scientist* (July 1969): 181.

3. David Hewitt, "Some Features of Leukemia Mortality," *British Journal of Preventative Social Medicine* 9 (1955): 81.

4. Kneale, "Excess Sensitivity of Pre-leukemics to Pneumonia."

5. Since there are three recessive genes to one dominant, the likelihood of recessive gene damage is three to one; recessive genes require two pairs to meet before the characteristic shows.

6. Abraham Bergman, *The 'Discovery' of Sudden Infant Death Syndrome: Lessons in the Practice of Political Medicine* (Seattle: University of Washington Press, 1986), xi, 8–10.

7. D. J. Weatherall et al., "Hemoglobin and Red Cell Enzyme Changes in Juvenile Myeloid Leukemia," *British Medical Journal* 1 (1968): 679; G. G. Guilian et al., "Elevated Fetal Hemoglobin Levels in Sudden Infant Death Syndrome," *New England Journal of Medicine* 316 (1987): 1122; D. G. Fagan and A. Walker, "Hemoglobin F Levels in Sudden Infant Deaths," *British Journal of Haematology* 82 (1992): 422. See also Holcomb B. Noble, "Clues in Sudden Infant Death Syndrome," *New York Times* 24 November 1998.

8. Abraham Bergman et al., *Sudden Infant Death Syndrome, Proceedings of the Second International Conference on the Causes of SIDS* (Seattle: University of Washington Press, 1970), 13, citing J. B. Beckwith, "Observations on the Pathological Anatomy of SIDS."

9. Alice Stewart, "Infant Leukemias and Cot Deaths," *British Medical Journal* 11 (1975): 605–7; "Cot Deaths, Immune Deficiencies and Leukemia," *Journal of Preventive Social Medicine* 29 (1975): 63; "Factors Affecting the Recognition of Childhood Cancers: Respiratory Infections, Cot Deaths, and Season of Birth," *Pediatrics Digest* (January 1977): 9–20; "Factors Affecting the Recognition of Childhood Cancers: Respiratory Infections, Cot Deaths and Season of Birth," *Pediatrics Digest,* 19, no. 1 (1979): 9–20; "Recent Theories on the Cause of Cot Deaths," *British Journal of Medicine* 296 (1988): 358; "Sudden Infant Death Syndrome: Faulty Maturation of Haemoglobin and Immunoglobulins," *British Medical Journal* 9, no. 2 (1989): 521–22.

10. Alice Stewart and George Kneale, "The Immune System and Cancers of Foetal Origin," *Cancer Immunology Immunotherapy* 14 (1982): 110–16; George Kneale, Alice Stewart, and Kinnear Wilson, "Immunisations against Infectious Diseases and Childhood Cancers," *Cancer Immunology Immunotherapy* 21 (1986): 129–32.

11. D. Dalldorf et al., "The Malignant Lymphomas of African Children: Burkitt's Lymphoma, Lymphoblastic Leukemia and Malaria," Fourteenth International Congress of Hematology, San Paulo, July 1972, 16–21; Alice Stewart et al., "Malignant Lymphomas of African Children," *Proceedings of the National Academy of Science,* USA 70, no. 1 (1973): 15–17.

12. B. Ramot et al., "Observations on Lymphatic Malignancies in Israel," *Pathogenesis of Leukaemias and Lymphomas: Environmental Influences,* ed. I. T. Magrath et al. (New York: Raven Press, 1984).

13. Alice Stewart, "Childhood Cancers and Competing Causes of Death," *Leukemia Research* 19, no. 2 (1995): 103–11.

14. See n. 2 in this chapter for the association of Down's syndrome with leukemia.

Chapter 16

1. Sheldon Samuels, Ramazzini awards ceremony, Carpi, Italy, November 1991; thanks to Boniske and Robinson, who conducted the interview for the documentary cited in note 24, chap. 14.

2. Interview with Rosalie Bertell (with Dr. Vicki Ratner), 30 April 1994.

3. Interview with Steve Wing, 17 December 1995. See Helen Longino, "Can There Be a Feminist Science?" *Feminism and Science,* ed. Nancy Tuana (Bloomington: Indiana University Press, 1989), 45–57, on incentives for conformity in science.

4. Personal communication, 30 October 1995.

5. Personal communication, 25 October 1995.

6. Evelyn Fox Keller, "The Gender/Science System: or, Is Sex to Gender as Nature Is to Science?" *Feminism and Science,* Tuana, 33–57.

7. See Lynn Bindman et al., eds. *Women Physiologists* (London: Portland

Press, 1993); and Shils and Blacker, *Cambridge Women,* 150–51, on scarcity of funding and space for women scientists.

8. Cf. Ruth Hubbard, "Thus science is made, by and large, by a self-perpetuating, self-reflexive group: by the chosen for the chosen"; "Science, Facts, Feminism," *Feminism and Science,* Tuana, 119–31. The quotation is on p. 120.

9. Hilary Rose, *Love, Power, and Knowledge: Towards a Feminist Transformation of the Sciences* (Cambridge, UK: Policy Press, 1994), 15.

10. Ibid. 14.

11. "A lot of women want to be honorary men. Margaret Thatcher is an honorary man. She hasn't lifted a finger for women." Christy, cited in my chapter 10, n. 40.

12. Personal communication, 30 October 1995. Sarah Darby copublishes regularly with Richard Doll and works with him at the Imperial Cancer Research Fund; Valerie Beral, director of the Imperial Cancer Research Fund's cancer epidemiology unit at Radcliffe Infirmary, Oxford, also works with him. Gilbert copublishes with Sidney Marks and has worked closely with him at Batelle. One scientist refers to these women, along with Shirley Fry, as "carefully groomed epidemiologists."

13. Personal communication, 23 January 1996; 5–20 February 1996.

14. Hilary Rose notes that "interventionism, so central to modern science, was written into the foundational texts of the new masculine knowledge" but that the observational sciences are less interventionist than the laboratory sciences (86–88, 99).

15. Keller, my chapter 14, n. 23, pp. 158–76.

16. Wing, "Sex and the Scientist," my chapter 14, n. 28.

17. Dr. John Bithell, statistician on the Oxford Survey: "The lateral thinking side of science, the creative, imaginative part of it, is so important. She would make connections other people didn't think to make, and that's where her contribution is most evident." Ibid.

18. Margaret A. Boden, *The Creative Mind: Myths and Mechanisms* (New York: Basic Books, 1990); Scott G. Isaksen, ed., *Frontiers of Creativity Research: Beyond the Basics* (New York: Bearly Limited, 1987); Robert W. Weisberg, *Creativity: Beyond the Myth of Genius* (New York: W. H. Freeman, 1993).

19. The metaphor has resonances for Hilary Rose, who describes the feminist scientist's task as not "hold[ing] up a mirror to either the social or the natural world. Her task is to go behind the mirror, to go behind the appearance of things," as Alice went through the looking glass (27). One aspect of the "mirror" that needs subverting is "the deeply rooted popular mythology that casts objectivity, reason, and mind as male, and subjectivity, feeling, and nature as female"—that is, the central assumption of the Baconian worldview that casts the scientist/knower as masculine and nature as female, according to which model, "understanding and competence . . . come to have value mainly to the extent that they serve to promote mastery or dominance" (Keller 6–7, 122–23). Alice's contributions to epidemiology as an observational science, her combination of analytical brilliance and intuitive leaps, may be seen as challenges to this dichotomy.

20. "Sex and the Scientist."

21. She notes that the number of x-rays a person has accumulated in his or her

life is not kept track of, so there is little attention to their long-term, cumulative effects—though what physicians do attend to, especially in the United States, is the ever-present threat of litigation, which may make them use x-rays more often than is medically necessary. She says of mammograms, "I'd no sooner have one than I'd x-ray a fetus."

22. On the advantages of a solar-hydrogen economy, see Tom Athanasiou, *Divided Planet: The Ecology of Rich and Poor* (New York: Little Brown, 1996), 246–47. Athanasiou cites Joan M. Ogden and Robert H. Williams, *Solar Hydrogen: Moving beyond Fossil Fuels* (Washington, DC: World Resources Institute, 1989) as the standard reference; and *Fossil Fuels in a Changing Climate: How to Protect the World's Climate by Ending the Use of Coal, Oil, and Gas* (Amsterdam: Greenpeace, 1993).

23. Ralph Moss, *The Cancer Industry* (New York: Paragon House, 1989), 429–30.

24. "Sex and the Scientist."

25. Ibid.

Chapter 17

1. Besides Wing, there is Ethel S. Gilbert and Shirley A. Fry et al., "Analyses of Combined Mortality Data on Workers at the Hanford Site, Oak Ridge, and Rocky Flats," *Radiation Research* 120 (1989): 19–35. Several researchers besides Morgan support Alice's view of a "supralinear" relationship of dose and response, that is, a greater cancer-causing effect per unit dose for low-dose than high-dose exposures. John Gofman, *Radiation-Induced Cancer from Low-Dose Exposure* (San Francisco: Committee for Nuclear Responsibility, 1990); idem, *Radiation and Human Health* (New York: Pantheon, 1983); R. H. Nussbaum, "New Data Inconsistent with Scientific Concerns on Low Level Radiation Cancer Risks," *Health Physics* 56 (1989): 961; idem, "The Linear No-Threshold Dose Effect Relation: Is It Relevant to Radiation Protection Regulation?" *Medical Physics* 25, no. 3 (March 1998): 291–99; and B. Modan, "Cancer and Leukemia Risks after Low Level Radiation—Controversy, Facts and Future," *Medical Oncology and Tumor Pharmacotherapy* 4 (1987): 452.

2. *Hiroshima and Nagasaki: The Physical, Medical and Social Effects of the Atomic Bombings*, "The Committee for the Compilation of Materials on Damage Caused by the Atomic Bombs in Hiroshima and Nagasaki," trans. Eisei Ishikawa and David L. Swain (New York: Basic Books, 1981).

3. I. Shigematsu et al., *Effects of A-Bomb Radiation on the Human Body* (Tokyo: Harwood Academic Publishers, 1995); and Y. Shimizu et al., "Cancer Risk among Atomic Bomb Survivors: The RERF Life Span Study," *Journal of the American Medical Association* 264 (1990): 601–4.

4. Interview with Sir Richard Doll, 9 October 1998.

5. Quoted in Rob Edwards, "Radiation Roulette," *New Scientist* (11 October 1997): 36–40.

6. B. I. Lord et al., "Tumour Induction by Methyl-Nitrose-Urea following Preconceptual Paternal Contamination with Plutonium-239," *British Journal of*

Geiger, H. Jack et al., eds. *Dead Reckoning: A Critical Review of the Department of Energy's Epidemiologic Research.* Physicians for Social Responsibility, Physicians Task Force on the Health Risks of Nuclear Weapons Production, 1992.
Gerson, Joseph. *With Hiroshima Eyes: Atomic War, Nuclear Extortion and Moral Imagination.* Philadelphia: New Society Publishers, 1995.
Gofman, John. *Radiation and Human Health.* New York: Pantheon, 1983.
———. *Radiation-Induced Cancer from Low-Dose Exposure.* San Francisco: Committee for Nuclear Responsibility, 1990.
Gofman, John, and Arthur R. Tamplin. *Poisoned Power: The Case against Nuclear Power Plants before and after Three Mile Island.* Emmaus, PA: Rodale Press, 1971.
Grossman, Karl. *Cover-up: What You* Are Not *Supposed to Know about Nuclear Power.* Sagaponack, NY: Permanent Press, 1980.
———. *Power Crazy.* New York: Grove Press, 1986.
Gyorgy, Anna and friends. *No Nukes: Everyone's Guide to Nuclear Power.* Boston: South End Press, 1979.
Hendee, William R., ed. *Health Effects of Low-Level Radiation.* Norwalk, CT: Appleton-Century Crofts, 1984.
Hilgartner, Stephen, Richard Bell, and Rory O'Connor. *Nukespeak: The Selling of Nuclear Technology in America.* San Francisco: Sierra Club Books, 1982.
Keller, Evelyn Fox. *Reflections on Gender and Science.* 1985. New Haven: Yale University Press, 1995.
———. *A Feeling for the Organism.* New York: W. H. Freeman, 1983.
Kohn, Howard. *Who Killed Karen Silkwood?* New York: Summit, 1981.
Leff, S. L. *Social Medicine.* London: Routledge and Kegan Paul, 1953.
Lifton, Jay, Robert Mitchell, and Greg Mitchell. *Hiroshima in America: A Half Century of Denial.* New York: Avon, 1995.
Lilienfeld, Abraham M., and David E. Lilienfeld. *Foundations of Epidemiology.* New York: Oxford, 1980.
Lindee, M. Susan. *Suffering Made Real: American Science and the Survivors of Hiroshima.* Chicago: University of Chicago Press, 1994.
Makhijani, Arjun, Howard Hu, and Katherine Yih, eds. *Nuclear Wastelands: A Global Guide to Nuclear Weapons Production and Its Health and Environmental Effects.* Special Commission of International Physicians for the Prevention of Nuclear War and the Institute for Energy and Environmental Research. Cambridge, MA: MIT Press, 1995.
McKewon, Thomas. *The Role of Medicine: Dream, Mirage, or Nemesis.* Oxford: Blackwell, 1979.
McSorley, Jean. *Living in the Shadow: The Story of the People of Sellafield.* London: Pan Books, 1990.
McWilliams-Tullberg, Rita. *Women at Cambridge: A Men's University—Though of a Mixed Type.* London: Gollancz, 1975.
Moss, Ralph. *The Cancer Industry.* New York: Paragon House, 1989.
Nader, Ralph. *The Menace of Atomic Energy.* 1977. New York: Norton, 1979.
Proctor, Robert. *Cancer Wars: How Politics Shapes What We Know and Don't Know about Cancer.* New York: Simon and Schuster, 1986.

Rhodes, Richard. *The Making of the Atomic Bomb.* New York: Simon and Schuster, 1986.

Roff, Sue Rabbit. *Hotspots: The Legacy of Hiroshima and Nagasaki.* London: Cassell, 1995.

Rose, Hilary. *Love, Power and Knowledge: Towards a Feminist Transformation of the Sciences.* Cambridge: Policy Press, 1994.

Rudig, Wolfgang. *Anti-nuclear Movements: A World Survey of Opposition to Nuclear Energy.* Harlow, UK: Longman Group, 1990.

Ryle, John A. *Changing Disciplines: Lectures on the History, Method and Motives of Social Pathology.* London: Oxford University Press, 1948.

Shapiro, Jacob. *Radiation Protection: A Guide for Scientists and Physicians.* Cambridge, MA: Harvard University Press, 1990.

Shigematsu, I. et al. *Effects of A-Bomb Radiation on the Human Body.* Tokyo: Harwood Academic, 1995.

Shils, Edward, and Carmen Blacker, eds. *Cambridge Women: Twelve Portraits.* Cambridge: Cambridge University Press, 1996.

Snow, C. P. *The Search.* New York: Charles Scribner's Sons, 1934.

Stauber, John, and Sheldon Rampton. *Toxic Sludge Is Good for You: Lies, Damn Lies and the Public Relations Industry.* Monroe, ME: Common Courage Press, 1995.

Sternglass, Ernest. *Secret Fallout: Low Level Radiation from Hiroshima to Three Mile Island.* New York: McGraw-Hill, 1981.

Stoler, Peter. *Decline and Fail: The Ailing Nuclear Power Industry.* New York: Dodd, Mead, 1985.

Timmreck, Thomas C. *An Introduction to Epidemiology.* Boston: Jones and Bartless, 1994.

Totten, Sam, and Martha Wescoat Totten. *Facing the Danger: Interviews with Twenty Anti-Nuclear Activists.* Trumansburg, NY: Crossing Press, 1984.

Tuana, Nancy, ed. *Feminism and Science.* Bloomington: Indiana University Press, 1989.

Wasserman, Harvey and Norman Solomon with Robert Alvarez and Eleanor Walters. *Killing Our Own: The Disaster of America's Experience with Atomic Radiation.* New York: Dell, 1982

Index

References to Alice Stewart in subheads are indicated by "AS." Family relationships identified parenthetically are in reference to Alice Stewart.

Abalone Alliance movement, 154
ABCC. *See* Atomic Bomb Casualty Commission (ABCC)
acne, incidence of, 73
AEC. *See* Atomic Energy Commission (AEC), U.S.
age: and cancer, 238; and effects of radiation, 121, 127, 139, 191, 192, 204, 222, 238, 268; and survival, 235
Agency for Toxic Substances and Disease Registry, 188
Aldermaston weapons research facility: anti-nuclear marches to, 86, 87; cancer cluster at, 197
Alice in Wonderland (Carroll), 7
Allington, Lavinia, 54
Allis-Smith, Janine, 174–75, 176
alpha particles, 44
alternative energy sources, 165
Alvarez, Robert ("Bob"), 169, 223; appointment to DOE, 145; congressional investigation of Mancuso dismissal instigated by, 150; at EPI, 123; on Hanford contamination, 184; interest in nuclear workers' records, 182; support for AS's work, 149, 219, 267
American Journal of Industrial Medicine, 191
Anderson, Elizabeth Garrett, 3, 21
Anglesey (Wales), 31–32, 33, 36–37
ankylosing spondylitis, 80, 88, 216–18
anoxia, 241, 242
Anti-Ballistic Missile (ABM) system, 156
antibiotics, effect of, 92, 233, 234, 246
anti-Communism, 87
antigas protection, 58–59
anti-nuclear movement: development of, 1, 11–12, 13, 86–88, 152–54; following Three Mile Island accident, 178; scientists as expert witnesses in, 154–58; women in, 262–63
aplastic anemia, 137, 141
arms race, 8, 86
Armstrong, John, 202
asbestos exposure, dangers of, 278n. 1
Association of Physicians, 59 Atomic Bomb Casualty Commission (ABCC): creation of, 130, 131; criticism of, 133–35; criticism of Sternglass's study by, 156; study of Hiroshima survivors by, 80, 116, 130, 131–43. *See also* Hiroshima, study of survivors
Atomic Energy Authority, U.K., 168, 169
Atomic Energy Commission (AEC), U.S.: bomb shelter instructions from, 86; criticism of, 87; nuclear R&D by, 12; pro-nuclear publicity from, 8; study of Hiroshima survivors funded by, 135, 142; study of nuclear weapons workers' safety funded by, 113, 114; ties to RERF, 9

atomic physics, Cambridge research in, 43–44
Atoms for Peace campaign, 152
Auden, W. H., 3, 42
autonomic nervous system, 239

background radiation: AS's study of, 194–97; effects of, 171, 181, 193, 213, 234, 263. *See also* cancer clusters, existence of
badge monitoring of workers, 120, 121
Ban, Ida, 226
Barber, Renate, 95
Bartlett, George, 166
Battelle-Pacific Northwest Laboratories, 147, 152
Bell, Julian, 42
Bellamy, David, 168
Ben Gurion University (Israel), 1996 conference at, 146, 268
Beral, Valerie, 256, 292n. 8
Berger, Daniel, 189
Bertell, Dr. Rosalie, 123, 144, 158–60, 167–68, 223, 248, 251
beta particles, 44
Billingham nuclear waste project, 168
Birmingham University: AS as professor at, 227; AS as senior honorary research fellow at, 99–100, 101–2, 113
birth defects, 87, 133, 238
birth notifications, 81
Bithell, Dr. John, 298n. 17
Black, Sir Douglas, 165
Blackmur, R. P., 106
Blackwell, Elizabeth, 21
blast cells, 238
Bloomsbury Group, 41
Blunt, Anthony, 42
BNFL. *See* British Nuclear Fuels (BNFL)
Bohr, Neils, 43
bomb shelters, construction of, 86
bone cancer, 240
bone marrow, damage to, 136–38, 141
Bowden, Ruth, 54
brain cancer, 206, 239
breast feeding, 30–31
Bristol, flood data from, 133, 140
Bristol University, honorary doctorate from, 268
British Atomic Test Veterans Association, 202
British College of Physicians, 2, 268–69
British Journal of Industrial Medicine, 59
British Journal of Social Medicine, 226
British Kennel Club, 22
British Medical Journal, 85, 88
British Nuclear Fuels (BNFL): compensations cases against, 197–99; nuclear facilities of, 163, 165, 168, 169
British Paediatric Association, 32
British Society of Social Medicine, 107–8
Bronowski, Jacob, 42
Brookhaven facility (NY), 12
Brooks, Cleanth, 106
Bross, Dr. Irwin, 90, 123, 158, 159, 161
Bulletin of Atomic Scientists, 125, 148, 155, 156
Burchett, Wilfred, 130
Bureau of Radiological Health, 96, 115
Burkitt's lymphoma, 232, 244–45
Busby, Chris, 199
Bush, George, 294n. 35
Byron, Lord, 33, 107

Califano, Joseph, 149
Cambridge University: AS as student at, 3, 39–45, 249; AS snubbed for job at, 60; medical education for women at, 21; women's admittance to, 39, 40–41
Campaign for Nuclear Disarmament (CND), 86, 166, 173
cancer: description of, 236–37; la-

tency period of, 115, 117, 193, 220–21. *See also specific cancers by name*
cancer clusters, existence of, 171, 181, 193–97, 269. *See also* background radiation
carbon tetrachloride poisoning, 58–59
cardiovascular disease, 78
Carson, Rachel, 159, 280n. 22
Carter, Jimmy, 152
Cartier-Bresson, Henri, 42
Cavendish Laboratory (Cambridge), 43–44
CDC. *See* Centers for Disease Control (CDC)
censorship, 10, 130, 135, 157, 158–61, 187, 250
Centers for Disease Control (CDC), 152, 226; National Center for Environmental Health, 188
Central Electricity Generating Board (CEGB), 166, 172, 174
Chadwick, James, 43
Chambers Encyclopedia, 20, 34
Changing Disciplines (Ryle), 68
Channel 4 documentary (British television), 16, 230, 264
chemical carcinogens, 234
chemical warfare, 180
Chernobyl nuclear accidents, 159, 166, 179
Child Health Survey, 71, 75
childhood cancer: AS's research about, 8, 70, 78–93 passim, 218–20, 232; and background radiation, 181; and delayed congenital defects, 238–40; Doll's research about, 102–3; geographical distribution of, 194; risks of contracting, 1, 2, 79, 83, 85–86, 89–90, 114
Childhood Cancer Research Institute (CCRI), 180, 182, 185, 243
children's welfare issues, 3, 19, 23–25
China Syndrome, The (film), 150
chloromatous myeloid leukemia, 245
cholera, etiology of, 23–24, 69, 214
Christmas Islands, nuclear testing at, 11, 155, 199–202
"Citizen Atom," 11
Civilian Medical Board records, 67, 72–73
Clamshell Alliance, 154
Clark, Sir Wilfred Legros, 107
climate effects, 195
clinical medicine, 2, 45–46, 51, 52, 69, 225
coal mining, worker safety standards in, 59–60
Cobb, Dr. John, 203–4
colitis, 218
Colleges of Physicians and Surgeons of England, 21
Columbia River, pollution in, 113
Committee for Nuclear Responsibility, 157
Committee on Medical Aspects of Radiation in the Environment (COMARE), 269
Committee to Bridge the Gap, 209, 268
compensation claim cases. *See* insurance
Conference on Low Level Radiation (Bristol, 1997), 269
Congress, U.S.: investigation of Mancuso's dismissal, 150–52, 153, 161; investigation of nuclear weapons facilities, 186
Consortium of Local Authorities (COLA), 172
control groups, 79–80, 81
Cooke, Alistair, 42
Cook-Mozaffari, Paula, 292n. 4
Copeland Borough Council, 164
Council for Energy Awareness, 272n. 23
Court-Brown, William, 80, 88, 89, 217
crib death. *See* Sudden Infant Death Syndrome (SIDS)
Crick, Francis, 16
Cumbrians Opposed to a Radioactive Environment (CORE), 174, 175
Cummins, Dr. Wally, 208

Daily Mirror (London), 164
Darby, Dr. Sarah, 173, 201–2, 249, 256
Darwin, Robert, 42
Dead Reckoning (Physicians for Social Responsibility), 187
DeBoskey, Bruce, 161, 204, 205–7
Denver (CO), cancer rates in, 291–92n. 1
Department of Energy (DOE), U.S.: compensation cases against, 202–8; criticism of, 1, 2, 179–81; development of, 12, 117; waste management and environmental restoration programs of, 272–73n. 25; workers' health studies contracted by, 147, 152, 186–87; workers' records controlled by, 186, 187–88
Department of Health and Human Services (HHS), U.S., 15, 149, 152, 180, 186, 187–88
Department of Veterans Affairs (VA), U.S., 208–9
Derwen cottage (Anglesey), 37
Diablo Canyon (CA) nuclear reactor, 154
disarmament. *See* anti-nuclear movement
disaster effects, 133, 140
DNA, damage of, 237
DOE. *See* Department of Energy (DOE), U.S.
dogmatism, criticism of, 142
Doll, Sir Richard: AS's criticism of, 216–18; on AS's work, 226–27, 230, 257, 266; ankylosing spondylitis research by, 80, 88, 89, 216–18; appointment at Oxford, 98–101; atomic test veterans research by, 201–2; as expert witness, 173, 249; lung cancer research by, 7, 78, 100–101, 220, 224; rebuttal of Gardner's study by, 198–99; women mentored by, 298n. 12
dose fractionation, 124–25
Dose Rate Effect of Radiation (DREF), 125
Dounreay nuclear facilities: cancer cluster at, 171, 193, 195–96, 197; investigation of, 169–72
Dow Chemical, 204
Down's syndrome, 246
Dunster, John, 163, 169
Du Pont Nemours, 10

Eagleton, Terry, 106
ear diseases, 67
eczema, incidence of, 73
Eisenhower, Dwight D., 11
Eliot, T. S., 42, 106
Elizabeth Garrett Anderson Hospital (London), 52–53, 71
Emery, John, 27, 32, 33, 34
Empson, Hester Henrietta Crouse ("Hetta"), 103, 105, 107, 108
Empson, Jacob, 105
Empson, Mogador, 4, 62, 105, 1087
Empson, Sir William: AS's affair with, 3–4, 7, 42, 47–48, 103–10; death of, 109–10; health of, 109; and literary criticism, 3, 47, 105–6
Energy Research and Development Administration (ERDA): critique of Mancuso/Stewart/Kneale research by, 122–23, 126–27, 162, 281n. 27; studies funded by, 12, 117, 122
English, S. E., 286n. 13
environmental movement, 148–49
Environmental Policy Institute (EPI), 123, 149, 180, 182
Environmental Protection Agency (EPA), 179
epidemiology: AS's interest in, 56, 58, 70–73; description of, 213; development of, 15, 23, 67–70, 74–75, 224–25; lacking in ICRP membership, 144; proper use of, 214–18; teamwork in, 223–24
ERDA. *See* Energy Research and Development Administration (ERDA)
Esquire, 156
external radiation (contamination), 120

Farr, William, 24, 214
Faulkner, Joan, 84
Fawler cottage (near Oxford), 4–5, 62, 268
Federal Privacy Act, U.S., 182
Federation of Atomic Scientists, 155
Feed Materials Production Center (Fernald, OH): class-action suit against, 183; closing of, 180; disregard of safety standards at, 179, 180, 183; workers' records released, 189, 190
Fell, Honor, 260–61
feminism, 7
feminist science, 257
Fernald (OH) nuclear facility. *See* Feed Materials Production Center (Fernald, OH)
fetal x-ray exposure, dangers of: AS's study of, 1, 2, 7, 8, 70, 83–86, 114, 196, 218–20, 232; acceptance of, 90–91; criticism of AS's research, 8, 88–90, 93; other studies of, 88–90
Financial Times, 264
Firth's Steel Works (Sheffield), 28
Firvale Hospital (Sheffield), 26
Firvale Workhouse (Sheffield), 25
FOIA. *See* Freedom of Information Act (FOIA), U.S.
Forster, E. M., 42
Francis, Dick, 54
Freedom of Information Act (FOIA), U.S.: access to DOE files under, 151, 187–90; access to ERDA critiques under, 126; access to workers' data under, 127, 182; passage of, 292n. 5
French, Barbara, 175
Friends of the Earth (FOE), 153, 164, 165, 166, 168, 173
Fry, Dr. Shirley, 182, 285n. 3, 298n. 12

Gabel, Don, 206–7
gamma particles, 44
Gardner, Martin, 198–99, 219, 267
Geiger, Dr. Jack, 180, 187, 188, 189, 229
gender issues, 7, 254–59
genetic damage, danger of, 14, 87, 140, 143, 157, 237, 238, 263, 267
Germany, anti-nuclear movement in, 86–87, 153–54, 300n. 8
Gilbert, Dr. Ethel, 164, 173, 184, 189–90, 205, 249, 256
Girton College, 39, 40, 41, 42. *See also* Cambridge University
Glenn, John, 180, 187, 188, 189
global warming, 13
Gofman, Dr. John, 123, 145, 156–58, 231
Goldsmith, Dr. John, 230, 268, 280n. 16
Gorleben (Germany) anti-nuclear demonstration, 154, 178, 253
Gould, Jay, 158
Government Code and Cypher School (Bletchley), 53
Great Ormond Street Hospital (London), 22
Green, Dr. (funding source), 81
Green, Patrick A., 145, 284n. 35
Greenberg, Dr. Morris, 143, 230
Greenham Common protest, 12, 166, 167–68, 269
Greenpeace, 168
Grossman, Karl, 11
Groves, Gen. Leslie, 130

Haffenden, John, 107, 108
Hall, Dr. Arthur, 26, 27–28
Hall, John, 200
Hanford Education Action League (HEAL), 180, 185
Hanford (WA) weapons complex: AS's review of Mancuso's study, 8–9, 102, 118–22, 123, 151; AS's study of, 13, 170, 178–79, 181–92, 222; Battelle-Pacific Northwest Laboratories' study of health effects at, 147, 152; closing of, 180; disregard of safety standards at, 179,

Hanford (WA) weapons complex: (*continued*) 184–85; Mancuso barred from releasing results of study at, 122–24; Mancuso's study of health effects at, 12, 113, 114–18, 162; plutonium manufacturing at, 113–14; Wing's study of, 228; workers' records released, 189, 190
Harrow, 49–50, 53, 61
Harvard University School of Public Health, 89
health physics, 90
Health Physics (journal), 123
Health Physics Society, 151
healthy survivor effect, 136
healthy worker effect, 121, 187, 281n. 33
heart disease, 67, 143
Heasman, Dr. Michael, 170
hernia, incidence of, 73
Herrington, John, 179
Hewitt, David, 78, 79, 80, 82, 97, 234, 240
Hijayama Hill research center (Japan), 135
Hildyard, Nicholas, 164–65
Hill, Sir Austin Bradford, 78, 84, 88, 101, 220
Himsworth, Sir Harold, 84
Hinkley Point nuclear facility, 172–74, 193
Hiroshima: bombing of, 128–29; commemoration of, 252; data on multiple acute injuries, 145–46, 268, 269; difficulty of data collection about, 134–36; study of survivors, 5, 9, 15–16, 80, 88, 92–93, 103, 116, 129–43, 181, 196, 250, 266; and theory of "silent forces," 138–40
Hodgkins, Dorothy, 226
Hodgkin's lymphoma, 240
Holroyd, Michael, 41
Hope, Vivien, 197
Horner, Arthur, 60
Hospital for Sick Children (London), 24
House Un-American Activities Committee (HUAC), U.S., 87
Hueper, Dr. Wilhelm, 116

Idaho National Engineering Laboratory, 185
immune system: boosting of, 244; functioning of, 235–36, 239; suppression of, 92, 136–38, 141, 143, 195, 232, 233–38, 241–42, 246
Imperial Chemical Industry (ICI), 168
Impressions of Factory Life (anthology), 58
industrial medicine, development of, 59, 70–71, 95, 224
industrial pollution, 7, 116, 179
Infant Life Protection Society, 24
infant mortality: from leukemia, 79; and nuclear accidents, 177; and nuclear testing, 156; rates of, 24, 25, 67, 68, 75, 114
Infant Welfare Clinic (Sheffield), 24–25, 29
infant welfare movement, 24–25
infections, susceptibility to. *See* immune system, suppression of
inoculations as cancer inhibitors, 181, 232, 244
insurance: expert witnesses in compensation claim cases, 161, 197–210; fear of liability and compensation claims, 9, 10, 124, 130, 151, 187, 207; setting standards for compensation, 267–68
intellectual property rights, 182
internal radiation (contamination), 120
International Association of Epidemiology, 226
International Association of Machinists and Aerospace Workers, 149
International Commission on Radiation Protection (ICRP), 9, 122, 125, 132, 143–45, 194
International Institute of Concern for Public Health, 159

international nuclear regulatory agencies, criticism of, 1, 9, 128, 143–46
International Physicians for the Prevention of Nuclear War, 286n. 23
International Radiobiological Society meeting (Pisa, 1986), 194
International Workshop on Radiation Exposures by Nuclear Facilities (Portsmouth, 1996), 269
interventionism, 298n. 14
in utero mutations, 234–35, 238. *See also* genetic damage, danger of
Irish Sea, pollution of, 163

Jablon, S., 93
Johnson, Dr. Carl, 123, 160–61, 185
Jones, Eleanor Davis, 54
Journal of Radiological Research, 285n. 45
Journal of the National Cancer Institute, 89

Kato, H., 93
Keller, Evelyn Fox, 7, 229, 257
Kemeny Commission, 177–78
Kendall, Dr. Henry, 157
Kerr-McGee plutonium processing plant (Crescent, OK), 149–50
Keynes, Geoffrey, 3, 41
Keynes, John Maynard, 3
Kinlin, Dr. Leo J., 199
Kleeman, David, 180
Kneale, George W.: on Oxford Survey of Childhood Cancer, 92, 97–98, 234, 262, 269; review of Mancuso's Hanford study by, 8–9, 102, 118–22, 123; as statistician, 223–24; study of cancer clusters by, 196–97; study of nuclear weapons workers' records by, 13, 148, 189–92, 222; testimony at Windscale hearing, 164; on Three Mile Island Public Health Fund's study, 178
Knox, George, 200, 227
K-25 nuclear plant (Oak Ridge, TN), 189
Krebs, Hans, 107
Krumback, Florence, 203, 204
Krumback, Leroy, 203, 204

Lady Tata Memorial Fund for Leukemia Research, 81
Lancet, 30, 84, 93, 170, 201, 279n. 20
Lashoff, Dr. Joyce, 2
lateral thinking, 258
Lawrence Livermore Laboratory (CA), 12, 156
Leavis, F. R., 105, 106
leukemia: AS's study of, 78–79, 214, 233–34; among British children, 165–66, 173, 174–75, 198; detection of, 244; etiology of, 136, 239, 240; geographical distribution of, 195; among Hiroshima survivors, 122; latent period for, 79; risks of, 89–90, 159, 276n. 1; among Scottish (Dounreay) children, 170, 171; in twins, 243; types of, 79, 92
London Hospital, 70–71
London School of Hygiene and Tropical Medicine, 49, 50, 51
London School of Medicine for Women, 21
Long Island Lighting Company, 153
Los Alamos (NM) National Laboratory: A-bomb construction at, 113; DOE's study of health effects at, 147; Mancuso's study of health effects at, 13, 114; nuclear research at, 12; workers records released, 189
low-dose radiation: dangers of, 1, 9, 13, 85–86, 103, 113, 124–25, 191; safe levels alleged for, 88, 116, 142–43, 144–45, 194. *See also* threshold hypothesis
Lowry, Malcolm, 42 lung cancer: among nuclear weapons industry workers, 122; and smoking, 7, 100, 101, 220; study of, 78
Lunsden, Louisa Innes, 38–39
Lushbaugh, Dr. Clarence, 147–48
lymphoma, 137, 239

MacArthur, Gen. Douglas, 130
MacMahon, Brian, 89
MacMichael, Anthony, 280n. 16
MacNalty, Sir Arthur, 68
MacNeice, Louis, 108
Magnuson, Ed, 289–90n. 8
malaria and Burkitt's lymphoma, 244–45
Mallincrodt Chemical Company (MO), workers records released, 189
Manchester *Guardian,* 219
Mancuso, Dr. Thomas, 229; congressional investigation of dismissal of, 150–52, 161; dismissed from Hanford study, 122–24, 126–27, 147, 150–52, 153, 158; Hanford study by, 12, 113, 114–18, 162; long-term impact of dismissal on, 178, 187; support to restore funding to, 148–49; as witness in Silkwood case, 150
Manhattan Project, 10, 12, 113, 130, 143, 155, 156
Mantel-Haenzel statistical procedures, 90
Marie Curie Cancer Foundation, 100
Marks, Dr. Sidney, 117, 147, 151, 298n. 12
Marshall, Sir Walter, 166
Marshall Plan, 96
mastoid diseases, 67
McCarthy, Sen. Joseph, 87
McClintock, Barbara, 7, 229
McGinley, Ken, 202
McKewon, Thomas, 99
McMasters, Jo and Stella, 175
Medical Research Council (MRC) funding: of ankylosing spondylitis, 217; of cancer epidemiology, 80, 84, 100, 102, 249; of carbon tetrachloride poisoning study, 58; of pneumoconiosis among Welsh coal miners study, 59–60; of Radiation and Genome Stability Unit, Harwell, 266–67; of smoking and lung cancer research, 101; of TNT poisoning study, 56, 57
Memoranda of Understanding, DOE, 188
Metropolitan Edison, 178
midwives, reporting of birth anomalies by, 134
Milham, Dr. Samuel, 117, 127
Mills, John (maternal great-great-grandfather), 19
Milne, Roger, 170
Morgan, J. P., 54
Morgan, Dr. Karl, 115, 123, 230, 269; censorship of, 158; criticism of ERDA/DOE research, 148; criticism of ICRP, 143–44; on radiation risks, 90–91, 124, 125, 205
Moss, Ralph, 264
Mound (OH) nuclear facility, workers records released, 189
MRC. *See* Medical Research Council (MRC) funding
Ms. Magazine, 16, 264
Muller, Hermann, 87
munitions industry, worker safety standards in, 56–58
Murdoch, Iris, 3
Murray, Dr. Flora, 26
myeloma, 122, 137, 241–42

Nader, Ralph, 153, 157
Nagasaki, bombing of, 114, 129
Naish, Dr. Albert Ernest (father), 3, 6, 19; career of, 22–34 passim, 254; death of, 34; health of, 34; marriage of, 22–23, 33
Naish, Anthony (a.k.a. David, brother), 31, 32, 35, 38
Naish, Charles (brother), 23, 31, 35, 39
Naish, Charlotte (sister), 35, 37
Naish, Ernest (brother), 31, 35
Naish, George (brother), 35, 36, 38
Naish, Jean (sister), 23, 35, 44
Naish, John (brother), 35, 39, 274n. 11
Naish, Dr. Lucy Wellburn (mother), 3, 6, 7, 19–34; background and upbringing of, 19–20; career of, 22–

34, 254–55; death of, 34; education of, 20–22; health of, 31, 33; marriage of, 22–23, 33; reforms introduced and supported by, 25–27
Naish, Michael (nephew), 34
Naish, Nora (former sister-in-law), 22, 27, 32
NAS. *See* National Academy of Sciences (NAS), U.S.
Natarajan, Nachimuthu, 90
National Academy of Sciences (NAS), U.S.: access to workers' records through, 186; committee on the Biological Effects of Ionizing Radiation (BEIR), 9, 132, 140, 144, 165, 191; creation of Atomic Bomb Casualty Commission by, 131; study of nuclear safety by, 179
National Cancer Institute (NCI), 96, 116, 159, 273n. 27
National Council of Radiation Protection (NCRP), 91
National Health Service, 67, 82, 217
National Institute for Occupational Safety and Health (NIOSH), 188, 228
National Institutes of Health (NIH), 56
National Organization for Women, 150
National Projectile Factory (Sheffield), 28–29
National Radiological Protection Board (NRPB), British, 9, 126, 132, 144, 163, 181, 194, 200, 201–2
National Research Council, 135
national security issues, 155, 180–81
National Union of Seamen, 168
Nationwide (BBC television), 200, 201
NATO, 167
Nelson, Sara, 150
neural tumors, 239
New Criticism, 3, 47, 105–6
Newhouse, Dr. Molly, 55, 95, 229
Newnham College, 41, 46
New Scientist, 114, 170
New York Times, 2, 7, 13, 16, 105, 130, 149, 179, 189, 190, 191, 198
Nichols, Faith, 253
Nienhuys, Klarissa, 197, 218, 246, 253–54, 256
no-threshold hypothesis, 144–45
nuclear accidents: at Chernobyl, 159, 166, 179; risk of, 178; at Rocketdyne division, Rockwell International, 208–10; at Three Mile Island, 150, 154, 163, 177–79; at Windscale (Sellafield), 163, 201
nuclear energy, commercial exploitation of, 10–11, 13–14
nuclear industry: development of, 10–12, 13–14; moratorium on development in Great Britain, 174; placement of personnel in, 120–21; public relations by, 152; secrecy in, 155, 180–81, 186, 187, 195, 209
Nuclear Industry Radioactive Waste Executive (NIREX), 168–69
nuclear installations: inspection of, 179–81; risks and benefits of, 9
nuclear medicine. *See* radiography
nuclear physics at Cambridge, 44
Nuclear Regulatory Commission (NRC), U.S., 9, 117, 178
nuclear terrorism: fear of, 170; threat of, 14
nuclear test bans, 87, 152, 155
nuclear testing: military personnel exposed during, 150, 199–202; postwar, 11, 86, 155; underground, 11, 152; undisclosed, 267
nuclear war: prevention of, 155, 156; survival after, 86
nuclear waste disposal, 168–69, 179
nuclear weapons industry: AS's study of, 13, 170, 178–79, 181–92; criticism of worker safety standards in, 1, 2, 8–9, 12, 113–14, 115–18,
nuclear weapons industry (*continued*) 187; development of, 11; low-dose exposure of workers in, 113; rise of, 8; urine testing of workers in, 119–20

Nuffield, Lord (donor), 76
Nuffield Department of Clinical Medicine/Hospital (Oxford), 55–56
Nuffield Provincial Hospitals Trust, 68
Nussbaum, Rudi, 208, 221, 251–52, 253, 264

Oak Ridge Associated Universities, 148, 182
Oak Ridge (TN) nuclear facility: AS's study of, 13, 191; Lushbaugh's study of health effects at, 147–48; Mancuso's study of health effects at, 12, 114, 122; uranium manufacturing at, 115; Wing's study of, 192, 221, 268; workers' records released, 189, 190
Observer, 189, 200
Occupational and Environmental Medicine, 191
occupational safety. *See* worker safety standards
Occupational Safety and Health Administration (OSHA), U.S., 179
Office of Population Censuses and Surveys, British, 214
Official Secret Act, British, 195
Oil, Chemical and Atomic Workers Union, 148
O'Leary, Hazel, 145, 150, 267
One Damned Thing after Another, 42
Oppenheimer, Robert, 44, 87, 113, 155
Orwell, George, 105, 106
Oster House (St. Albans), 53–54
O'Toole, Dr. Tara, 267–68
Oxford Study of Background Radiation, 194–97
Oxford Survey of Childhood Cancer: AS's commitment to, 8, 61, 94, 98, 100; criticism of, 84–85, 88–90, 93; data collection for, 76, 81–83, 91–92, 100, 218–20; end of, 91–92, 98, 100, 101, 250, 259; findings of, 83–86, 116, 153, 232–47; long-term implications of, 86–88, 90–93, 246–47; research design of, 78–81
Oxford University, medical education for women at, 21
Oxford University, Imperial Cancer Research Fund Cancer Epidemiology and Clinical Trials Unit, 201–2
Oxford University, Institute of Social Medicine (Social Medicine Unit): AS as assistant to Dr. Ryle at, 68, 70–73; AS as head of, 2, 3, 60, 74–77, 94; established, 67–70; funding and staffing of, 78, 81, 94–97, 101

pancreatic cancer, 122
Pankhurst, Christabel, 26
Pankhurst, Emmeline, 26
Pantex (TX) nuclear facility, workers records released, 189
Parfit, Jesse, 95
Parker, Michael, 164–65
Partial Test-Ban Treaty, 152, 155
Paterson Institute for Cancer Research (Manchester), 267
Pattie, Geoffrey, 201
Pauling, Linus, 87, 155
pediatrics, 19, 22, 24
Pennybacker, Winifred, 95
Physicians for Social Responsibility (PSR), 155, 180, 182, 187, 209, 268
Physician's Task Force on the Health Risk of Nuclear Weapons Production, 180
"plutonium economy," creation of, 170
pneumoconiosis, among coal minters, 59–60
pneumonia, 92, 136, 234
poliomyelitis, 78
Poor Laws, administration of, 25, 26, 274n. 9
prenatal x-ray exposure. *See* fetal x-ray exposure, dangers of
preventive medicine, 98–99, 155, 224, 278n. 4
Proctor, Robert, 224

prospective research, support for, 80, 88, 89, 220
public health: AS's questionnaires distributed by, 82; development of, 23–24, 67; and social medicine (epidemiology), 70, 74, 75
Public Health Act, British, 24
public interest groups, 149, 180
Pugwash, 155

Queen Charlotte's Hospital (London), 22
Quigley, Dianne, 183, 184, 185

Radcliffe Infirmary (Oxford), 55
Radford, Edward, 165
radiation effects, types of, 237
Radiation Effects Research Foundation (RERF), 9, 131–43, 252–53
radiation poisoning, alleged effects in Hiroshima, 129–30
radiation sickness, identification of, 135
radioactive dust, collection of, 200
radioactive fallout, exposure to, 86, 87, 273n. 27
radioactive waste: classification of, 9; dangers of, 14; disposal of, 9–10; stockpiles of, 14
radiography: hazards studied, 80, 159; medical applications of, 8
radionuclides, 44
Raine, Kathleen, 43, 47, 104, 106
Ramazzini, Bernardino, 272n. 20
Ramazzini Prize, 13, 248
Ransom, John Crowe, 106
Ratner, Dr. Vicki, 1–2
Ready Reckoner, 203, 204, 207, 268
Reagan, Ronald, 96, 101
Reay, Dorothy, 197
recall bias, 80, 85, 295–96n. 20
Redgrave, Michael, 42
Reich, Robert, 294n. 35
respiratory diseases, 67, 195
RES (reticulo-endothelial system) tissue, cancer of, 122, 136–37, 201, 232, 239
retrospective research, 80, 85, 90, 220
Richards, I. A., 48, 105, 106
Richardson, David, 222, 228, 268
Ricks, Christopher, 109–10
Right Livelihood Prize/Award, 12–13, 159, 170, 208, 229, 250–51
Roberts, Dr. (Sheffield Medical Officer of Health), 81–82
Rocketdyne Clean-up Coalition, 209
Rocketdyne division, Rockwell International, 208–10, 268, 269
Rocky Flats Health Advisory Panel, 268
Rocky Flats (CO) nuclear facility: closing of, 180, 203; compensation cases against, 185, 202–4, 206; conference on radiation effects near, 268; disregard of safety standards at, 160–61, 179, 202–3; Mancuso's study of health effects at, 114; workers' records released, 189, 190
Roff, Sue Rabbit, 142, 250, 255–56
Rogers, Paul, 149, 151, 152
Roman, E., 292–93n. 10
Rose, Hilary, 254, 255, 298n. 19
Roswell Park Memorial Research Institute (Buffalo, NY), 158
Rotblat, Dr. Joseph, 148, 155, 282n. 9, 283n. 29
Royal College of Physicians, 28, 32, 51, 59
Royal College of Physicians for Public Health and Social Medicine, 226, 227
Royal Free Hospital (London), 21, 22, 45, 51–52, 62
Royal Hospital (Sheffield), 29
rubber industry, worker safety standards in, 116
Rudig, Wolfgang, 288n. 10
Rutherford, Ernest, 43–44
Ryle, Dr. John A., 15, 67–70, 74, 83, 224

Sagan, Leonard, 280n. 6
St. Leonards School (St. Andrews, Scotland), 38–39

Samuels, Sheldon, 248
Sanger, Eugene, 205
Savannah River (SC) nuclear facility: AS's study of, 13; closing of, 180; disregard of safety standards at, 179, 185; Mancuso's study of health effects at, 12, 114; workers records released, 189
Schlesinger, James, 148
Science, 156
Scottish Campaign to Resist the Atomic Menace (SCRAM), 164, 169
Scottish Conservation Society, 170
Scurfield, Dr. (Sheffield Medical Officer of Health), 25
Seabrook (NH) anti-nuclear demonstration, 154
Seaview cottage (Anglesey), 37
Sellafield cancers case, 197–99
Sellafield nuclear facility. *See* Windscale (Sellafield) nuclear facility
Severnside Campaign against Radiation (SCAR), 175, 176
Sheffield Board of Guardians of the Poor Law, 25
Sheffield Girls' High School, 38
Sheffield Hospital, 45
Sheffield Medical School, 29
Sheffield University, 31
Shigematsu, I., 266
shoe industry, worker safety standards in, 71–72
Shook, Larry, 184
Shoreham (NY) nuclear reactor, 153
SIDS. *See* Sudden Infant Death Syndrome
Sierra Club, 180
"silent forces," theory of, 138–40
Silkwood, Karen, 149–50, 159
Silsoe, Lord, 173
Simi Valley, California, 209, 268
60 Minutes (television), 16
Sizewell nuclear facilities, 165–66, 193
Skaggs, David, 294n. 35
Smith, Honor, 54, 76
Smith-Wilson, Maggie Kinnear, 95, 100
smoking: and lung cancer, 7, 100, 101, 220; parental, and risk of childhood cancer, 198, 219, 247
Snow, C. P., 43
Snow, Dr. John, 23–24, 69, 214
social class and disease, 67–68
socialism, 6
socialized medicine, 69
social medicine. *See* epidemiology; public health
social responsibility among physicians, 15, 19, 68–69
Social Security records, 115–16
Society of Social Medicine, 226
solar energy, 10
Sorahan, Tom, 199, 200–201, 202, 219, 247, 257
South Pacific, nuclear testing in, 11, 155, 199–202
Soviet Union, radioactive contamination in, 14. *See also* Chernobyl nuclear accidents
Spanish influenza, 31, 234
"Special Orders Concerning Poor Law Institutions," 26
species death, risk of, 159
spinal curvature, 73
Standardized Mortality Ratio (SMR), 120, 127
STAR (Standing for Truth About Radiation) Foundation, 269
statistical significance as proof, 295n. 4
Stein, Gertrude, 42
Sternglass, Dr. Ernest, 114, 155–56, 158, 160
Stevenson, Adlai, 86
Stewart, Dr. Alice Naish: as assistant to Dr. Ryle, 68, 70–73; awards and honors, 12–13, 59, 159, 170, 176, 208, 227, 229, 248, 250–51, 268, 269; background and upbringing of, 3, 6, 19–34; biographies of, 3, 15; birth of, 3, 23, 35; career of, 2, 49, 50, 51–60, 224–29, 258–59,

260–62; children of, 2, 4, 7, 49, 50, 55, 61, 255; credibility of, 1, 88; defense of Greenham Common protesters, 167–68; divorce of, 61; education of, 3, 21, 38–48; grandchildren of, 62, 63; health of, 31; as independent scientist, 1, 13, 228, 248–49, 259; interviews with, 2–3, 14; love of research, 6–7; marginalization of, 7, 226, 249–54; marriage of, 45, 46–47, 48, 49; mentors of, 255–56; ongoing invisibilizing of, 193–210; personality of, 2, 5–6, 19, 256, 257–58; radicalizing of, 262–63; reputation of, 1, 58; retirement of, 99, 113, 249–50; siblings of, 6, 35, 37–38; testimony at anti-nuclear hearings, demonstrations, and conferences, 164, 166, 168, 170–72, 173, 175–76, 182–84, 186, 195, 204–5, 206, 269; wartime experiences of, 53–60; works by, 8–10, 16
Stewart, Anne (daughter), 5, 37, 49, 62
Stewart, Charles (grandson), 61, 62, 63
Stewart, Christabel (granddaughter), 61, 62, 63
Stewart, Hughie (son), 4, 50, 52, 61–62
Stewart, Jeanette Johnston (daughter-in-law), 4, 61–62
Stewart, Jessie (mother-in-law), 46
Stewart, Katherine (sister-in-law), 46
Stewart, Ludovick (husband): career of, 48, 49, 60; divorce of, 61; marriage of, 46–47, 49; as musician, 50; war service by, 53
Still, Dr. George Frederic, 22
Stock Hinkley Expansion (SHE), 173
Stocks, Dr. Percy, 101
Storey, Emily P., 39
Sudden Infant Death Syndrome (SIDS), 92, 181, 232, 240–44
Suffragist movement, 26–27
suicide, 6
Superfund cleanup, 180
Sutcliffe, Dr. Jill, 176, 291n. 36
Szilard, Leo, 44, 155

Taft, William H., 271n. 8
Tamplin, Dr. Arthur, 156–57, 158
Tate, Allen, 106
Taylor, Dr. Everley, 20
Taylor, Peter, 168
Thackeray, William Makepeace, 6
Thatcher, Margaret, 166, 298n. 11
thermal oxide reprocessing plant (THORP), 163, 165
thermonuclear war. *See* nuclear war
Three Mile Island nuclear facility: class-action suit against, 13, 178; design of, 166; nuclear accident at, 150, 154, 163, 177–79
Three Mile Island Public Health Fund: grant to AS, 13, 178–79, 181–92; workers records released to, 187–90
threshold hypothesis: challenges to, 90, 116, 122, 124–25, 132, 142, 157, 171–72, 221, 237–38; dosage for, 122, 143, 153; and effective dose calculations, 203, 204; and proof, 215–16. *See also* low-dose radiation
thyroid cancer, 184
Times (London), 44, 71, 72, 109, 219
Times Higher Education Supplement (London), 16, 264, 269
TNT (trinitrotoluene) poisoning, 56–58
tobacco industry, 108
Tokyo Radio, 129
Totter, John, 280n. 6
Town and Country Planning Association, 164, 173
Trevelyan, Julian, 47
Trinity nuclear test, 114
Tri-State Leukemia Survey, 89–90, 96, 158–59, 161
Truman, Harry, 10
Tsuzuki, Masao, 131
tuberculosis among shoe industry workers, 71–72

Tucker, Kitty, 149, 150
tumors, 238, 239

ulcers, 67, 73
ultrasound, effects of, 246–47
Union Carbide, 10
Union of Concerned Scientists, 153, 157
United Kingdom Childhood Cancer Survey, 102
United Kingdom Co-Ordinating Committee on Cancer Research, 102
United Nations: "Atoms for Peace" speech (Eisenhower), 11; Pauling's anti-nuclear petition presented to, 87, 155
United Nations Scientific Committee on the Effects of Atomic Radiation (UNSCEAR), 9, 132
United Steelworkers Union, 149
University College Hospital, 21
uranium, 44
Urey, Harold, 157
urine testing, 119–20
U.S. News and World Report, 284n. 31

varicose veins, 73
viral contagion, 199
visible tape, 75–76
vital statistics: collection of, 24, 73, 214; confidentiality of, 81
Voelz, George, 206
von Uexkull, Jakob, 250–51

Wald, George, 157, 282n. 9
Wales, coal mining in, 59–60
Walker, Anne, 227
water pollution, 163, 168
Watkins, James, 186, 188
Watkins, Ronnie, 50–51, 53
Watson, Dr. James, 16, 153, 157
Webb, Josephine, 78, 80, 81, 82
Webster, John, 265
Wellburn, Anne Matilda (maternal grandmother), 19–20, 21
Wellburn, Henry (maternal grandfather), 19
Westropp, Celia, 95
Whyl (Germany), anti-nuclear demonstration at, 153–54
Wilms' tumors, 239
Wimsatt, W. K., 106
Windscale (Sellafield) nuclear facility: cancer cluster at, 193, 195, 197, 198–99; investigation of, 126, 162–65; nuclear accident at, 163, 201; plutonium manufacturing at, 11
"Windscale—the Nuclear Laundry" (television documentary), 165
Wing, Steve, 192, 221–22, 228–29, 248, 257–58, 264, 268
Wirth, Timothy, 186, 187
Witts, Dr. Leslie, 55, 56, 59, 67, 71, 78
Wolffian ridge, 239
Women's Social and Political Union, 27
Woolf, Virginia, 42
worker safety standards: in coal mining, 59–60; in munitions industry, 56–58; in nuclear weapons industry, 1, 2, 8–9, 12; in shoe industry, 71–72
workers records, struggle for access to, 12, 13, 15, 181–90
World War I, 28–29, 37
World War II, 31, 53–54
Wright, Eric, 267

x-rays. *See* fetal x-ray exposure; radiography

Yorkshire News, 251
Yorkshire Television, 165

Acknowledgments

Being Alice Stewart's biographer has been a bit like becoming part of a large, complex, and very interesting family. I found myself plugged into an extensive network, reaping the good will built up by Alice in the course of her long life. This book has many benefactors.

Rudi and Laureen Nussbaum, professors emeriti at Portland State University, helped by finding me support in the form of a grant from the Oregon Community Foundation, Portland State University, and by sending dozens of tapes of interviews and seminars made the term Alice was on sabbatical at Portland State University, fall 1984. My thanks to those who did the interviews—Dr. Karen Steingart, Fern Shen, Karen Dorn-Steele, Tom Graham, Nancy Salomon—and to Ruth Bergman, who transcribed the tapes for me. Virginia Myers, who came to this project already an expert on Alice Stewart, having had a major role in producing the Channel 4 documentary, *Sex and the Scientist: Our Brilliant Careers,* has been an immensely knowledgeable, imaginative, and sensitive researcher. Klarissa Nienhuys, a longtime friend of Alice's who has thought long and deeply about her work, was more than generous in sharing her ideas.

Jill Boniske and Susan Robinson, who have been at work for several years on a documentary, "Second Opinion: Alice Stewart and Low Level Radiation," kindly sent me transcriptions of their interviews. A few dozen more tapes of talks and interviews in possession of the Children's Cancer Research Institute (CCRI, Concord, MA) came to me from Diane Quigley, director of the institute. CCRI was founded by David Kleeman, who met Alice when he was ten, the summer his parents, vacationing in the Cotswolds, rented her cottage at Fawler. Years later, when funding for the Oxford Survey dried up, he and his wife Palmer established this institute, which works with grassroots groups and health agencies—federal, state, and local—to help communities respond to nuclear contamination. CCRI sponsored several speaking tours for Alice in the eighties, the years just after she'd been awarded $2 million from the Three Mile Island Fund to study the nuclear workers, as part of its campaign, along with public interest groups such as Physicians for Social Responsibility, to get the workers' records away from the Department of Energy. Jean Kasperson at CCRI helped me locate media coverage of these public appearances.

I wish to thank Alice's family for their hospitality: Anne Marshall, her daughter; Charles Stewart, her grandson; brothers and sisters-in-law, Ernest and Joan Naish, John and Barbara Naish; and Nora Naish, an extraordinary woman who turned to writing novels (best-sellers!) in her seventies, after a career as a physician and raising four children. Her biography of Alice's mother, "Dr. Lucy" (recently published as a novel, *A Time to Learn*), gave me a way into the lives of Alice's mother and father.

Thanks also to Jane Gould, who gave such an enthusiastic account of Alice Stewart that I made a special effort to interview her in May 1994, with Dr. Vicki Ratner; and to Jay Gould, who, along with Jane, provided Vicki and me with hospitality and a weekend-long interview with Dr. Ernest Sternglass. Thanks to Renate Barber, whose sketch "That's Alice" was enormously helpful; to Faith Nichols, Alice's devoted friend, who provided meals, tea, and Alice anecdotes through one of our longest and most arduous work sessions, the summer of 1995; and to Anne Walker, Alice and George's secretary; Keith Webber; and Dr. Vicki Ratner, who helped with interviews and advice.

Warm thanks to all those who took the time and trouble of being interviewed by Virginia Myers or me. Several remained in correspondence with me in the years it took to complete the book: Bob Alvarez, Senior Policy Advisor to Secretary of Energy Bill Richardson; Dr. Wally Cummins, who continues work on the study of Hanford veterans he and Alice began in the mid-eighties; Bruce DeBoskey, the lawyer who brought Alice in on several Rocky Flats cases; Dr. John Goldsmith of Ben Gurion University; David Kleeman and Diane Quigley of CCRI; and Dr. Dave Richardson and Dr. Steve Wing, Department of Epidemiology, University of North Carolina. Others interviewed were Janine Allis-Smith of CORE (Cumbrians Opposed to a Radioactive Environment); Dr. Rosalie Bertell of the International Institute of Concern for Public Health, Toronto; Peter Bunyard, former editor of the *Ecologist;* Sir Richard Doll, Imperial Cancer Research Fund, Cancer Studies Unit; Dr. John Gofman of Committee for Nuclear Responsibility, San Francisco; Patrick Green of FOE (Friends of the Earth); Dr. Morris Greenberg; Dr. George Kneale; Dr. Karl Morgan; William Peden of CND (Campaign for Nuclear Disarmament); Larry Shook of Handford Education and Action League; Dr. Tom Sorahan, Department of Public Health and Epidemiology, University of Birmingham; Dr. Ernest Sternglass, professor emeritus, University of Pittsburgh;

Dr. Jill Sutcliffe, environmental scientist. There are others whose correspondence was helpful: Frank Barnaby of FOE; Professor John Haffenden, Sheffield University, who has been at work on a biography of William Empson; Dr. Wolfgang Kohnlein; Dr. Philip Landrigan; Susan Preston-Martin; Sue Rabbit Roff, University of Dundee; Dr. Inge Schmitz-Feuerhake of University of Bremen; and Dr. David Sumner.

I wish to thank those who have read the manuscript: Dr. Devra Lee Davis of World Resources Institute; Professor Rena Fraden, Pomona College; Dr. John Goldsmith; Bob Jourdain; Virginia Myers; Professors Emeriti Rudi and Laureen Nussbaum; Dr. Bob Tager; Mike Zeller. And thanks to those who have read parts of the manuscript: Professor Elizabeth Abel, University of California, Berkeley; Professor Janet Adelman, University of California at Berkeley; Peter Bunyard; Professor Marilyn Fabe, University of California, Berkeley; Professor Claire Kahane, State University of New York at Buffalo; Faith Nichols; Dr. Vicki Ratner; Dr. Jill Sutcliffe; Dr. Myron Wollin of PSR. Continued gratitude to Scripps College for research support through the years and thanks to the Right Livelihood Foundation for a stipend that helped launch the project. Special appreciation goes to my student Lisa Sharihari, who spent hours tracking down references and compiling a twenty-five-page bibliography, which, unfortunately, had to be mostly cut.

Coming in with splendid photographs at the last moment were Carole Gallagher, author of *America Ground Zero,* and Robert del Tredici, author of *At Work in the Fields of the Bomb,* both of whom I was honored to have associated with the book. Most of the other photos were donated by Alice's family and friends. Thanks to Susan Griffin for permission to use as an epigraph to the book her poem "Curve" from *Bending Home* (Port Townsend, WA: Copper Canyon Press, 1998), written for Alice Stewart when Griffin heard her lecture at Berkeley. And finally, special mention of my editor, LeAnn Fields, for her warm and imaginative response to this project.

As always and ever, to Bob Jourdain, who read, discussed, analyzed, argued and suffered with and through drafts, and was there with comfort, advice, airport transportation, dinner, dog-sitting, and much more. And finally to Alice, scientist, physician, activist, inspiration, teacher, friend, who—among many other things—has shown me the way to grow old.